Advances in Chromatographic Analysis of Bioactive Compounds

Advances in Chromatographic Analysis of Bioactive Compounds

Editor

Liangliang Liu

MDPI

Basel • Beijing • Wuhan • Barcelona • Belgrade • Novi Sad • Cluj • Manchester

Editor
Liangliang Liu
Institute of Bast Fiber Crops,
Chinese Academy of
Agricultural Sciences,
Changsha
China

Editorial Office
MDPI AG
Grosspeteranlage 5
4052 Basel, Switzerland

This is a reprint of articles from the Special Issue published online in the open access journal *Separations* (ISSN 2297-8739) (available at: https://www.mdpi.com/journal/separations/special_issues/Chromatographic_Bioactive).

For citation purposes, cite each article independently as indicated on the article page online and as indicated below:

Lastname, A.A.; Lastname, B.B. Article Title. *Journal Name* **Year**, *Volume Number*, Page Range.

ISBN 978-3-7258-1987-4 (Hbk)
ISBN 978-3-7258-1988-1 (PDF)
doi.org/10.3390/books978-3-7258-1988-1

Contents

About the Editor

Liangliang Liu

Associate researcher, is currently working at the Analysis and Testing Center, Institute of Bast Fiber Crops, Chinese Academy of Agricultural Sciences. He holds a Ph.D. in Analytical Chemistry from Central South University in Changsha, Hunan, China. His research interests include separation and analysis of natural products, molecularly imprinted materials, and modification of natural fiber materials. With more than a decade of experience in academia, he has published around 45 papers in peer-reviewed journals.

separations MDPI

Review

Factors Influencing the Prediction Accuracy of Model Peptides in Reversed-Phase Liquid Chromatography

Othman Al Musaimi *, Oscar M. Mercado-Valenzo and Daryl R. Williams

Department of Chemical Engineering, Imperial College London, London SW7 2AZ, UK
* Correspondence: o.al-musaimi@imperial.ac.uk

Abstract: Hydrophobicity is an important physicochemical property of peptides in solution. As well as being strongly associated with peptide stability and aggregation, hydrophobicity governs the solution based chromatographic separation processes, specifically reversed-phase liquid chromatography (RPLC). In addition, hydrophobicity is a major physicochemical property of peptides in comparison to H-bonding, electrostatic, and aromatic properties in intermolecular interactions. However, a wide range of molecular factors can influence peptide hydrophobicity, with accurate predictions depending on specific peptide amino acid compositions, structure, and conformation. It is noticeable that peptide composition, the position of the amino acid, and its neighbouring groups play a crucial role in the elution process. In light of this, the same amino acid behaved differently depending on its position and neighbouring amino acid in the peptide chain. Extra attention should be paid to the denaturation process during the course of elution, as it has been shown to complicate and alter the elution pattern. This paper reports on the key peptide properties that can alter hydrophobicity and, consequently, the RPLC elution behaviour of the peptides, and it will conclude by proposing improved prediction algorithms for peptide elution in RPLC.

Keywords: denaturation; peptides; separation; retention prediction

Citation: Al Musaimi, O.; Mercado-Valenzo, O.M.; Williams, D.R. Factors Influencing the Prediction Accuracy of Model Peptides in Reversed-Phase Liquid Chromatography. *Separations* **2023**, *10*, 81. https://doi.org/10.3390/separations10020081

Academic Editor: Liangliang Liu

Received: 4 January 2023
Revised: 18 January 2023
Accepted: 23 January 2023
Published: 24 January 2023

1. Introduction

Peptides are consolidating their importance in the pharmaceutical arena with a total FDA approvals of 22 peptide-based therapeutics during the last six years [1]. Peptides also exist in other pharmaceutical therapeutics such as the antibody drug conjugates, where they can be found as linkers or therapeutic payload, or both [2]. This success is ascribed to their high specificity and tolerable safety profile. Furthermore, the advancements in the main synthetic methodology called solid-phase peptide synthesis (SPPS) have enabled the synthesis of new families of peptides with both high purity and yield [3]. However, though SPPS produces relatively pure peptides, these purities are insufficient for therapeutic use by humans. Therefore, peptide separation and purification is a core facet for the manufacture of therapeutics products that are safe for human consumption. Reversed-phase liquid chromatography (RPLC) is the industry standard method to achieve such purity [4,5]. Peptide hydrophobicity is an important physicochemical property the governs the RPLC separation process of peptides and determines how they interact with the chromatographic packings [6,7]. Due to the intrinsic structural characteristics of peptides, synthetic peptide impurities often have very similar structures, complicating their separation and purification [8]. Hence, developing a reliable and accurate tool for predicting the chromatographic retention behaviour of structurally similar peptides is of the utmost importance for optimising industrial RPLC processes.

Previous studies have discussed the prediction of peptide retention times based on oligomer structures [7]. In fact, such structures do not represent the real "therapeutic", or model peptides or even those obtained from the tryptic digest. Hence, data generated from these oligomers-based peptides have only a limited use. However, when it comes to predicting the retention behaviour of real or model peptides, their validity is questionable. As will be discussed in this review, several groups have considered model peptides for the retention time prediction, but these overall efforts have only had limited success.

2. Retention Time Prediction Based on Model Peptides

Guo et al. [9] have reported on predicting the retention time of peptides by estimating the influence of the individual amino acid residues in a "model peptide" containing the following sequence: (Ac-Gly-X-X-$(Leu)_3$-$(Lys)_2$-Amide). All the proteinogenic amino acids are placed independently in position X. This work incorporates two amino acids of the same type at a time to amplify their influence on the final retention behaviour in the model peptide. Mobile phases investigated included pH 2 (A: 0.1%trifluoroacetic acid (TFA) in water; B: 0.1%TFA in ACN) and pH 7 (A: 10 mM $(NH_4)_2HPO_4$—0.1 M $NaClO_4$; B: 0.1 A4 $NaClO_4$ in 60% aq. ACN), which were evaluated in the study with the following columns: Beckman Ultrapore RPSC C_3, 5 µm, 4.6 × 75 mm; SynChropak RP-4 (C_4), 6.5 µm, 4.1 × 250 mm; SynChropak RP-8 (C_8), 6.5 µm, 4.1 × 250 mm; Whatman Partisil 5 (C_8), 6.5 µm, 4.6 × 250 mm; and three SynChropak RP-18 (C_{18}) columns: 6.5 µm, 4.1 × 250 mm, 6.5 µm, 4.1 × 50 mm, and 6.5 µm, 10 × 250 mm. All of these columns have a 300 Å pore size, except the Whatman Partisil, which has a size of 60 Å [9].

The most pronounced retention coefficient changes were noticed in the following residues: Glu, Asp, His, Arg, and Lys; basically, those are charged residues. At pH 7, the acidic residues Glu and Asp are completely ionised and hence elute earlier. At pH 7, His is neutral, which leads to it being retained more strongly. For Arg and Lys, the authors proposed that some kind of ionic interaction is taking place between their positive charge and the silanol groups of the stationary phase [9].

The retention coefficient of each amino acid is determined based on the retention time of the peptide containing the corresponding amino acid. Retention coefficient of the termini was determined using the following peptide: Y-Gly-$(Leu)_3$-$(Lys)_2$-Z, where Y: acetylated (Ac) or non-Ac *N*-terminus; Z: amide or acid *C*-terminus. The non-Ac peptides at pH 2.1 revealed a clear reduction in their retention time due to the positive charge associated with the protonation of the *N*-terminus. On the other hand, changing the amide *C*-terminus to an acid one had a minimal effect in decreasing the retention time, because it would be protonated with no charge at pH 2.1. On the contrary, the effect of pH 7.4 on the *N*-terminus (Ac or non-Ac) would have a small effect, whereas changing the *C*-terminus from amide to acid would show a pronounced reduction in the retention time due to the deprotonation of the carboxylic *C*-terminus (Table 1) [9].

Differences between the retention times among various groups might be attributed either to the absence of certain amino acids in one of the studies under comparison, or to the neighbouring effect or chain length. This research group also investigated the effect of various organic modifiers added to the mobile phase that can be used to enhance peptide separation. Acetonitrile proved to be superior in comparison with 2-propanol and methanol. However, for highly hydrophobic peptides, the 2-propanol, which is more non-polar, was more beneficial [9]. Similarly for a highly hydrophilic peptide, methanol, which is more polar, performed better. These findings are in agreement with the work of Meek and Rossetti [10]. Interestingly, when C_8 and C_{18} columns with different lengths were used, different elution patterns for alkylphenones were observed, while with peptides, the elution pattern was almost the same. This observation could lead to some insight into the mechanism of the elution. In the case of alkylphenones, it is mainly attributed to the partitioning effect, which explains the late elution when using C_8, which has almost double the ligand density compared to the C_{18}. Whereas adsorption/desorption is the driving force for the separation process in the case of peptides, it is to be noted that partitioning contributes as a retention mechanism for peptides, albeit to a lesser extent, and this is mainly dependent on the composition of the peptide being investigated. Thus, using an internal standard could help in decreasing the differences. The internal standard has to be peptide based to fulfil the prerequisite of being as close as possible to the nature of the material being separated. In a study using alkylphenones as an internal standard, the authors highlighted a different retention mechanism of separation [9].

Table 1. Retention time for pH 2.0 and 7.0 [9].

Amino Acid Residue	Retention Time Coefficient	
	pH 2.0	**pH 7.0**
Tryptophan	8.8	9.5
Phenylalanine	8.1	9.0
Isoleucine	7.4	8.3
Leucine	8.1	9.0
Tyrosine	4.5	4.6
Methionine	5.5	6.0
Valine	5.0	5.7
Proline	2.0	2.2
Threonine	0.6	0.3
Histidine	−2.1	+2.2
Alanine	2.0	2.2
Glutamine	0.0	0.0
Glutamic acid	1.1	−1.3
Glycine	−0.2	−0.2
Serine	−0.2	−0.5
Arginine	−0.6	+0.9
Aspartic acid	0.2	−2.6
Asparagine	−0.6	−0.8
Lysine	−2.1	−0.2
α-Amino	−6.9, −3.0 *	−2.4, 0.0 *
α-COOH	−0.8	−5.2

* The charged α-amino group had a smaller effect on an *N*-terminal Arg residue than an *N*-terminal residue with an uncharged sidechain. Mobile phases: pH 2 (A: 0.1%TFA in H_2O; B: 0.1%TFA in ACN) and pH 7 (A: 10 mM $(NH_4)_2HPO_4$—0.1 M $NaClO_4$; B: 0.1 A4 $NaClO_4$ in 60% aq. ACN). Columns: Beckman Ultrapore RPSC C_3, 5 μm, 4.6 × 75 mm; SynChropak RP-4 (C_4), 6.5 μm, 4.1 × 250 mm; SynChropak RP-8 (C_8), 6.5 μm, 4.1 × 250 mm; Whatman Partisil 5 (C_8), 6.5 μm, 4.6 × 250 mm; and three SynChropak RP-18 (C_{18}) columns: 6.5 μm, 4.1 × 250 mm, 6.5 μm, 4.1 × 50 mm, and 6.5 μm, 10 × 250 mm. All data are in minutes.

The predicted retention time (τ) of a peptide is calculated as follows:

$$\tau = \sum R_c + t_0 + t_s \tag{1}$$

where $\sum R_c$ represents the sum of the retention coefficients for all amino acid residues and the termini, t_0 is the estimated dead volume of the column using the retention time of the TFA peak, and t_s is the time correction for the peptide standard. This last value is previously calculating using the same equation in the form $t_s = t_{R,std} - (\sum R_{c,std} + t_0)$, where $t_{R,std}$ and $\sum R_{c,std}$ are observed values for the standard that is run under the same chromatographic conditions as the peptides to be predicted.

Determining the retention values for standard peptide and unretained compounds is beneficial, as for any HPLC system where the column's specifications such as length, particle size, or packing are known, this data can then be used to predict the retention time of peptides with a known structure [9].

Guo and co-workers also carried out another study to investigate further factors that may influence the peptide's retention time [11]. A total of 58 peptides in the range of 2–16 residues were investigated in this study. An excellent correlation was obtained between predicted and measured retention times, r = 0.98, which proved the dependence of the retention time of peptides on their amino acid composition [11].

Several findings were observed as a result of changing various condition of the chromatographic system. Increasing the flow rate and decreasing the elution gradient steepness had a positive effect on the resolution of peptide mixture. This behaviour is ascribed to the fact that the retention time is not strongly affected by the flow rate, whereas the peptide tendency to diffuse decreases with faster flow rates, which leads to a decrease in the peak width and consequently enhances the resolution, provided that the resolution is calculated by dividing the difference in the retention times of the two peaks over their total widths. For the gradient steepness, the increase in the retention time difference is more pronounced than the increase in the peak width, and hence, a higher resolution is obtained [11].

Increasing the column temperature decreased the peptide retention time and enhanced the resolution. Interestingly, for chromatographed homologous peptide pairs (the same amino acids with different order), those pairs were eluted as a single peak, and in the extreme case where a complete rearrangement of the sequence was considered, a near doublet peak was observed. These findings reaffirmed the accuracy of their developed retention time predictivity model, which depends primarily on the amino acid composition in the peptide, rather than their order. Increasing the temperature increased the peptide solubility in the mobile phase, which translated to decreasing the retention time and enhancing the resolution. That is, the peptide had a stronger preference for the mobile phase compared to the stationary phase. This study considered temperatures in the range of 26–66 °C. The molecular weight of the peptide has a key role for peptides up to 20 residues, but beyond 20 residues, a molecular weight correction must be incorporated [11]. The fundamental function of a prediction tool is to narrow the retention time/volume in which the peptide of interest will be eluted, thus saving, time, chemicals, and effort during processing [11].

Hodges and co-workers (including Guo) also introduced another hydrophobicity scale [12] by normalizing the coefficients previously determined in their study (Table 1) [9]. Reporting a relationship between hydrophilicity and antigenicity, this study aimed to investigate which amino acid is antigenic based on its hydrophilicity value. Determination of retention time coefficients was carried out by assigning a value of −10 to the amino acid with the highest retention coefficient from their previous study (shown in Table 1) [9], whereas the amino acid with the lowest retention coefficient was assigned with a +10 value. Other amino acids will be scaled accordingly [12].

Dealing with real peptides showed that several factors play a role in the hydrophobicity of amino acids.

3. Factors Affecting Amino Acid Hydrophobicity

3.1. Stationary Phase

In 1991, Wilce et al. carried out an extensive study using 2106 peptides from various studies to extract the retention coefficients of the amino acids [13]. A multiple regression matrix approach was utilized for this purpose. This statistical analysis revealed that at least 100 peptides are required for accurate retention coefficients determination. Various studies (Section 3.4) proved that as the peptide chain becomes larger, more deviation is expected from the linear summations of the hydrophobic contributions of the individual amino acids. This limitation is ascribed to the developed secondary, tertiary, and quaternary structures by the peptides in solution. These conformational changes result in changing the overall interaction patterns of the peptide molecule while in the chromatographic system, with either mobile or stationary phases. The main idea of this study is to assess the influence of various chromatographic conditions on the amino acid hydrophobicity coefficients [13].

Assuming that the retention mechanism is being governed solely by the hydrophobic interactions between the solute, mobile phase, and stationary phase, with the absence of any other electrostatic or H bonding interactions, this could be translated in the expression:

$$k'_{hydrophobic} = \phi K_{hydrophobic} \quad (2)$$

where $k'_{hydrophobic}$ and $K_{hydrophobic}$ are the retention factor and the equilibrium association constant of a hydrophobic solute, and ϕ is the phase ratio of the column measured as the ratio of the volumes of the stationary phase and the mobile phase.

Since the selectivity ($\alpha_{i,j}$) of two peptide entities (P_i and P_j) under a defined chromatographic condition can be measured as the ratio of k'_i and k'_j, the difference in retention coefficient of two peptides with only one different amino acid would be:

$$\tau = \ln \alpha_{i,j} = \ln k'_i - \ln k'_j \quad (3)$$

Here, τ can be understood as the difference in the energy due to the transfer of the peptide *i* from the mobile phase to the stationary phase, with respect to that of peptide *j*.

According to the solvophobic theory [14], where the retention coefficients represent the interaction of a certain portion of the analyte with the hydrophobic stationary phase, and including the ligand immobilized on it, the authors predict that a linear relationship should also be present, provided that the retention coefficients were derived from experimental chromatographic retention data. So, this means that the approach is also useful to understand the relationship between the amino acids and the ligands immobilized on the stationary phase as well as their densities.

Casal et al. studied the elution profile of 25 different peptides using four columns [15]. In this study, multiple linear regression (MLR) and partial least square (PLS) regression analyses were used. The main idea of this study is to evaluate the influence of different stationary phases on the retention coefficients of short peptides as well as on their retention times. The following columns were incorporated: (C_8—Ultrasphere, 5 µm, 4.6 × 250 mm), (C_{18}—Ultrasphere, 5 µm, 4.6 × 250 mm), (polymeric RP—PLRP-S, 8 µm, 300 A, 4.6 × 150 mm), and (C_{18}—Nova-pak, 4 µm, 60 A, 3.9 × 150 mm). The following mobile phases were used: A: 0.1 TFA in water, B: 0,1% TFA in acetonitrile [15].

The MLR and PLS regression models assumed the following equation to predict the retention times of peptides based on their amino acid compositions:

$$t_R = b_0 + \sum R_{c,i} n_i \quad (4)$$

where b_0 is the intercept of the linear model, $R_{c,i}$ is the amino acid *i* retention coefficient, and n_i is the number of times the same amino acid repeats in the peptide sequence.

Computer-aided programming was carried out afterwards to predict the retention times of peptides.

Several short peptides were chromatographed using these four columns, and their retention times of were used establish the retention coefficient of each amino acid using MLR and PLS models (Table 2).

From the retention data in Table 2, a limited influence of the stationary phase on the retention behaviour of the amino acids was observed. Furthermore, the higher pore diameter of PLRP-S column had little effect on the behaviour of short peptides. This confirms the independence of the retention behaviour of peptides on the column length, packing, or even the length of the alkyl chain attached to the stationary phase. These observations were also previously noted by Meek and Rossetti [10] and Guo et al. [9]). However, the authors stated that the effect of alkyl chain length could be more significant with the long peptides, which is ascribed to fact that one of the investigated peptides (DRVYIHPFHLLVYS) exhibited an overestimated predicted retention time by 18 min [15].

The suitability of this study in predicting the retention times of peptides was exemplified by comparing the predicted and the observed retention times of peptides recruited in this study as well as those were not used to estimate the retention coefficients. Good agreements were obtained where the correlations were r = 0.999 and 0.941, respectively. In conclusion, this study proved the ability to predict the retention times of short peptides [15].

Field et al. studied the effect of 38 different stationary phases on the elution pattern of peptides. Interestingly, various peptide analogues were recruited to have most of the potential structural modifications. Thus, oxidation, racemisation, and an increase and decrease in the charge were included in the study [16]. The main motivation of this study is the failure of the small molecule databases to correlate with the chromatographic behaviour of peptides. The study offers a selection of novel stationary phases for enhanced selectivity and peak shape [16].

Table 2. Retention time coefficients of amino acids [15].

Amino Acid	Columns							
	Ultrasphere C_8		Ultrasphere C_{18}		PLRP-S		C_{18}-Nova-Pak	
	MLR	—PLS	MLR	PLS	MLR	PLS	MLR	PLS
Trp	22.520	22.711	20.436	20.619	21.597	21.692	18.653	18.650
Phe	18.338	18.950	16.759	17.277	17.114	17.501	15.960	16.309
Leu	14.188	13.735	12.500	12.120	11.503	10.989	10.843	10.468
Met	9.924	9.668	9.638	9.440	9.551	9.180	8.578	8.313
Val	8.772	8.298	8.317	7.940	7.230	6.599	6.928	6.422
Tyr	7.483	8.005	6.291	6.791	8.241	8.789	7.028	7.370
Ile	6.119	5.700	6.838	6.620	5.938	5.221	8.134	7.763
Asp	4.880	4.200	4.233	3.610	2.255	1.651	1.207	0.769
Pro	2.428	1.928	1.852	1.494	2.871	2.302	2.569	2.195
Thr	1.157	2.563	1.873	2.985	1.527	3.036	3.110	4.215
Arg	0.740	1.186	−0.058	0.255	−1.086	−0.892	−0.169	−0.013
Glu	0.105	0.178	−0.173	−0.120	0.814	1.051	−0.019	0.081
Gly	−0.081	−0.367	0.152	−0.061	0.145	−0.227	0.013	−0.251
Ala	−1.448	−1.307	−1.534	−1.387	−0.789	−0.702	−0.778	−0.672
Lys	−3.001	−2.934	−3.488	−3.380	−4.092	−4.105	−3.103	−3.201
His	−6.703	−6.211	−7.448	−7.098	−5.269	−4.469	−3.546	−2.995
Ser	−7.358	−6.162	−6.550	−5.513	−3.966	−3.066	−2.943	−2.181
Intercept (b_o)	12.973	12.803	13.080	12.846	6.443	6.671	5.439	5.578
R^2	0.993	0.999	0.990	0.999	0.994	0.999	0.994	0.999

Mobile phases: A: 0.1 TFA in H_2O, mobile phase B: 0,1% TFA in acetonitrile. Columns: (C_8—Ultrasphere, 5 µm, 4.6 × 250 mm), (C_{18}—Ultrasphere, 5 µm, 4.6 × 250 mm), (polymeric RP—PLRP-S, 8 µm, 300 A, 4.6 × 150 mm), and (C_{18}—Nova-pak, 4 µm, 60 A, 3.9 × 150 mm). MLR: multiple linear regression; PLS: partial least square. All data are in minutes.

3.2. Mobile Phase Composition and Alkyl Chain Length

Wilce et al. investigated the effect of different alkyl chain lengths as well as organic modifiers on the retention coefficients [13]. A total of 44 group sets were assigned for the incorporated peptides in this study. In this context, C_{18}, C_8, and C_4 columns, and TFA–acetonitrile–water, TFA–1-propanol–acetonitrile–water mobile phases were investigated for their influence on the retention coefficients. The study proved a linear relationship between the retention time and the amount of organic modifier in the mobile phase. A multiple regression matrix approach was considered to calculate the retention coefficients, in addition to an alternative computational-based approach (multiple linear analysis with forcing) being considered. Comparable data were obtained almost to a certain extent. However, as the matrix approach is performed via statistical means, it is considered superior to the computational approach, and it can also provide more information about the individual amino acids [13].

Wilce et al. extracted the retention coefficients of individual amino acids from two different mobile phases using experimental data from 2106 peptides via a complex multi-linear regression analysis approach [13]. Moving from a TFA–ACN mobile phase to TFA–1-propanol–ACN, certain amino acids exhibited a significant difference in their retention coefficients (F, L, I, Y, C and A). Specifically, F, L, Y, and A interacted more strongly with the C_{18} stationary phase for TFA–ACN. The other amino acids, I and C, interacted more with the TFA–1-propanol–ACN mobile phase. This confirmed the different selectivity based on the organic modifier used in the mobile phase [13]. Specifically, differences in alkyl chain length also resulted in differences among the retention coefficients, while changing from C_{18} to C_8 stationary phases affected the amino acids F, L, W, Q, M, A and D. Specifically F, L, and W showed longer retention times for C_8 versus C_{18}. On the other hand, Q, M, and D eluted earlier with C_8- than C_{18}-based columns. Significant differences were also observed by going from C_{18} to C_4 alkyl chain length stationary phases for the amino acids

F, L, C and H. The hydrophobic residues F and L eluted at a longer retention time with C_{18} columns compared to C_4 columns, while C and H residues behaved in the opposite manner [13]. The authors also referred to other factors that could play a role in the overall separation process. An NMR study revealed that the solution conformation of the alkyl chain themselves could vary, hence affecting the separation process. Furthermore, the molecular mobility of alkyl chains bonded to the stationary phase could be increased as a result of increasing the polarity of the mobile phase. In conclusion, the variety of retention coefficients reported using different stationary phases demonstrated the direct influence of the alkyl chain length on the separation process and the way the peptide is interacting with the surface of the stationary phase. Finally, the hydrophobic character of a specific amino acid could vary depending on the organic modifier used in the mobile phase and/or the length of the alkyl chain of the column [13].

While Wilce and co-workers previously used data reported in the literature [13], later, they experimentally measured the retention coefficients for 118 peptides selected as heptamers related to the primary sequence of the myohemerythrin protein [17]. A multiple linear regression approach was again used to calculate the retention constants for the constituent amino acids. The obtained retention coefficients were then compared with the previously determined values for 2106 peptides [13]. Five chromatographic mobile phase conditions were included in this study: ACN, methanol, 2-propanol as alternate organic modifiers, TFA or potassium phosphate-based mobile phase, in addition to different silica-based stationary phases (octadecyl or phenyl). The authors also investigated the effect of the peptide chain length on the prediction capability. The five solution/column conditions evaluated were:

Condition 1: mobile phases A: 0.1% aqueous TFA, B: 0.09% TFA-50% in ACN; Zorbax C_{18};

Condition 2: mobile phases A: 0.1% aqueous TFA, B: 0.09% TFA-50% in methanol; Zorbax C_{18};

Condition 3: mobile phases A: 0.1% aqueous TFA, B: 0.09% TFA-50% in 2-propanol; Zorbax C_{18};

Condition 4: mobile phases A: 25 mM KH_2PO_4 B: 35 mM KH_2PO_4-50% in ACN; Zorbax C_{18};

Condition 5: mobile phases A: 0.1% aqueous TFA, B: 0.09% TFA-50% in ACN; Zorbax phenyl silane.

A DuPont Zorbax C_{18}, 5 μm, 4.4 × 150 mm column was considered with all four conditions, and Condition 5 was the DuPont Zorbax phenyl silane; 5 μm, 4.6 × 150 mm with Condition 1 mobile conditions; both columns have a 75 nm pore size. Hydrophobicity coefficients for the amino acids were then extracted for these chromatographic conditions (see Table 3) [17].

Similar behaviours were observed in all conditions, where the highest correlation was between mobile phases 1 and 3 (r = 1.00), and the lowest was between mobile phase 4 and 2 (r = 0.800). Some differences between this study and the previous one [13] are ascribed to the differences in the frequency of the amino acids distribution within the peptide sequence. In addition, the number of the peptides used in the data analysis were 112 in this study [17] versus 2106 previously [13]. The specific peptide sequence of both studies could also have a role in the observed discrepancies. In the previous study [13], the origin of the peptides were from enzymatic and chemical cleavage of a wide range of proteins, whereas in this study, they are from only a single protein [17]. Given that, the first peptide represents the first seven residues of myohemerythrin sequence, and the second one comprises the residues from 2 to 8. In summary, the local environment around the amino acid residues influences the extent of the interaction of the peptides with the stationary phase in chromatographically [17].

Recently, Field and co-workers investigated the factors that could influence the robustness of the method that includes DoE as well as the robustness of mobile phase switching. The study addressed the mitigation strategies for the impact of gradient variation as well as the sample load and its influence on switching between low and intermediate pH values [18]. In a very recent study, the authors have also investigated a total of 51 mobile phase with different pH values on the selectivity of peptide separation process [19]. In this study, the authors compared mobile phases with various salts, ion pairs, pH, stationary phases, and hence, it is a quite comprehensive study [19]. The study concluded that different mobile phases would allow a vast selectivity difference if applied at a correct pH. Hence, this study, along with their previous work [16], will help in the development process of RPC process [19].

Table 3. Estimated amino acid retention coefficients in five different chromatographic conditions [17].

Amino Acid	Condition 1	Condition 2	Condition 3	Condition 4	Condition 5
Alanine	1.70	1.76	1.36	2.91	1.24
Cysteine	0.49	1.66	0.33	0.19	~1.41
Aspartic acid	1.10	0.31	0.71	−1.47	1.01
Glutamic acid	0.79	0.03	0.24	0.00	0.69
Phenylalanine	8.79	10.66	7.21	9.14	9.18
Glycine	0.39	0.00	0.00	0.36	0.00
Histidine	0.62	−0.46	0.12	2.21	0.23
Isoleucine	8.35	11.76	6.97	9.80	8.29
Lysine	0.00	−1.41	−0.54	3.36	−0.24
Leucine	9.51	14.69	7.56	7.88	8.88
Methionine	2.60	1.63	1.92	3.90	2.31
Asparagine	−0.02	−2.01	−0.40	1.31	−0.26
Proline	2.79	4.21	2.08	3.50	2.09
Glutamine	−0.66	−7.52	−0.76	3.09	−1.28
Arginine	2.36	1.62	1.66	4.61	3.19
Serine	0.27	2.64	−0.41	1.66	0.10
Threonine	1.80	2.10	0.97	2.40	1.40
Valine	4.93	6.03	3.80	4.97	4.72
Tryptophan	9.75	13.30	7.47	9.99	10.54
Tyrosine	6.14	9.01	4.06	4.93	5.95

Condition 1: mobile phase A: 0.1% aqueous TFA, mobile phase B: 0.09% TFA-50% in acetonitrile. Condition 2: mobile phase A: 0.1% aqueous TFA, mobile phase B: 0.09% TFA-50% in methanol. Condition 3: mobile phase A: 0.1% aqueous TFA, mobile phase B: 0.09% TFA-50% in 2-propanol. Condition 4: mobile phase A: 25 mM KH_2PO_4, mobile phase B: 35 mM KH_2PO_4-50% in acetonitrile. Columns: DuPont Zorbax C_{18}, 5 μm, 4.4 × 150 mm was considered with all four conditions, and DuPont Zorbax phenyl silane, 5 μm, 4.6 × 150 mm was considered with Condition 1. Both columns have a 75 nm pore size. All data are in minutes.

Ion-Pairing Reagent

In 1987, Guo and co-workers studied the ion-pairing effect on the prediction of peptide retention time [20]. As the TFA is a hydrophobic ion-pairing reagent, it interacts with the basic sites of the peptide, leading to an increase in the retention time and thus affecting the prediction accuracy. The hydrophobic ion-pairing reagent is not only capable of interacting with the analyte to form the ion pair, but it is also able to result in an increased affinity of the peptide with the stationary phase, leading to an increased retention time. On the other hand, a hydrophilic ion-pairing reagent, after forming the ion pair with the peptide, is unlikely to interact with the non-polar stationary phase. Phosphoric acid (H_3PO_4) can be used as a hydrophilic ion-pairing reagent. Guo and co-workers compared the influence of the three ion-pairing regents on the peptide retention: TFA, HFBA, and H_3PO_4. The model peptide (Ac-Gly-X-X-(Leu)$_3$-(Lys)$_2$-Amide) was studied, which was also used to establish the retention time coefficient in a previous study by the same researchers (Table 1) [9]. Columns considered were SynChropak C_{18}, 6.5 μm, 4.1 × 250 mm and Aquapore C_8 10 μm, 4.6 × 220 mm, both with 300 Å pore sizes [20].

This study plotted the retention times of the peptides/number of the positive charge on each peptide for each of the three ion-pairing reagents versus the values obtained for the other two. Excellent correlations using linear least square fit were obtained: HFBA and TFA r = 0.999, H_3PO_4 and TFA r = 0.998, H_3PO_4 and HFBA r = 0.997. These results suggested that each positive charge contributes equally to the retention time shifts, in addition to the fact that only positive charges can influence the retention mechanism. Moreover, the negligible change in the retention behaviour of neutral peptides supports those findings. The authors proposed an equation to predict the retention times and examined a mixture of peptides with various numbers of positively charged groups. The results showed the largest retention time was for HFBA (the most hydrophobic), while the lowest retention time was for H_3PO_4 (the most hydrophilic). Changing the ion-pairing reagent is beneficial for separating peptides with similar hydrophobicity, but with a different number of positive charges. This approach is more advantageous than searching for different columns and specifications. The results showed accuracy between the predicted and the observed retention times, though sometimes discrepancies can arise due to the fact that not all residues are involved in the interaction; this is especially the case if there are two charged residues in a close proximity to each other, for example, a charged residue at the *N*-terminal of the peptide chain. Overall, this study reported an excellent tool to predict the retention times and to evaluate the effect of various ion pairs on the peptide separation process [20].

3.3. Sample Size

Wilce et al. included seven different sample sizes, where each set contained randomly selected peptides [13]. The separation was performed on a C_{18} column using TFA–ACN–water as the mobile phase. Afterwards, a multiple linear regression analysis was performed to generate the retention coefficients. The study revealed that the smaller the sample size, the more accurate the retention coefficients, in which deviation would be highly expected in case of any other peptide out of the set that was considered to generate the coefficients. Thus, the larger the sample size, the more universal the retention coefficients will be. It is worth mentioning that the correlation between the tested peptide sets decreased as the sample size increased. However, what really matters is the overall influence on the retention coefficient rather than the correlation. To explain the last point regarding the low correlation in the large sample size; assuming the nearest neighbour factor, obviously, this effect would be significantly pronounced in the large-sized sets as the possibility of having various amino acids is more likely to be high in comparison with the small sized ones. In turn, this would affect the final correlation among the sets under investigation [13].

3.4. Peptide Chain Length

Wilce et al. investigated several peptide sets, each with 100 peptides and a peptide chain length of 4–15 residues. The obtained correlation of 0.58 to 0.66 proved that in the selected peptide chain length range, there is no substantial effect associated with chain length on the retention coefficients [13]. The study considered eight peptide groups, in which the chain lengths were as follows: 2–30, 220, 2–15, 2–10, 2–8, 2–7, 2–5, and 2–4. In fact, the highest correlation was with the peptide group of 2–15 residues, which compromises an average chain length of 7.2 residues. Thus, it is not surprising to have a good correlation considering the heptamer peptide fragments that were used in this study [17]. On the other hand, a lower correlation was observed with 2–10, 2–8, 2–7, 2–5, and 2–4 amino acid residues. The low correlation with the latter groups could be circumvented by incorporating the coefficients for *N*- and C-termini. As the peptide chain reaches 19 residues, a poor correlation ($r = 0.38$) with the previous study [13] was obtained. This confirms the effect of the chain length on the predicted retention time. Excluding the peptides of more than seven residues in length resulted in an enhanced correlation of $r = 0.82$. Again, this emphasizes that other factors are influencing the retention behaviour of peptides. The conformational flexibility of the peptide has an important influence on the retention time as it directly controls the way the peptide will interact with the stationary phase [17].

The ability to predict the retention time using the scales in the previous study [13] and the current one (Table 3) [17] was exemplified by the good correlation values. A total of 118 peptides were examined and showed a good correlation ($r = 0.98$) between the predicted and observed retention times, according to the scales obtained from this study. Moreover, using the previously estimated scale, the correlation between the predicted and the observed retention times was r = 0.91. It is worth mentioning that some adjustments were needed while using the previous study to account for the differences in the column configurations. This study proved the general usefulness of using retention time constants to predict the behaviour of new peptides other than those used to establish the scale [17].

Mant et al. investigated the effect of the chain length on the retention behaviour of peptides [21]. The authors agree with the key assumption that the amino acid composition is driving the retention process, but not for long peptides. The previous rule is valid for up to 15 amino acid residues, after which, the retention time starts to become shorter and deviates from the predicted retention times. In this study, the authors investigated the elution behaviour of four peptides (5–50 residues), which in turn resulted in extending the utility of their retention time prediction for up to 50-residue peptides. In fact, other factors must be considered when it comes to peptide separation. The neighbouring groups do contribute and could even reduce the retention behaviour of the primary amino acid. In another words, the retention coefficient of certain amino acids might change in the case of having another adjacent amino acid, and the extent of this change depends on the type of the amino acid in the close proximity. Moreover, the conformation of the peptide structure also plays a significant role in the elution process. Conformation can reduce the overall hydrophobicity in comparison to a random coiled structure, leading to the retention time being shorter, which is mainly ascribed to some amino acid surface residues being masked and not in a direct contact with the stationary phase. The preferred binding domain (will be discussed later, Section 3.9) also has a clear influence on the separation process [21].

These peptides were designed to have similar chain length but with different hydrophobic constituents. The study considered the chromatographic conditions and the retention coefficients from the Guo et al. study [9]. A correlation was obtained in this study with the penta- and decapeptides, confirming the validity of the Guo et al. model. However, this behaviour was only true for up to 10 amino acid peptides with considerable hydrophobicity, and the model does not hold true when moving to 50 amino acid residues. Some decapeptides had also deviated from Guo's prediction model, and this was in the case of highly hydrophobic peptides. Thus, the study showed that the higher the hydrophobic character, the more likely was deviations from Guo's model. The non-linear relationship between the predicted retention time and the observed one was confirmed in this study using three columns with different alkyl chains: SynChropak C_4, 6.5 μm, 4.1 × 250 mm, Aquapore C_8 7 μm, 4.6 × 220 mm, and SynChropak C_{18}, 6.5 μm, 4.6 × 250 mm, all with a 300 Å pore size. The phenomenon was confirmed by plotting the observed retention time versus the number of amino acid residues (N) or versus ln N, and in both cases, a linear relationship was attained with various slopes depending on the hydrophobicity of each peptide under investigation [21]. The core problem that arose in this work is even if linearization is achievable with respect to the chain length, the hydrophobicity of the various resides will cause correlation divergence, depending on the hydrophobicity extent. A correlation between the discrepancy between the predicted and observed retention times with the chain length and the hydrophobicity was drawn using the linear least-squares fitting and showed a high correlation of almost r = 1.00 for the C_{18} column and r = 0.99 for the other C_4 and C_8 columns. These data reemphasised the importance of considering both the chain length as well as hydrophobicity. Excluding the latter from the final equation resulted in a non-linear relationship. With respect to the stationary phase, the alkyl chain length of the stationary phase has almost no influence on the separation process of the peptide molecules, which was previously noted by Meek and Rossetti [10] and Guo et al. [9].

Based on the above findings, a modification to Guo's predicting equation was proposed by incorporating a correction factor to account for the chain length as well as the hydrophobicity, hence enhancing the prediction capability of the model:

$$\tau = \sum \mathrm{R_c} + t_0 + t_s - \left(\mathrm{m}\sum \mathrm{R_c} \ln \mathrm{N} + \mathrm{b}\right) \quad (5)$$

where the first part is the same as in Equation (1), and the second part is a correction based on the sum of the retention coefficients for all amino acid residues and the termini ($\sum R_c$), the number of times that amino acid repeats on the peptide sequence (ln N), the slope (m), and the intercept (b) of a linear model.

Using the above equation, the prediction accuracy was enhanced with a difference between the predicted and observed retention time of not more than 1.9 min on average, with a high correlation as well (r = 0.99) [21]. It is worth highlighting that using molecular weight instead of the chain length with hydrophobicity did not exhibit a high correlation [21].

In 1989, Mant et al. extended their findings to large proteins of up to 300 amino acids [22]. The authors examined 23 proteins with a known sequence using RP-HPLC and employed columns with different hydrophobicities and ligand densities. They concluded that their model from their previous study [21] is also valid for large proteins; however, an understanding of the three-dimensional structures of proteins upon interacting with the stationary phase is important for a better accuracy.

Chabanet and Yvon predicted of the retention time of the peptides based on the relative hydrophobicity contribution of each amino acid [23]. However, as this prediction may overestimate the retention time of longer peptides of 15 residues and more, they proposed considering the contribution of each amino acid as a decreasing function of the peptide length. The study used 104 peptides with a non-linear multiple regression analysis. The main assumption in this study is that the amino acid residues in large peptides may be less accessible to adsorption on the stationary phase, so some amino acid residues are analysed as being "hidden", which means their contribution is less [23]. Mobile phase A was 0.11% aqueous TFA and mobile phase B was 0.1% TFA in acetonitrile, with a Waters μBondapak C_{18}, 10 μm, 4.6 × 250 mm column. In this study, the retention time was expressed as the percentage of acetonitrile at the elution point, which is calculated by multiplying the peak retention time (subtracted from the gradient elapsed time) by the percentage of acetonitrile in the gradient program.

For short peptides, the following linear model was adopted for predicting peptides retention times:

$$t_R = \sum_{j=1}^{19} n_{i,j} a_j + b_0 + \varepsilon_i \quad (6)$$

where $n_{i,j}$ is the number of times amino acid j in the peptide i, a_j is the retention coefficient of the amino acid j, b_0 is the retention coefficient for both *N*- and *C*-termini, and ε_i is the independent error.

The authors tried to simulate the decrease in the retention time as a result of chain length. Thus, they suggested a new model that considers the contribution of each amino acid residue (A_i) to the retention time as a decreasing function of peptide length (l_i). It was assumed that the slope equals zero when $l_i = 0$.

$$t_R = \sum_{j=1}^{19} n_{i,j} A_{i,j} l_i + b_0 + \varepsilon_i \quad (7)$$

Here, the contribution of each amino acid ($A_{i,j}$) is calculated using $A_{i,j}(l) = (a_j - l_j)e^{-b_{j,l}^2} + l_j$ and $l_j = a_j/k_j$.

In small peptides, the contribution of each residue (A_j) is very close to the retention coefficient of each residue (a_j), while in the long peptides, this contribution (l_j) is proportional to aj ($I_j = a_j/k_j$), and b_j represents the curve's slope. Based on this model, they considered two scenarios or sub-models: (i) considering the chain length will have an effect irrelative to the amino acid composition, and thus, k_j and b_j would be similar for all residues; and (ii) taking each amino acid residue into consideration when applying the decreasing function. To make the evaluation process easier, and to decrease the number of parameters that need to be estimated, the authors have classified the amino acids under various groups: non-polar (Gly, Ala, Val, Met, Ile, Leu, Phe, Trp), polar (Asp, Asn, Thr,

Ser, Glu, Gln, Pro, Tyr, His), and charged residues (Lys and Arg). Then, the same k_j was assigned for all residues of the same group. Consequently, the residues of the same group would have an identical accessible surface area, hence the same decreasing effect (b_j). The following equations would be generated:

Model 1—all residues:

$$A_j(l) = (a_j - l_j)e^{-bl^2} + l_j\,; l_j = a_j/k \tag{8}$$

Model 2—non polar residues:

$$A_j(l) = (a_j - l_j)e^{-b_1 l^2} + l_j\,; l_j = a_j/k_1 \tag{9}$$

Model 3—polar residues:

$$A_j(l) = (a_j - l_j)e^{-b_2 l^2} + l_j\,; l_j = a_j/k_2 \tag{10}$$

Model 4—charged residues:

$$A_j(l) = (a_j - l_j)e^{-b_3 l^2} + l_j\,; l_j = a_j/k_3 \tag{11}$$

A total of 104 peptides with various chain lengths were investigated, and the retention time prediction was estimated using the forementioned models [23].

Based on Guo et al., Equation (1) [9], the decrease in the retention time as a function of chain length was simulated in a way that makes it applicable even for short peptides [23]. Using peptides of different lengths helped in having accurate data for each amino acid. As discussed earlier, some long chains could result in decreasing the accessibility to certain residues, leading to an unexpected elution pattern (earlier than expected). Thus, the retention time coefficients for each amino acid were determined using the linear model and utilizing short peptides; 67 (decapeptides) or 55 (heptapeptides). The correlation was higher in the case of heptapeptides. This work confirms the effect of the chain length on the retention time, which also reaffirms the findings of Mant et al. [21]. For long peptides, the new Models 1 and 2, which considered the chain length effects, were used. It should be noted that using Model 2 for some cases, the authors had to pre-set some values in advance to overcome some difficulties in obtaining all the required parameters. Specifically, they set the parameters for the charged amino acid, as they were only two residues (Arg and Lys), in addition, their retention coefficients were small [23]. These authors also determined the retention times for 19 amino acids using the small peptides (seven residues and fewer) and applying the linear model (Table 4) [23].

The estimated retention coefficients using either Model 1 or 2 correlated well with those obtained from the linear model. Where the difference with respect to the linear model were observed, these were less in the case of Model 2 than Model 1. The correlation between the predicted and the observed retention times in both Models 1 and 2 were $r = 0.98$ and 0.99, respectively [23].

Using Model 1, the retention times were underestimated for peptides in the range of 4 to 10 amino acids and for those of more than 20 amino acids. Furthermore, they were overestimated for the very short peptides of less than 3 amino acids as well as those in the range of 10 to 20 amino acids. Overall, the results obtained from Model 2 were more satisfactory than those from Model 1. The authors pointed out that having only three groups to classify the amino acids with may not be sufficient; in addition, the distribution of the amino acids might not be performed accurately [23]. The effect of polar residues on the peptide retention time decreased dramatically as the chain length increased. For example, the effect of polar residues is more pronounced in dipeptides than pentapeptides. Thus, the predicted retention time as a result of polar residues in peptides up to seven residues is usually lower than observed. As a result, peptides that were predicted based on the linear model reported underestimated retention times, unless there are more than three residues [23]. The authors proposed that this phenomenon is due to the development of secondary structure in solution, and this topic would be discussed later in this review [24]. The estimated retention constants were compared with other studies of similar chromatographic conditions. Satisfactory correlations were obtained, for example, comparing the data from this study with that of Sasagawa et al. [25], giving a correlation of r = 0.93 and 0.94 for Model 1 and 2, respectively.

Table 4. Retention time coefficients for amino acids and termini using linear and modified models [23].

Amino Acid	Retention Coefficient		
	Short Peptides—Linear Model	**All Peptides—Linear Model**	**All Peptides—MODIFIED Model**
Trp	10.24 (11)	9.8 (25)	10.64 (25)
Phe	8.81 (22)	8.15 (64)	8.65 (64)
Leu	6.91 (27)	5.93 (134)	6.51 (134)
Tyr	4.45 (16)	4.10 (71)	6.16 (71)
Ile	6.16 (11)	5.50 (63)	5.90 (63)
Met	5.15 (11)	5.13 (35)	5.54 (35)
Pro	2.39 (21)	2.26 (178)	3.58 (178)
Val	2.55 (9)	2.03 (91)	2.56 (91)
Ser	0.58 (5)	0.64 (85)	1.14 (85)
Gln	−0.41 (8)	0.30 (102)	1.11 (102)
Arg	0.74 (9)	0.84 (30)	1.06 (30)
Glu	0.24 (12)	0.56 (94)	1.03 (94)
Asn	−1.10 (2)	1.02 (43)	1.03 (43)
Thr	0.73 (3)	0.31 (55)	0.65 (55)
Gly	−0.05 (15)	0.12 (63)	0.50 (63)
Asp	1.30 (2)	−0.40 (32)	0.31 (32)
Lys	−1.35 (18)	−0.55 (70)	−0.18 (70)
Ala	−0.39 (7)	−0.61 (57)	−0.27 (57)
His	−0.96 (6)	−1.35 (31)	−1.24 (31)
α-amino + α-COOH	−4.66 (55)	−3.45 (104)	−5.25 (104)

Mobile phase A: 0.11% TFA in H_2O; B: 0.1% TFA in ACN. Column: Waters μBondapak C_{18}, 10 μm, 4.6 × 250 mm. Numbers in brackets are the number of amino acids in the sequence. All data are in minutes.

For Model 1, the effect of chain length on the predicted retention times was related to the nature of each amino acid. Hence, it is expected that some conformational constraints would mask some interactions between certain residues and the stationary phase, thus decreasing the lability of certain residues to adsorb on the stationary phase [23].

To assess the accuracy of the developed models the authors challenged them using 47 new peptides, which were not used in the original retention coefficient estimation work. The same chromatographic conditions used for estimating the retention coefficients were adopted. The selected peptides comprised chain lengths from 2 to 58 amino acids. A satisfactory correlation of r = 0.97 was obtained. Some peptides with a negative predicted retention coefficient showed a zero actual retention time [23]. The authors also applied their models to other group's work, in which the chromatographic conditions are different. A quite good correlation of r = 0.93 was obtained. They predicted the retention time for 71 peptides out of 100 peptides tested by Sasagawa and co-workers [25]. The other 29 peptides were not checked as they have some amino acid residues whose their retention coefficients were not determined in this study [23].

The discrepancies did exist, which could be attributed to effects such as neighbouring amino acid effects or certain sequence-specific conformations. However, the model did show a good overall predicting capability [23].

3.5. Alpha-Amino Group

In 1993, Hodges and co-workers studied the effect of the α-amino group on the retention time of peptides, in addition, they determined the pKa for the α-amino groups in 19 peptides [26]. They considered two peptide analogues, acetylated and non-acetylated *N*-terminal, where the latter represents the α-amino group. The idea of the study was to compare the retention times of the two analogues with that of a Gly-containing analogue. Studies of simple organic molecules presented that the effect of the substituents next to the ionizable terminal groups (end groups) could affect the dissociation constants of those groups. Nevertheless, this study confirmed the difference in hydrophobicity between the presence or absence of the α-amino groups at the *N*-terminus. This effect was also proved to be sequence dependent. Increasing the pH led an increase in the retention time in the case of non-acetylated analogues as a result of amino group deprotonation, which resulted in a neutral charge and hence more retention. In this study, the 20 proteinogenic amino acids were investigated using the following analogues [26]:

1st analogue: Ac-X-Leu-Gly-Ala-Lys-Gly-Ala-Gly-Val-Gly-Amide.

2nd analogue: H-X-Leu-Gly-Ala-Lys-Gly-Ala-Gly-Val-Gly-Amide.

Core peptide: Ac-Leu-Gly-Ala-Lys-Gly-Ala-Gly-Val-Gly-Amide.

The rationale for the core peptide compositions investigated is to represent an ideal model peptide which lacks structural factors which are known to contribute to deviations from the expected retention behaviour, for example, the presence of an amphipathic helix with preferred site of binding [24]. A decapeptide chain length was chosen because it is the most common average length for peptides following proteolytic digest, and also to avoid any effects from the chain length as previously reported by Mant et al. [21]. The hydrophobicity of the amino acids that compose the core peptide sequence would cause the peptide to be eluted at around 15–40% of acetonitrile, where the optimum resolution could be achieved. The presence of the Lys residue is to confer good solubility to the peptide [26]. The two model peptides (acetylated versus non-acetylated) with the same amino acid substitution were separated using HPLC over a pH range from 2 to 6.8. pH had little effect on the majority of the acetylated analogues. Anomalous behaviour was observed in the case of Leu residue in which, as the pH increased from 2 to 6, a decrease in the retention time of the acetylated analogue was observed, whereas an increase in the retention time was observed with the non-acetylated analogue which is ascribed to the deprotonation of the *N*-terminal. The increase in the pH also led to an inversion of the elution order of the Leu and Ile residues [26].

Interestingly, the deprotonation explanation was not the case with all analogues. Furthermore, the α-amino group in the non-acetylated analogue not only influenced the hydrophobicity, but it was also sequence dependent. Five peptide analogues were investigated (acetylated and non-acetylated) on a C_{18}/C_2, 5 μm, 4 × 250 mm column at pH 2. The effect was mainly shifts to the shorter retention time of the non-acetylated analogues as a result of the positive charge that was developed on the *N*-terminus. Nevertheless, it cannot be concluded that the effect of α-amino group is to decrease the retention time as despite the decreased retention times, the elution pattern had also changed. For example, some acetylated analogues were baseline resolved; however, this was not the case with the non-acetylated analogues, in which a coeluted elution profile was obtained instead. Some non-acetylated analogues were well resolved at pH 2 but coeluted at pH 6.8, and the opposite is true for other analogues as well. So, each analogue has shown a distinct elution profile, and this suggests that besides affecting the hydrophobicity of the non-acetylated analogues, the α-amino group is also sequence dependent with respect to the amino acid residue at the *N*-terminal [26].

In order to quantify the effect of α-amino group, pairs of acetylated and non-acetylated peptides were chromatographed using polystyrene-based columns (PLRP-S, 5 μm, 4.6 × 250 mm) to allow the use of high pH elutions (where silica-based columns have silanol activity concerns and they could be negated at high pHs) and determining the pKa values of the α-amino group and the basic sidechains simultaneously [26]. The researchers proposed several equations to quantify factors that are believed to affect the elution process of peptides:

To determine the effect of α-amino group:

$$a = t_{R,\,H-Gly} - t_{R,Ac-Gly} \quad (12)$$

To determine the hydrophobicity in the absence of α-amino group.

$$H = t_{R\ Ac-X} - t_{R\ Ac-Gly} \quad (13)$$

$t_{R\,H-X} - t_{R\ Ac-Gly}\ a + h = t_{R,\,H-X} - t_{R\ Ac-Gly}$ represents a combination of both effects (α-amino group (a) and the hydrophobicity of the sidechain of the *N*-terminal in the presence of α-amino group (h)), then:

$$h = \left[t_{R,\,H-X} - t_{R,Ac-Gly}\right] - a \quad (14)$$

To determine the effect of α-amino group (a) and the hydrophobicity of the sidechain of *N*-terminal in the presence of α-amino group

$$a = t_{R,\,H-X} - t_{R,Ac-Gly} - t_{R,\,H-Gly} + t_{R,Ac-Gly} = t_{R,\,H-X} - t_{R,H-Gly} \quad (15)$$

To determine the effect of the α-amino group on the hydrophobicity

$$s = h - H = [t_{R,\,H-X} - t_{R,Ac-X}] - a \quad (16)$$

It is to be noted that the obtained results were comparable to those obtained by Guo et al. [9]. If the α-amino group has no effect on the hydrophobicity of the peptide, this would result in zero value of the *s*, which was not the case. Plus, the difference in the retention time between the acetylated and non-acetylated analogues was not the same among various analogues [26].

Interestingly, plotting the difference in the retention time between the acetylated and non-acetylated analogues versus the pH over the range from 2–9, helped determine the pKa of the α-amino group. Polystyrene-based columns are the best choice for a high-pH mobile phase, whereas silica-based columns might decompose. The authors determined the pKa of the α-amino group by plotting the difference in the retention time between the acetylated and non-acetylated analogues [$t_{R\,H-X} - t_{R\ Ac-X}$] versus the pH from 2 to 9. The obtained pKa values were higher than in the case of free amino acids, and this is in line with the fact the acidic amino acids have higher pKa in proteins than in free state (Table 5). Furthermore, it was reported that the hydrophobic environment could also affect the dissociation of the ionizable groups [27]. Increasing the percentage of the organic solvent has led to a decrease in the dissociation of the α-carboxyl group of Gly (increase in the pKa from 2.35 to 3.96), and increase dissociation of α-amino group (decrease in the pKa from 9.78 to 7.42) [27].

Interestingly, the plot of each amino acid resembled a titration curve, and for each amino acid it was different and unique. This suggests that the deprotonation of the α-amino group varies based on the substituted amino acid at the *N*-terminus [26].

The authors were also able to establish the pKa of the ionizable sidechain (with the absence of α-amino group effect) by plotting the difference in the retention time between the acetylated analogues and acetylated core peptide over the same pH range of 2–9. The study showed that the pKa of the sidechain of the acidic amino acid is significantly higher whilst part of a protein than in its free state (Table 6). The opposite was observed with the amino acids of a basic sidechain. Nevertheless, the values obtained in this study are similar to those in protein [26].

Table 5. pKa values for various α-amino group of peptide analogues, 19 *N*-terminus amino acids [26].

Peptide Analogue	This Study	In Free State [28]
Pro	7.1	10.6
Gly	7.0	9.8
Asp	6.8	9.6
Ala	6.8	9.7
Glu	6.6	9.7
Val	6.5	9.6
Ile	6.4	9.7
Gln	6.4	9.1
Trp	6.3	9.4
Ser	6.3	9.2
Thr	6.3	9.1
Leu	6.3	9.6
His	6.3	9.2
Lys	6.2	9.2
Asn	6.1	8.8
Arg	6.1	9.0
Tyr	6.1	9.1
Met	6.1	9.2
Phe	6.0	9.2
Cys	-	10.5

Mobile phases: pH 2: A: 20 mM H_3PO_4 in H_2O, containing 2% ACN; B: 20 mM H_3PO_4 in ACN-H_2O (1:1); pH 4–7: A: 20 mM triethylammonium phosphate (TEAP) in H_2O, containing 2% ACN; B: 20 mM TEAP in ACN-H_2O (1:1); pH 7–9: A: 10 mM $(NH_4)_2HPO_4$ in H_2O, containing 2% ACN; B: 10 mM $(NH_4)_2HPO_4$ in ACN-H_2O (1:1); both eluents contain 100 mM sodium perchlorate. Column: polystyrene-based (PLRP-S, 5 μm, 4.6 × 250 mm).

Table 6. pKa values for sidechain of peptide analogues (absence of α-amino group effect) [26].

Peptide Analogue	This Study	In Free State [28]	In Protein
Asp	7.5	3.65	6.7/6.8/10 *
Glu	7.4	4.25	6.0/6.5/8.0–8.5 *
Arg	7.3	12.48	11.6–12.6 *
Lys	7.4	10.79	9.11/5.9 *
His	5.8	6.0	5.0–8.0 *

* Different values depending on the protein. Mobile phases: pH 2: A: 20 mM H_3PO_4 in H_2O, containing 2% ACN; B: 20 mM H_3PO_4 in ACN-H_2O (1:1); pH 4–7: A: 20 mM triethylammonium phosphate (TEAP) in H_2O, containing 2% ACN; B: 20 mM TEAP in ACN-H_2O (1:1); pH 7–9: A: 10 mM $(NH_4)_2HPO_4$ in H_2O, containing 2% CAN; B: 10 mM $(NH_4)_2HPO_4$ in ACN-H_2O (1:1); both eluents contain 100 mM sodium perchlorate. Column: polystyrene-based (PLRP-S, 5 μm, 4.6 × 250 mm).

These findings suggest that the stationary phase could mimic the hydrophobic environment as found in the proteins, provided that these pKa values were comparable to those reported for proteins [26].

3.6. Sidechain Amino Acid in the Absence of Nearest Neighbour Effect

Kovacs et al. [29] have studied the sidechain hydrophobicity of 20 proteinogenic amino acids as well as norleucine, norvaline, and ornithine as non-proteinogenic amino acids in the absence of the nearest neighbour group effect. The following decapeptide was considered: Ac−X−Gly−Ala−Lys−Gly−Ala−Gly−Val−Gly−Leu−amide. It is noticeable that X position has Gly as a nearest neighbour group. Gly has only H as sidechain, which is known not to have any steric effect. This ensures the unrestricted rotation around the peptide (amide) bond from either side between the substitution site and the residue next to it. To demonstrate the free rotation, all the 23 amino acids were substituted in their L and D isomers, while the adjacent amino acid is the Gly. Having an unrestricted rotation means that both diastereomers should elute at the same retention time. Taking into account that the overall composition of the two peptides is identical, whether the adjacent amino acid is Gly or Leu (Table 7) [29].

Table 7. Comparison between the retention times of (L and D)-peptides, absence of the nearest group effect (Ac–XGAKGAGVGL–amide) [29].

Amino Acid Substitution	Retention Time		Difference
	L-Isomer	D-Isomer	
Trp	67.5	67.8	0.0
Phe	64.3	64.3	0.0
norLeu	59.8	59.8	0.0
Leu	58.6	58.6	0.0
Ile	56.5	56.5	0.0
Met	51.3	51.3	0.0
norVal	50.6	50.6	0.0
Tyr	50.6	50.6	0.0
Val	49.0	49.0	0.0
Pro	44.6	44.6	0.0
Cys	43.3	43.3	0.0
Ala	38.8	38.8	0.0
Glu	38.8	38.8	0.0
Thr	38.0	38.0	0.0
Asp	37.4	37.9	0.5
Gln	35.7	35.7	0.0
Ser	35.2	35.2	0.0
Asn	35.2	35.2	0.0
Gly	35.2	35.2	0.0
Arg	30.2	30.2	0.0
His	28.2	28.2	0.0
Lys	28.2	28.2	0.0
Orn	27.6	27.6	0.0

Mobile phase: A: 20 mM H_3PO_4 in H_2O; B: 20 mM H_3PO_4 in ACN. Column: Kromasil C_{18}, 5 μm, 2.1 × 150 mm. All data are in minutes.

When the adjacent residue is Leu, well-resolved diastereomers were obtained at pH 2. On the other hand, when the Gly was the adjacent residue, these diastereomers were inseparable. However, the gradient was very shallow (0.25% ACN). The exception was for two amino acids out of 23 (Asp and Trp). Using a more standard gradient condition (1% ACN), even these two pairs were not separable anymore. The *N*-terminal was acetylated, and the *C*-terminal was amidated, to eliminate any potential effect from the charges that might develop during various pH environments [29]. Having proven the above concept, the L- amino acid peptides were investigated using six mobile phases with various pH values: 2, 5 and 7. Different ion-pairing reagents were also considered as well as the presence and the absence of different salts: 20 mM H_3PO_4 or 20 mM TFA at pH 2; 10 mM PO_4 buffer at pH 5; and 10 mM PO_4 buffer at pH 7, containing no salt, containing 50 mM NaCl, or containing 50 mM $NaClO_4$ (Table 8). The following columns were used: for mobile phase of pH 2: Kromasil C_{18}, 5 μm, 2.1 × 150 mm; for that of pH 5 and 7: Zorbax Eclipse XDB-C_8, 5 μm, 2.1 × 150 mm [29].

The results confirmed that the hydrophobicity is independent of the mobile phase's pH, buffer conditions, and alkyl chain length in the stationary phase for 17 amino acid residues. Conversely, for the potentially charged residues (His, Asp, Glu, Arg, Lys, Orn), the pH was proven to play a crucial role in the hydrophobicity coefficients of those residues (Table 8) [29].

Peptides are generally highly retained in case of the hydrophobic TFA than the hydrophilic H_3PO_4. At pH 5 and 7 without added salt, the retention times of all peptides have decreased except for Orn, Lys, His, and Arg. This effect is ascribed to the deprotonation of their sidechain, resulting in a neutral charge and enhancing their hydrophobic character and hence their retention. It should be noted that the pKa of the highly basic residues are decreased in the hydrophobic environment such as protein or RP stationary phase. Adding 0.05 M of $NClO_4$ to the mobile phase of pH 7 increased the retention time of the peptides more than when the NaCl was added [29].

Table 8. Peptide retention times in various mobile phases (absence of the nearest group effect) [29].

Amino Acid Substitution	pH 2		pH 5	pH 7 (10 mM PO_4)		
	20 mM H_3PO_4	20 mM TFA	10 mM PO_4	No Salt	+50 mM NaCl	+50 mM $NaClO_4$
Trp	67.5	73.5	73.0	72.0	71.3	79.9
Tyr	64.3	70.2	69.9	69.0	68.4	77.0
Glu	38.8	43.9	39.3	38.2	37.9	39.1
Asp	37.4	42.7	38.8	38.2	37.5	38.6
Gln	35.7	41.7	40.4	39.6	39.9	46.2
Asn	35.2	40.5	39.8	39.6	39.3	45.4
Orn	27.6	40.5	33.0	35.5	36.3	48.3

Mobile phases in H_2O as mobile phase A or in ACN as B; 20 mM H_3PO_4 or 20 mM TFA at pH 2; 10 mM PO_4 buffer at pH 5; and 10 mM PO_4 buffer at pH 7, containing no salt, containing 50 mM NaCl, or containing 50 mM $NaClO_4$. Columns: for pH 2: Kromasil C_{18}, 5 µm, 2.1 × 150 mm; for pH 5 and 7: Zorbax Eclipse XDB-C_8, 5 µm, 2.1 × 150 mm. All data are in minutes.

Comparing the retention times of the 19 amino acids, except the positively charged ones, in TFA versus in H_3PO_4, there was a high correlation of r = 0.999. This reflects that the relative hydrophobicity is not affected by the type of ion pair, considering that TFA can make ions pair with the Lys residue, leading to a longer retention time. As for the positively charged amino acids Orn, Lys, His, and Arg, their hydrophobicity indexes have increased in the TFA-containing mobile phase in comparison with the H_3PO_4 one. Comparing the retention time of the 17 amino acids with neutral sidechains in mobile phase of pH 2 and 7 with no added salt, a good correlation of r = 0.999 was obtained, reflecting the independence of the hydrophobicity on the pH. On the other hand, the hydrophobicity indexes of the positively charged amino acids have increased as a result of their deprotonation, and their positive charge has diminished. For the negatively charged amino acids Asp and Glu, at pH 7, their hydrophobicities have decreased due to the deprotonation effect and developing of the negative charge, the same happened at pH 5. The elution profile for the amino acids with neutral sidechain was similar either in pH 2 (H_3PO_4) or pH 7 with no added salt. The major changes in hydrophobicity were noticed with the charged amino acids, in which higher hydrophobicity was observed with peptides having residues Orn, His, Lys, and Arg as the pH was raised from 2 to 7. On the contrary, a decrease was observed with the Asp- and Glu-containing peptides [29].

Comparing pH 7 with 5 in the absence of added salt, a good correlation of r = 0.999 was obtained, with the exception of Orn, His, Lys, and Arg. The non-linear relationship of the charged residues could be ascribed to the deprotonation effect of these residues, leading to a longer retention time. As for Asp and Glu, they were both deprotonated and held a negative charge at both investigated pHs [29].

The effectiveness of Cl^- vs. ClO_4^- was evaluated at pH 7. The comparison between the mobile phases with and without Cl^-, showed a good correlation (r = 0.998) for all 23 amino acids. Also included were the positively charged residues, but with little effect so far. This reflects the ineffectiveness of the Cl^- ion as an ion-pairing reagent. On the other hand, ClO_4^- affected the positively (except His) and negatively charged residues, reflecting the effectiveness of this ion-pairing reagent. All the neutral residues (in addition to the charged His) showed a high correlation of r = 0.999. The positively charged residues Orn, Lys, and Arg were eluted at a longer retention time due to the ion-pairing effect. As for His, it is deprotonated at pH 7, so it became neutral and thus behaved such that it showed no response to the addition of $NaClO_4$. The poor correlation with Asp and Glu could be ascribed to the decreased ion-paring capability of ClO_4^- with the positively charged residues, as indicated by the net charge on the peptide of zero [29].

The retention time difference between the X-substituted peptides (22 amino acids other than Gly) and the Gly-substituted peptide represents the hydrophobicity of the sidechain in the absence of the nearest group effect (Table 9) [29].

Table 9. Hydrophobicity coefficients in five different mobile phases (absence of the nearest group effect) [29].

Amino Acid Substitution	pH 2 20 mM H_3PO_4 Δt_{RGly}	pH 2 20 mM TFA Δt_{RGly}	pH 5 10 mM PO_4 Δt_{RGly}	pH 7 (10 mM PO_4) No Salt Δt_{RGly}	pH 7 (10 mM PO_4) +50 mM NaCl Δt_{RGly}	pH 7 (10 mM PO_4) +50 mM $NaClO_4$ Δt_{RGly}
Trp	32.3	32.4	33.2	32.9	33.0	33.7
Phe	29.1	29.1	30.1	29.9	30.1	30.8
norLeu	24.6	24.6	25.6	25.6	25.9	26.6
Leu	23.4	23.3	24.1	24.2	24.6	25.1
Ile	21.3	21.4	22.2	22.4	22.8	23.0
Met	16.2	15.7	16.4	16.3	17.3	16.8
norVal	15.4	15.2	15.9	16.3	16.9	16.8
Tyr	15.4	14.7	15.2	15.4	16.0	15.1
Val	13.8	13.4	14.0	14.4	15.0	14.6
Pro	9.4	9.0	9.4	9.7	10.4	9.9
Cys	8.1	7.6	7.9	8.3	9.1	8.2
Ala	3.6	2.8	3.3	3.9	4.1	3.4
Glu	3.6	2.8	−0.5	−0.9	−0.4	−7.1
Thr	2.8	2.3	2.8	3.9	4.1	2.5
Asp	2.2	1.6	−1.0	−0.9	−0.8	−7.6
Gln	0.5	0.6	0.6	0.5	1.6	0.0
Ser	0.0	0.0	0.0	0.5	1.2	−0.5
Asn	0.0	−0.6	0.0	0.5	1.0	−0.8
Gly	0.0	0.0	0.0	0.0	0.0	0.0
Arg	−5.0	0.6	−3.7	3.9	4.1	6.4
His	−7.0	0.0	−5.1	3.4	4.7	3.4
Lys	−7.0	2.8	−3.7	−1.1	−2.0	3.4
Orn	−7.6	−0.6	−6.8	−3.6	−2.0	2.1

H_2O as mobile phase A or in ACN as B; 20 mM H_3PO_4 or 20 mM TFA at pH 2; 10 mM PO_4 buffer at pH 5; and 10 mM PO_4 buffer at pH 7, containing no salt, containing 50 mM NaCl, or containing 50 mM $NaClO_4$. Columns: for mobile phase of pH 2: Kromasil C_{18}, 5 μm, 2.1 × 150 mm; for pH 5 and 7: Zorbax Eclipse XDB-C_8, 5 μm, 2.1 × 150 mm. Δt_{RGly}: retention time difference relative to Gly-peptide. All data are in minutes.

It was observed that the norVal is more hydrophobic than Pro, despite both of them having the same number of carbon atoms. This behaviour is ascribed to the fact that the cyclisation in case of Pro makes it less exposed to the stationary phase, thus meaning less retention and less hydrophobicity. Comparing the hydrophobicity between Ala, Val, Ile, norVal, norLeu, and Gly showed that the greater the distance between the added carbon and the peptide backbone will result in higher hydrophobicity. It can be noted that in case of Ala, the methyl group is at the β-carbon, whereas in case of Ile, it is at the δ-carbon, and in norLeu, it is the at ϵ-carbon. Interestingly, the addition of a methyl group at β-carbon resulted in increased hydrophobicity, like in the case of Gly to Ala or Ser to Thr. However, the increase was more pronounced in the former than the latter. This is ascribed to the fact that the OH is also attached to the β-carbon which shields the methyl group from interacting with the stationary phase, hence decreasing its hydrophobicity expression [29].

All uncharged amino acids in all mobile phases showed a good correlation of r = 0.997, where the retention time difference (with Gly) in various mobile phases was plotted. This highlights the independence of those residues on the pH, nor the ion pair composition of the mobile phase. Ionizable residues showed higher hydrophobicity at pH 2, due to the protonation of their sidechain, and hydrophobicity in TFA was higher than in the H_3PO_4 mobile phase [29]. At pH 5 and 7, the Asp and Glu are considered hydrophilic due to the deprotonation of their sidechain, leading to the development of a negative charge; it was also noticed that their hydrophilicity increases even more in the presence of ClO_4^- [29]. Plotting the retention time of peptides versus the net charge (considering the Lys that is already included in all peptides), showed that the increased hydrophobicity is ascribed to the ion-pairing effect of the ClO_4^-. On the other hand, the ineffectiveness of Cl^- was clear from the independence of the retention behaviour in this mobile phase on the net charge [29].

For charged amino acids, the pKa is important, as is their protonated or deprotonated state. In addition, the concentration of the HPO_4^{3-} is also important as it can efficiently neutralize the positively charged sidechains in the peptide [29].

3.7. *Sidechain Amino Acid That Restricts Conformation (Presence of Nearest Neighbour Effect)*

Kovacs et al. quantified the effect of the nearest group effect (i to i + 1) using RP-HPLC [30]. They noticed that no nearest group effect is observed to i (any amino acid) when i + 1 is Gly [29]. However, after replacing the Gly with a bulky and hydrophobic amino acid such as Leu, a pronounced effect started to show depending on the nature of i amino acid. They calculated the minimum energy conformations for certain peptides with L and D amino acid next to L-Leu and proved that the most energetically favourable configuration is dependent on its orientation with respect to the Leu. Consequently, this translates in the way they will be interacting with the stationary phase [30]. The authors believe that the restriction of conformational space of the protein could start from the backbone of the polypeptide itself, depending on the type of amino acids in close proximity to each other [30].

3.7.1. Rationale of Selecting the Model Peptide

The following model peptide was considered: Ac-X-Leu-Gly-Ala-Lys-Gly-Ala-Gly-Val-Gly-amide, where X is substituted by the 19 L and D- amino acids. The presence of the Gly residues suppresses any secondary structure formation. In order to eliminate the potential effect of any possible positive charge at the α-amino group or negative charge at the C-terminus, both termini were acetylated and amidated, respectively. One Lys residue was incorporated into the structure to enhance the solubility over a wide pH range. Four hydrophobic residues were included to ensure satisfactory retention, and they were distributed in a way that guaranteed no hydrophobic interactions among them and, hence, no preferred domain of interaction which could alter important interactions. To supress any effect due to chain length, a decapeptide length was considered [21,23]. Leu was considered as a bulky, hydrophobic, and neutral amino acid, thus eliminating any interactions due to charges, and also any polar interactions such as H bonding [30].

The X-amino acid is placed at the *N*-terminal to study the effect of being next to only one Leu residue; in addition, the Gly was placed at position 3, to avoid any other nearest-group interactions with the Leu from the other side, provided that Gly has no effect, as shown previously (Table 7) [29]. It is to be noted that such a neighbour effect is considered as a peptide-based approach that can be used to evaluate the protein folding as well as stability [30].

3.7.2. Nearest Neighbour Effect on the Elution of L and D Diastereomers

In the previous study (Table 7) [29], the same researchers studied the L and D amino acids at the *N*-terminal position, where the Gly adjoined. The study stated that the coelution of both diastereomers reflects a completed freedom of rotation about the peptide bond between the L-/D-amino acid and the Gly residue. Meaning therefore that the Gly has no nearest group effect. Therefore, any variation in the retention behaviour in this study would definitely be ascribed to the substituted amino acid being next to the bulky Leu residue [30]. The study considered five different mobile phases (the same as in their previous study (Table 7), except the pH 5 mobile phase) [29]; a shallow gradient (0.25%) was considered to magnify any difference between the L-/D-peptide diastereomers. The same columns as in the previous study were considered [29]. A model peptide was tested using the 20 proteinogenic amino acids under these conditions. Unlike the previous study (Table 7) [29], it is noticeable that the presence of the bulky Leu group next to the substitution site led to significant separation of L and D diastereomers (Table 10) [30].

Table 10. Retention times in mobile phases (presence of nearest neighbour effect) [30].

Amino Acid Substitution Diastereomer	pH 2				pH 7 (10 mM PO_4)					
	20 mM H_3PO_4		20 mM TFA		No Salt		+50 mM NaCl		+50mM $NaClO_4$	
	D	L	D	L	D	L	D	L	D	L
Trp	69.9	66.5	78.4	74.7	71.1	67.9	73.1	69.6	83.3	79.9
Phe	71.4	63.6	79.0	71.5	72.0	64.5	74.0	65.5	84.4	77.1
Leu	64.1	57.4	73.2	66.6	66.3	59.9	68.8	62.2	79.3	72.9
Ile	64.4	56.6	72.9	65.1	65.7	58.4	68.1	60.8	78.3	71.1
Met	55.1	49.1	63.3	57.3	57.3	51.4	59.3	53.6	68.8	63.2
Val	54.6	47.9	62.8	55.8	56.8	50.5	58.8	52.2	68.1	61.5
Tyr	46.6	47.6	54.6	55.5	49.1	50.0	50.9	51.8	58.9	60.2
Pro	45.4	42.0	54.6	51.3	48.9	45.9	51.0	47.8	59.9	57.2
Cys	45.7	41.3	53.4	48.8	48.4	44.4	52.6	48.2	58.5	54.2
Glu	37.7	36.3	44.4	43.4	36.5	35.5	38.6	37.8	40.2	40.1
Ala	38.4	35.5	46.0	42.8	41.4	38.6	45.4	42.3	51.2	48.2
Asp	34.7	34.7	41.6	41.6	36.1	37.0	37.9	38.9	37.7	38.4
Thr	35.6	34.3	42.8	41.4	38.5	37.1	40.4	39.1	47.8	46.3
Gly	33.3	33.3	40.1	40.1	36.0	36.0	37.6	37.6	45.1	45.1
Ser	31.6	33.2	38.7	40.1	34.7	36.2	36.4	37.8	43.3	44.6
Gln	32.8	33.0	39.5	39.9	35.6	35.9	39.2	39.5	44.1	44.4
Asn	31.9	32.8	38.7	39.7	35.0	35.9	38.3	39.3	43.3	44.1
Arg	25.8	27.1	40.9	42.2	30.1	31.4	35.3	36.8	48.8	50.5
His	23.9	27.0	38.3	41.6	28.5	32.8	31.9	37.3	46.1	49.3
Lys	23.5	25.5	38.4	39.8	30.6	30.0	34.4	33.4	45.5	47.6

H_2O as mobile phase A or in ACN as B; 20 mM H_3PO_4 or 20 mM TFA at pH 2; and 10 mM PO_4 buffer at pH 7, containing no salt, containing 50 mM NaCl, or containing 50 mM $NaClO_4$. Columns: for mobile phase of pH 2: Kromasil C_{18}, 5 μm, 2.1 × 150 mm; for that of pH 5 and 7: Zorbax Eclipse XDB-C_8, 5 μm, 2.1 × 150 mm. All data are in minutes.

The retention times of the peptides in various mobile phases was compared by the mean of subtraction to estimate any effect due to the mobile phase [30]. At pH 2, the negatively charged counter ion of the TFA and/or H_3PO_4 would form an ion with the positively charged amino acids. The study showed that because TFA is more hydrophobic, it enhanced the retention of the peptides more than the H_3PO_4, which is more hydrophilic. The increase was significant in those with a positive charge at pH 2 Lys, His, and Arg. It is worth noting that this higher hydrophobicity effect of TFA over H_3PO_4 was only pronounced with the charged amino acids and became almost the same in the case of neutral amino acids. However, in the selected model peptide, there would be little effect due to the presence of Lys residue, which is positively charged at pH 2 and it will interact with counter ion. Thus, little increase in the retention time was also observed, especially, when using a TFA-containing mobile phase [30].

At pH 7, the positively charged amino acids were not affected as they were deprotonated at that pH. The negatively charged amino acids Asp and Glu, became deprotonated and exhibited a negative charge. NaCl as an added salt in the mobile phase did not affect the retention, as the Cl^- is ineffective as an ion-pairing reagent. On the other hand, $NClO_4$ had a dramatic effect on increasing the retention times of all the amino acids except the Asp and Glu. The lesser effect on Glu and Asp might be ascribed to the ineffective ion-pairing capability of ClO_4^- in the presence of negatively charged residues. Almost the same enhancement that was observed earlier in the case of TFA with respect to H_3PO_4 was observed. Here, the retention times of the positively charged amino acids containing peptides His, Arg, and Lys increased in comparison to the other 17 amino acids [30].

In conclusion, the presence of the hydrophobic TFA (pH 2) and the $NClO_4^-$ (pH 7) increased the retention times of the peptides with positively charged amino acids, compared to the other peptides with other residues. These observations were ascribed to the higher hydrophobicity of TFA over H_3PO_4 and the higher pairing efficacy of the $NClO_4^-$ over the Cl^-, respectively. Away from the charged amino acids, the effect of the added salt, decreased with decreasing the hydrophobicity of the substituted residue [30].

3.7.3. Evaluating the Nearest Neighbour Effect

The difference in the retention times between the L-/D- diastereomers was almost negligible, which means that the nearest group effect is independent of the mobile phase composition or pH. Some differences were observed in the case of Asp, Glu, and His among the various mobile phases (Table 11). In this study, the difference in the retention times is ascribed to the different hydrophobicity of the individual amino acid as well as its final configuration when it is adjacent to the bulky Leu [30].

Table 11. Retention times differences in mobile phases (presence of nearest neighbour effect) [30].

Amino Acid Substitution	pH 2 20 mM H_3PO_4 $\Delta t_{R\ D\text{-}L}$	 20 mM TFA $\Delta t_{R\ D\text{-}L}$	pH 7 (10 mM PO_4) No Salt $\Delta t_{R\ D\text{-}L}$	 +50 mM NaCl $\Delta t_{R\ D\text{-}L}$	 +50 mM $NaClO_4$ $\Delta t_{R\ D\text{-}L}$
Trp	3.4	3.7	3.2	3.5	3.4
Phe	7.8	7.5	7.5	8.5	7.3
Leu	6.7	6.6	6.4	6.6	6.4
Ile	7.8	7.8	7.3	7.3	7.2
Met	6.0	6.0	5.9	5.7	5.6
Val	6.7	7.0	6.3	6.6	6.6
Tyr	−1.0	−0.9	−0.9	−0.9	−1.3
Pro	3.4	3.3	3.0	3.2	2.7
Cys	4.4	4.6	4.0	4.4	4.3
Glu	1.4	1.0	1.0	0.8	0.1
Ala	2.9	3.2	2.8	3.1	3.0
Asp	0.0	0.0	−0.9	−1.0	−0.7
Thr	1.3	1.4	1.4	1.3	1.5
Gly	0.0	0.0	0.0	0.0	0.0
Ser	−1.6	−1.4	−1.5	−1.4	−1.3
Gln	−0.2	−0.4	−0.3	−0.3	−0.3
Asn	−0.9	−1.0	−0.9	−1.0	−0.8
Arg	−1.3	−1.3	−1.3	−1.5	−1.7
His	−3.1	−3.3	−4.3	−5.4	−3.2
Lys	−2.0	−1.4	0.6	1.0	−2.1

H_2O as mobile phase A or in ACN as B; 20 mM H_3PO_4 or 20 mM TFA at pH 2; and 10 mM PO_4 buffer at pH 7, containing no salt, containing 50 mM NaCl, or containing 50 mM $NaClO_4$. Columns: for mobile phase of pH 2: Kromasil C_{18}, 5 µm, 2.1 × 150 mm; for that of pH 5 and 7: Zorbax Eclipse XDB-C_8, 5 µm, 2.1 × 150 mm. $\Delta t_{R\ D\text{-}L}$: retention time difference in D- relative to L-diastereomer. All data are in minutes.

Using computational analysis, the authors speculated that the most favourable energetic conformation is dependent on the orientation of the substituted amino acid with respect to the adjacent Leu residue. For example, substituting a large hydrophobic residue D-Leu next to the L-Leu, showed a difference in the retention time of 6.7 min. This high hydrophobicity reflects that the most favourable energetic configuration of the D-Leu is that when it is interacting with the hydrophobic stationary phase more than the L-Leu. Indeed, it seems that in the D-configuration, both bulky groups are being directed to the one side of the molecule and hence facilitating their interaction with the stationary face. As for the L-Leu, the opposite scenario is more likely to drive the separation process [30]. The same explanation could also be adopted for the polar and charged residues. For example, the most favourable configuration of D-Lys was when adjacent to L-Leu and would bring the positive charge in close proximity to the stationary phase more than the L-Lys, which means that the D-Lys diastereomer will be eluted faster than the L-Lys diastereomer (represented by negative value of $t_{R\ D\text{-}L}$) (Table 11). In this case, the L-Lys is more hydrophobic than the D-Lys [30].

Interestingly, the hydrophilic character for D-Ser was predominant in all mobile phases ($\Delta t_{R\ D\text{-}L}$ = −1.6, −1.4, −1.5, −1.4, and −1.3) (Table 11), which could be ascribed to having the polar hydroxyl group directed towards the stationary phase in the D-Ser diastereomer than in the L-Ser. On the other hand, as for the Thr residue, the hydrophobic character was predominant in all mobile phases in the D-Thr diastereomer than the L-Thr ($\Delta t_{R\ D\text{-}L}$ = 1.3, 1.4, 1.4, 1.3, and 1.5) (Table 11). The reason could be the masking effect of the hydroxyl group on β-carbon by the methyl group on the same carbon, which led to a favoured hydrophobic interaction with the stationary phase. In addition, in the L-Ser, the methyl group of the acetyl at the *N*-terminal is at the same side of the sidechain at the adjacent L-Leu, leading to more preferred interaction with the stationary phase. Thus, it will be eluted later than the D-Ser diastereomer. In contrast, in the D-Ser, the polar hydroxyl group is directed on the same side with the sidechain of the adjacent L-Leu, leading to less affinity to interact with the stationary phase, and hence, it will be eluted faster. As Trp is highly hydrophobic, it is expected that the retention time of Trp-containing peptides would be high: at pH 2, 20 mM H_3PO_4, they were 66.5 and 69.9 for L- and D-Trp, respectively (Table 10). Unexpectedly, the nearest neighbour effect in the case of the highly hydrophobic Trp residue was close to the moderately hydrophobic Ala residue: at pH 2, 20 mM H_3PO_4, $\Delta t_{R\ D\text{-}L}$ = 3.4 and $\Delta t_{R\ D\text{-}L}$ = 2.9, respectively (Table 11). It was confirmed that the conformational structure of D-Trp, which is adjacent to the L-Leu, did have an interaction with the nonpolar surface with respect to L-Trp, but the difference is not significant enough for such a hydrophobic amino acid. The explanation could be ascribed to the overall hydrophobicity of Trp-containing peptides with either the enantiomers (L or D). Thus, the interaction with the stationary phase is mainly governed by the hydrophobic component. Furthermore, it was computationally predicted that the ring of D-Trp is directing towards the stationary phase and masking the full interaction with the stationary phase that takes place normally via H-bonding [30]. Generally, the hydrophobic environment enhances the hydrophilic character of the polar sidechain. Given that D-Tyr conformation is directed towards the stationary phase when it is adjacent to the bulky Leu, this would likely increase the hydrophilicity of the OH with respect to the L-Tyr, leading to an overall lower hydrophobicity of the D-Tyr and to it being eluted earlier ($\Delta t_{R\ D\text{-}L}$ = −1.0) at pH 2/H_3PO_4 (Table 11) [30].

Examining the $\Delta t_{R\ D\text{-}L}$ for these 20 amino acids in five mobile phases at pH 2 (H_3PO_4) versus the TFA mobile phases, and at pH 7 (in the absence of the salt versus the presence of 50 mM NaCl or 50 mM $NaClO_4$), the effect of nearest neighbour effect proved to be independent of the relative hydrophobicities of the counter ions in the mobile phase (TFA versus H_3PO_4). So, for the mobile phases of pH 2, a satisfactory correlation of r = 0.997 was obtained. The same observation was noticed with the mobile phases at pH 7, where the correlation was r = 0.995, which confirmed the independence of the nearest neighbour effect on the effectiveness of the anion of the added salt (Cl^- vs. ClO_4^-). It is worth mentioning that some anomalous behaviour in the His residue was observed in case of the mobile phase with the added NaCl [30]. A good linear correlation of r = 0.972 was obtained between the observed retention times (nearest neighbour effect, $\Delta t_{R\ D\text{-}L}$) and the sidechain hydrophobicity of D-substituted peptides ($\Delta t_{R\ Gly}$). As the hydrophobicity of the sidechain increased, the effect of the neighbour group also increased. Additionally, a good correlation of r = 0.999 was obtained with three amino acids Trp, Tyr, and His, whereas a smaller nearest neighbour effect was observed [30].

3.7.4. Retention Coefficients Determination

The determination of the sidechain hydrophobicity was performed by comparing the retention times of the X-substituted (L and D) diastereomers with that of the Gly-substituted peptide (Table 12).

Table 12. Retention coefficients of amino acids [30].

Amino Acid Substitution	pH 2				pH 7 (10 mM PO_4)					
	20 mM H_3PO_4		20 mM TFA $\Delta t_{R\ L\text{-}Gly}$		No Salt $\Delta t_{R\ D\text{-}Gly}$		+50 mM NaCl $\Delta t_{R\ D\text{-}L}$		+50 mM $NaClO_4$ $\Delta t_{R\ D\text{-}L}$	
	$\Delta t_{R\ D\text{-}Gly}$	$\Delta t_{R\ L\text{-}Gly}$	$\Delta t_{R\ D\text{-}Gly}$	$\Delta t_{R\ L\text{-}Gly}$	$\Delta t_{R\ D\text{-}Gly}$	$\Delta t_{R\ L\text{-}Gly}$	$\Delta t_{R\ D\text{-}Gly}$	$\Delta t_{R\ L\text{-}Gly}$	$\Delta t_{R\ D\text{-}Gly}$	$\Delta t_{R\ L\text{-}Gly}$
Trp	36.6	33.2	38.3	34.6	35.0	31.9	35.5	32.1	38.2	34.9
Phe	38.0	30.3	38.9	31.3	35.9	28.5	36.4	27.9	39.3	32.0
Leu	30.8	24.1	33.1	26.4	30.3	23.9	31.2	24.7	34.2	27.9
Ile	31.1	23.3	32.8	25.0	39.7	22.4	30.5	23.2	33.3	26.0
Met	21.8	15.7	23.2	17.2	21.3	15.4	21.7	16.0	23.7	18.1
Val	21.3	14.6	22.7	15.7	20.8	14.5	21.2	14.6	23.0	16.4
Tyr	13.2	14.2	14.5	15.3	13.0	14.0	13.3	14.2	13.8	15.2
Pro	12.1	8.6	14.5	11.2	12.8	9.9	13.4	10.2	14.8	12.1
Cys	12.4	7.9	13.3	8.7	12.4	8.3	15.0	10.6	13.5	9.1
Glu	4.3	3.0	4.3	3.2	0.5	−0.5	1.1	0.3	−4.9	−5.0
Ala	5.0	2.2	5.9	2.7	5.4	2.5	7.8	4.7	6.2	3.1
Asp	1.4	1.4	1.4	1.5	0.1	1.0	0.3	1.3	−7.4	−6.7
Thr	2.2	1.0	2.7	1.3	2.5	1.1	2.8	1.5	2.7	1.2
Gly	0.0	0.0	0.0	0.0	0.0	0.0	0.0	0.0	0.0	0.0
Ser	−1.8	−0.2	−1.4	−0.1	−1.3	0.2	−1.2	0.3	−1.8	−0.5
Gln	−0.6	−0.3	−0.6	−0.3	−0.4	−0.1	1.6	1.9	−1.0	−0.6
Asn	−1.5	−0.6	−1.4	−0.5	−1.0	−0.1	0.7	1.7	−1.8	−0.9
Arg	−7.6	−6.2	0.7	2.1	−6.0	−4.7	−2.3	−0.7	5.4	3.8
Lys	−9.4	−6.3	−1.8	1.5	−7.5	−3.2	−5.7	−0.3	1.0	4.2
His	−9.8	−7.9	−1.7	−0.3	−5.4	−6.1	−3.2	−4.2	0.4	2.6

H_2O as mobile phase A or in ACN as B; 20 mM H_3PO_4 or 20 mM TFA at pH 2; and 10 mM PO_4 buffer at pH 7, containing no salt, containing 50 mM NaCl, or containing 50 mM $NaClO_4$. Columns: for mobile phase of pH 2: Kromasil C_{18}, 5 µm, 2.1 × 150 mm; for that of pH 5 and 7: Zorbax Eclipse XDB-C_8, 5 µm, 2.1 × 150 mm. $\Delta t_{R\ D\text{-}Gly}$: retention time difference in D- relative to Gly-peptide; $\Delta t_{R\ L\text{-}Gly}$: retention time difference in L- relative to Gly-peptide. All data are in minutes.

Hydrophobicity for the mobile phases pH 2/H_3PO_4 and pH 7/$NClO_4$ showed a good correlation: r = 0.994 and 0.997, respectively. Only the charged residues deviated as a result of adopting different charges at different pHs. Nevertheless, the behaviour of the neutral residues proved the independence of the retention coefficients on either pH or the type of the anionic counter ion (PO_4^{3-} vs. ClO_4^-) [30]. Comparing two mobile phases at pH 2 with different counter ions (H_3PO_4 versus TFA) showed a good correlation for all the included amino acids in the study: r = 0.982 and 0.976, respectively. This means that the sidechain hydrophobicity coefficient is independent of the counter ion in the mobile phase. A good correlation was noted for all amino acids when comparing the three mobile phases at pH 7: no added salt, with added NaCl, or with added $NClO_4$; r = 0.977, 0.972, and 0.979, respectively [30].

In conclusion, the study confirmed that the sidechain coefficient is independent on either the absence or presence of the anion, or even its nature. Furthermore, it is independent of having the ineffective Cl^- or the effective $NClO_4^-$ as ion-pairing reagents in the mobile phase [30].

3.8. N and C Termini Position

Tripet et al. [31] studied the hydrophilicity/hydrophobicity when substituting 20 amino acids at the *N* and C-termini. The following models were considered: N^{α}-acetyl-X- and N^{α}-amino-X- at the *N*-terminus and –X-C^{α}-carboxyl and –X-C^{α}-amide at the *C*-terminus. The retention time coefficients were determined for each amino acid at these positions, and these were also compared with the sidechain coefficients and the internal coefficients determined at the centre of the peptide chain. Thus, the accuracy of retention times prediction would be enhanced (Figure 1).

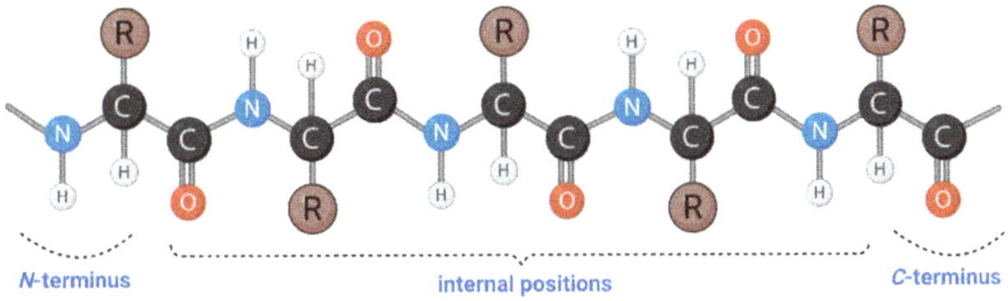

Figure 1. Schematic representation of the three different positions of an amino acid in a peptide chain. Created with Biorender.com.

A TFA-containing mobile phase was considered in this study: mobile phase A: 20 mM TFA in water (containing 2% v/v of ACN); mobile phase B: 0.2% TFA in ACN. Kromasil C_{18}, 5 μm, 2.1 × 150 mm was used (Table 13).

Table 13. Retention times of peptide models [31].

Substituted Amino Acid at X Position	-G-X-OH	-G-X-NH_2	Ac-X-G-	NH_2-X-G-	-G-X-G-
Trp	76.3	68.9	68.0	57.1	96.9
Phe	73.3	65.3	64.4	51.5	94.4
Leu	68.5	58.4	58.5	45.0	90.6
Ile	66.8	57.6	56.6	43.4	89.3
Met	57.5	50.4	51.3	41.0	84.9
Tyr	55.2	48.8	50.4	42.0	82.1
Val	56.3	47.8	48.6	37.3	82.4
Pro	48.5	39.9	45.0	33.7	77.5
Cys	47.1	41.0	42.9	33.5	79.7
Ala	41.3	35.4	38.2	30.7	76.6
Glu	38.4	36.2	37.9	30.6	76.3
Thr	39.9	35.7	37.6	31.1	75.2
Arg	38.8	36.1	37.2	32.2	72.7
Asp	37.7	34.3	36.5	30.6	75.4
Gln	36.3	34.0	35.5	30.6	74.7
Gly	36.3	32.4	34.8	29.2	73.8
His	36.3	33.6	34.8	30.6	71.6
Ser	35.5	32.4	34.8	29.2	74.4
Lys	35.3	32.4	34.5	30.5	71.4
Asn	34.0	32.4	34.3	29.2	73.4

Mobile phase A: 20 mM TFA in H_2O (containing 2% v/v of ACN); mobile phase B: 0.2% TFA in acetonitrile. Column: Kromasil C_{18}, 5 μm, 2.1 × 150 mm. All data are in minutes.

The predicted retention time (τ) was determined as follows:

$$\tau = \sum R_c + t_g \tag{17}$$

where the gradient time (t_g), the time for the gradient to reach the detector, is calculated from the summation of the dwell or gradient delay time (t_d), the time for the gradient to reach the top of the column from the proportioning valve via the pump, solvent mixer, and injection loop, and the void volume (t_0) estimated from the elution time of an unretained peak. Thus, $t_g = t_d + t_0$.

3.8.1. Rationale of Selecting the Model Peptides for Investigation

Peptides were selected not to have more than 10 residues. Furthermore, they must not have any propensity to form secondary structures which could block certain residues from interacting with the stationary phase leading to misleading results. The peptide must have an overall hydrophobicity to maintain column retention behaviour. Having Gly distributed in the peptide sequence helps in ensuring that no secondary structures will form. Gly was placed next to the substitution position to allow a free rotation around the peptide bond between the substitution position and the adjacent residue. When the substitution is at the *N*-terminal, the *N*-terminal was either acetylated or left free; the same applies when the substitution is at the *C*-terminal, where it was amidated or left free. This would help in demonstrating the effect of these end groups on hydrophobicity, e.g., by preparing the same peptides with different *N*- or *C*-termini. When the substitution is at the internal position, a Gly residue was placed on both sides. All the selected peptides should have at least one positive charge (at pH 2) to enhance their solution solubility [31].

The retention time coefficients were considered from Lau et al., and their work [32] was conducted using a TFA mobile phase, as it is the most common ion-pairing reagent for reverse-phase peptide separation. A low gradient (0.35%) was considered to enhance the separation among peptides and magnify the differences in the hydrophobic character [31].

3.8.2. Hydrophobicity Determination

For the sake of hydrophobicity determination, a Gly-substituted peptide was used as a reference, in which the retention time difference was recorded for the 19 proteinogenic amino acids other than the Gly (Table 14) [31].

$$\Delta t_{R,X} = t_{R,X-\text{substituted peptide}} - t_{R,\text{ Gly}-\text{substituted peptide}} \quad (18)$$

Table 14. Hydrophobicity of peptide models [31].

Substituted Amino Acid at X Position	-G-X-OH (1) $\Delta t_{R,X-Gly}$	-G-X-NH$_2$ (2) $\Delta t_{R,X-Gly}$	Ac-X-G- (3) $\Delta t_{R,X-Gly}$	NH$_2$-X-G- (4) $\Delta t_{R,X-Gly}$	-G-X-G- (5) $\Delta t_{R,X-Gly}$
Trp	40.0	36.5	33.2	27.9	23.1
Phe	37.0	32.9	29.6	22.3	20.6
Leu	32.2	26.0	23.7	15.8	16.8
Ile	30.5	25.2	21.8	14.2	15.5
Met	21.2	18.0	16.5	11.8	11.1
Tyr	18.9	16.4	15.6	12.8	8.3
Val	20.0	15.4	13.8	8.1	8.6
Pro	12.2	7.5	10.2	4.5	3.7
Cys	10.8	8.6	8.1	4.3	5.9
Ala	5.0	3.0	3.4	1.5	2.8
Glu	2.1	3.8	3.1	1.4	2.5
Thr	3.6	3.3	2.8	1.9	1.4
Arg	2.5	3.7	2.4	3.0	−1.1
Asp	1.4	1.9	1.7	1.4	1.6
Gln	0.0	1.6	0.7	1.4	0.9
Gly	0.0	0.0	0.0	0.0	0.0
His	0.0	1.2	0.0	1.4	−2.2
Ser	−0.8	0.0	0.0	0.0	0.6
Lys	−1.0	0.0	−0.3	1.3	−2.4
Asn	−2.3	0.0	−0.5	0.0	−0.4

Mobile phase A: 20 mM TFA in H_2O (containing 2% v/v of ACN); mobile phase B: 0.2% TFA in acetonitrile. Column: Kromasil C_{18}, 5 μm, 2.1 × 150 mm. The more positive the Δt_R, the more hydrophobic the peptide. $\Delta t_{R,X—Gly}$: retention time difference in X-peptide relative to Gly-peptide. All data are in minutes.

3.8.3. Comparison of the Different *C*-Terminal

The $\Delta\Delta t_R$ of the obtained hydrophobicity coefficients, from Table 14, for the 20 peptides with an acid terminal versus those of the amide terminal, peptides 1 and 2, are reported in Table 15. Plotting the obtained values gave a correlation of r = 0.993, which means that the relative difference in the hydrophobicity for both acidic and amide *C*-terminus is almost the same regardless of what the *C*-terminus is (Table 15), taking into consideration that several sidechain values varied quantitatively between both termini, especially for the hydrophobic residues (Pro, Val, Met, Ile, Leu, Phe, Tyr, and Trp). On the other hand, the hydrophilic residues showed little variations, including the polar (Asp, Glu, Asn, Gln, Ser, and Thr) and the charged residues (His, Lys, and Arg) [31].

Table 15. Hydrophobicity differences among the peptide models [31].

Substituted Amino Acid at X Position	$\Delta\Delta t_R$ (1–2)	$\Delta\Delta t_R$ (3–4)	$\Delta\Delta t_R$ (1–4)	$\Delta\Delta t_R$ (1–5)	$\Delta\Delta t_R$ (4–5)
Trp	3.5	5.3	12.1	16.9	4.8
Phe	4.1	7.3	14.7	16.4	1.7
Leu	6.2	7.9	16.4	15.4	−1.0
Ile	5.3	7.6	16.3	15.0	−1.3
Met	3.2	4.7	9.4	10.1	0.7
Tyr	2.5	2.8	6.1	10.6	4.5
Val	4.6	5.7	11.9	11.4	−0.5
Pro	4.7	5.7	7.7	8.5	0.8
Cys	2.2	3.8	6.5	4.9	−1.6
Ala	2.0	1.9	3.5	2.2	−1.3
Glu	−1.7	1.7	0.7	−0.4	−1.1
Thr	0.3	0.9	1.7	2.2	0.5
Arg	−1.2	−0.6	−0.5	3.6	4.1
Asp	−0.5	0.3	0.0	−0.2	−0.2
Gln	−1.6	−0.7	−1.4	−0.9	0.5
Gly	0.0	0.0	0.0	0.0	0.0
His	−1.2	−1.4	−1.4	2.2	3.6
Ser	−0.8	0.0	−0.8	−1.4	−0.6
Lys	−1.0	−1.6	−2.3	1.4	3.7
Asn	−2.3	−0.5	−2.3	−1.9	0.4

Reference (green); hydrophobic (yellow); hydrophilic (blue)-polar; hydrophilic (orange)-positively charged; slightly hydrophobic (grey). Mobile phase A: 20 mM TFA in H_2O (containing 2% *v*/*v* of ACN); and mobile phase B: 0.2% TFA in ACN. Column: Kromasil C_{18}, 5 µm, 2.1 × 150 mm. $\Delta\Delta t_R$ (1–2): hydrophobicity difference between acid and amide peptide; $\Delta\Delta t_R$ (3–4): hydrophobicity difference between acetylated and free *N*-terminal peptide; $\Delta\Delta t_R$ (1–4): hydrophobicity difference when the residue is placed at the *C*-terminus or at the *N*-terminus; $\Delta\Delta t_R$ (1–5): hydrophobicity difference when the residue is placed at the *C*-terminus or at an internal position; $\Delta\Delta t_R$ (4–5): hydrophobicity difference when the residue is placed at the *N*-terminus or at an internal position. All data are in minutes.

Residues that differ by 2.5 min or more are considered hydrophobic, including Pro, Val, Met, Ile, Leu, Phe, Tyr and Trp—shown in yellow. Residues that showed little variation are considered hydrophilic, such as the polar residues Asp, Glu, Asn, Gln, Cys, and Thr (shown in blue) and the positively charged residues His, Lys, and Arg (shown in orange), and the small hydrophobic residues are Ala and Cys (shown in grey) [31].

3.8.4. Comparison of the Different *N*-Terminal (Difference from 0.0 to 7.9)

The $\Delta\Delta t_R$ of the 20 peptides with the free *N*-terminal were compared with those of acetylated *N*-terminal, peptides 3 and 4, respectively (Table 15). Plotting those values showed a high correlation r = 0.982. Thus, the relative differences in the hydrophobicity for both *N*-terminal situations are the same regardless of what is the *N*-terminus. Unsurprisingly, several sidechains showed quantitative differences. The hydrophobic residues that differ by 2 min or more are the same as those in *C*-terminal study. In addition to Cys, the hydrophilic residues are the same as those of *C*-terminal study: polar and positively charged residues. The small hydrophobic residue here is only Ala. Acetylation is one of the major post-translational modifications; thus, determining their hydrophobicity coefficients are useful for predicting the retention time of such *N*-terminal modified peptides [31].

3.8.5. Comparison of Terminals with the Largest Difference (from 0.0 to 16.4 min)

The $\Delta\Delta t_R$ of the 20 peptides where the residue was placed at the *C*-terminus were compared with the *N*-terminus locations: peptides 1 and 4, respectively (Table 15). Plotting the differences resulted in a good correlation of r = 0.967, which suggests a similar relative difference in the hydrophobicity regardless of the amino acid position (*N*- or *C*-terminus). There are quantitative differences between amino acids. The hydrophobic residues that differ by 3 min or more are as follows: Ala, Cys, Pro, Val, Tyr, Met, Ile, Leu, Phe, and Trp. Meanwhile, the hydrophilic residues are the polar residues: Asp, Glu, Asn, Gln, Ser, and Thr in addition to the charged ones His, Lys, and Arg [31].

3.8.6. Relative Hydrophobicity of *N*- and *C*-Termini

This model highlighted the absence of any effects related to *N*- or *C* termini. So, comparing the Gly-substituted peptide at the *N*-terminal with the acetylated *N*-terminal, demonstrated that the acetylated analogue is more hydrophobic. This behaviour is ascribed to the fact the free *N*-terminal is protonated at pH 2. Depending on the concentration of the ion pairing (in this case the TFA), the hydrophobicity of the free *N*-terminal could be increased proportionally with the ion-pairing concentration. Consequently, the lower the concentration of TFA, the lower the hydrophobicity of the free *N*-terminus, and the greater the difference would be [31]. The acid *C*-terminal is more hydrophobic than the amide one. The reason here is the carboxyl group is fully protonated at pH 2 and thus more hydrophobic than the amide one. Any ionization that could happen to the carboxyl group (e.g., at high pH) would increase its hydrophilicity [31]. In conclusion, a single value is not possible to be assigned for any of the termini as they are basically amino acid dependent [31].

3.8.7. Comparison of the *C*- and *N*-Termini with the Internal Coefficients

When the amino acid residue is place at the centre of the chain with two Gly residues around it, this peptide, with the 20 amino acids, showed smaller hydrophobicity differences than having the amino acid at the C-terminus; see peptides 5 and 1, respectively in (Table 15). Plotting $\Delta\Delta t_R$ of these 20 peptides resulted in good correlation (r = 0.984), which means having the same relative differences in the hydrophobicity. Several residues showed considerable differences (from 17.1 to 4.8 min) (Table 15). The only non-hydrophobic (hydrophilic) residue in this study was the Arg. Arg, when at the centre of the chain, is much more hydrophilic than when at the termini when it is always hydrophobic (Table 14). The other charged residues such as His and Lys also became more hydrophilic at the centre than at the C-terminus (Table 14). In summary, the hydrophilic amino acids showed little difference, including the charged amino acids, than the hydrophobic ones [31]. This phenomenon is less pronounced when comparing the residues at the centre of the chain with those at the *N*-terminus; peptides 4 and 5, respectively. The charged residues His, Lys, and Arg became more hydrophobic at the *N*-terminal than at the centre. Interestingly, the aromatic amino acids Tyr and Trp became more hydrophobic at the *N*-terminus than the centre; however, the other hydrophobic amino acids remained almost the same at both locations: Phe, Leu, Ile, Met, Val [31].

Acidic *C*-terminus showed the most pronounced difference when compared to internal coefficients for Amide *C*-terminus, free *N*-terminus and acetylated *N*-terminus. For example, Trp, when located at the *N*-terminus, has a retention coefficient of 27.9 min with respect to the Gly; when it was located at the centre, it dropped to 23.1 min, while it showed the maximum value when it was located at the *C*-terminus 40.0 min (Table 14). In summary, the value of the -Trp-COOH-substituted peptide showed a 3.4-fold higher hydrophobicity coefficient than for NH_2-Trp- (Table 14) [31].

3.8.8. Retention Time Prediction

The importance of this work was demonstrated by its ability to predict the retention times for the peptides, first by considering the coefficients from the model with the largest influence based on this study (which is the peptide with an acidic *C*-terminus, and the substitution is at the *C*-terminal). The large overestimation in the predicted retention times observed resulted on insufficient weighting of the residues at the centre of the chain (Column 3, Table 16) [31].

Table 16. Comparison of predicted and observed retention times [31].

Column # Amino Acid Substitution	1 Peptide 1 Observed t_R [a]	2 Pred [b]	3 Diff [c]	4 Pred [d]	5 Diff [e]	6 Pred [f]	7 Diff [g]	8 Pred [h]	9 Diff [i]	10 Pred [j]	11 Diff [k]
Trp	76.3	107.9	31.6	56.8	−19.5	76.8	0.5	59.6	−16.7	76.7	0.4
Phe	73.3	104.9	31.6	55.3	−18.0	73.8	0.5	57.3	−16.0	73.7	0.4
Leu	68.5	100.1	31.6	52.9	−15.6	69.0	0.5	53.5	−15.0	68.9	0.4
Ile	66.8	98.4	31.6	52.1	−14.7	67.3	0.5	52.0	−14.8	67.2	0.4
Met	57.5	89.1	31.6	47.4	−10.1	58.0	0.5	47.9	−9.6	57.9	0.4
Tyr	55.2	86.8	31.6	46.3	−8.9	55.7	0.5	44.9	−10.3	55.6	0.4
Val	56.3	87.9	31.6	46.8	−9.5	6.8	0.5	45.3	−11.0	56.7	0.4
Pro	48.5	80.1	31.6	42.9	−5.6	49.0	0.5	40.3	−8.2	48.9	0.4
Cys	47.1	78.7	31.6	42.2	−4.9	47.6	0.5	42.7	−4.4	47.5	0.4
Ala	41.3	72.9	31.6	39.3	−2.0	41.8	0.5	39.5	−1.8	41.7	0.4
Glu	38.4	70.0	31.6	37.9	−0.5	38.9	0.5	39.0	0.6	38.8	0.4
Thr	39.9	71.5	31.6	38.6	−1.3	40.4	0.5	38.2	−1.7	40.3	0.4
Arg	38.8	70.4	31.6	38.1	−0.7	39.3	0.5	35.6	−3.2	39.2	0.4
Asp	37.7	69.3	31.6	37.5	−0.2	38.2	0.5	38.2	0.5	38.1	0.4
Gln	36.3	67.9	31.6	36.8	0.5	36.8	0.5	37.5	1.2	36.7	0.4
Gly	36.3	67.9	31.6	36.8	0.5	36.8	0.5	36.7	0.4	36.7	0.4
His	36.3	67.9	31.6	36.8	0.5	36.8	0.5	34.3	−2.0	36.7	0.4
Ser	35.5	67.1	31.6	36.4	0.9	36.0	0.5	37.3	1.8	35.9	0.4
Lys	35.3	66.9	31.6	36.3	1.0	35.8	0.5	34.4	−0.9	35.7	0.4
Asn	34.0	65.6	31.6	35.6	1.6	34.5	0.5	36.2	2.2	34.4	0.4

[a] from Table 13; [b] by applying the retention coefficient estimated at C-terminal for all amino acids; [c] difference between the predicted 2 and the observed 1; [d] by applying the weighed retention coefficients estimated at C-terminal for all amino acids including at C-terminus position; [e] difference between the predicted 4 and the observed 1; [f] by applying the weighed retention coefficients estimated at C-terminal for all amino acids except at C-terminus position; [g] difference between the predicted 6 and the observed 1; [h] by applying the internal coefficients for all amino acids including at C-terminus position; [i] difference between the predicted 8 and the observed 1; [j] by applying a combination of the internal coefficients and C-terminus ones accordingly; [k] difference between the predicted 8 and the observed 1. Mobile phase A: 20 mM TFA in water (containing 2% *v*/*v* of acetonitrile); mobile phase B: 0.2% TFA in acetonitrile. Column: Kromasil C_{18}, 5 µm, 2.1 × 150 mm. Pred: predicted retention time; Diff: Difference between the predicted and observed retention time. All data are in minutes.

Secondly, they performed the prediction based on the coefficients estimated from the model peptide where the substitution is at the C-terminus and considering a weighing factor of 0.5 (Table 17) [31].

These results were better than the first un-weighted protocol. However, an underestimation in the predicted retention time was observed especially with the hydrophobic containing peptides of up to 19.5 min (Column #5, Table 16). This difference is ascribed to the fact that the hydrophobicity at the C-terminus is different to any other locations. Thus, the weighted coefficients must be applied to all residues except the C-terminus one. Thirdly, when using a weighing factor of 0.5 for all residues except those at the C-terminus, more concordant values, within a 0.5 min difference, were obtained (Column #7, Table 16). Fourthly, they predicted the retention times by considering the internal retention coefficients. An underestimation was observed, especially with the hydrophobic residues of up to −16.7 min (Column #9, Table 16). Again, the same reason as in the second scenario happened, it applies here (using the weighed coefficients from the C-terminus peptide and applying them to all residues including at C-terminus position). Unsurprisingly, the values were also similar to those obtained in the second scenario. Finally, they used a combination of the internal hydrophobicity coefficients and the C-terminus ones accordingly. These results proved to be the most accurate ones, where the maximum difference between the predicted and the observed retention times was less than 0.4 min for all peptides (Column #11, Table 16) [31].

Table 17. The weighed retention coefficients of peptide 1 [31].

Substituted Amino Acid at X Position	-G-X-OH $\Delta t_{R,X-Gly}$	-G-X-OH $\Delta t_{R,X-Gly}$ Weighted
Trp	40.0	20.0
Phe	37.0	18.5
Leu	32.2	16.1
Ile	30.5	15.3
Met	21.2	10.6
Tyr	18.9	9.5
Val	20.0	10.0
Pro	12.2	6.1
Cys	10.8	5.4
Ala	5.0	2.5
Glu	2.1	1.1
Thr	3.6	1.8
Arg	2.5	1.3
Asp	1.4	0.7
Gln	0.0	0.0
Gly	0.0	0.0
His	0.0	0.0
Ser	−0.8	−0.4
Lys	−1.0	−0.5
Asn	−2.3	−1.2

Mobile phase A: 20 mM TFA in H_2O (containing 2% *v*/*v* of ACN); mobile phase B: 0.2% TFA in ACN. Column: Kromasil C_{18}, 5 μm, 2.1 × 150 mm. $\Delta t_{R,X-Gly}$: retention time difference in X-peptide relative to Gly-peptide. All data are in minutes.

It is worth mentioning that plotting the over- and/or underestimated values versus the observed retention time in the three scenarios resulted in linear relationship with high correlation. Even though there were differences, this suggests the presence of a systematic error which they were able to circumvent by introducing the weighting factor. Thus, the C-terminal coefficients are not representative to the internal values and vice versa [31].

In conclusion, the hydrophobicity coefficients of the amino acids based on their position in the chain, is of utmost importance to accurately predict the peptide retention times. Especially important are those located at the C-terminus, as they showed a dramatic difference with respect to the other positions within the chain. In fact, such an approach seems to be greatly assisting and improving the prediction capability of the chromatographic method. However, additional factors could be included to ensure the robustness of this approach [31].

3.9. Preferred Domain of Binding

Zhou and co-workers studied the effect of the preferred domain of binding of the peptide with the non-polar stationary phase during the course of HPLC work [24]. If the molecule/peptide becomes helical on binding and has a preferred domain of binding, like the case with the amphipathic helix, this would affect the overall hydrophobicity of the peptide as some residues will not be contributing/interacting like in their primary structure (Figure 2).

The following mobile phases were investigated: A: 0.1% TFA in H_2O; B: 0.1% TFA in ACN. The following columns were also investigated: SynChropak C_4, 6.5 μm, 4.1 × 250 mm; Aquapore C_8 7 μm, 4.6 × 220 mm; SynChropak C_{18}, 6.5 μm, 4.6 × 250 mm; all columns had 300 Å pore size [24].

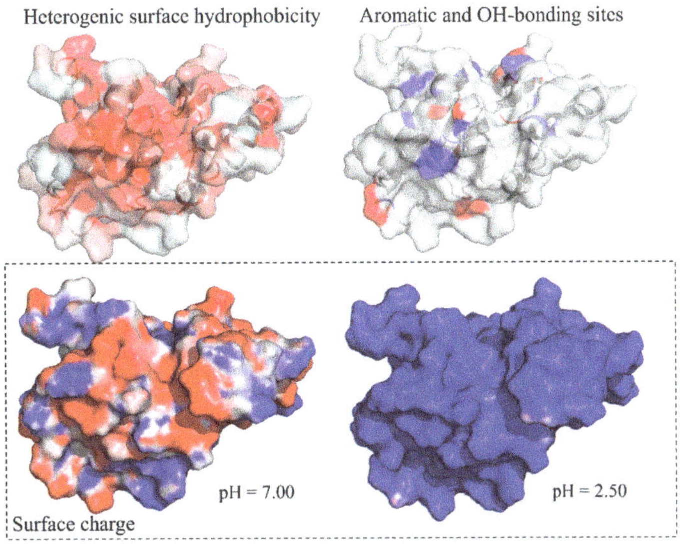

Figure 2. Peptide surface properties. Three-dimensional structure of a hydrolysing peptide represented by its accessible surface area. (**Top-left**) Red-coloured areas depict higher hydrophobicity. (**Top-right**) Blue and red show sites where π–π interactions and OH-bonding are possible, respectively.

Circular dichroism studies proved that both model peptides showed high propensity to form α-helical structures in the non-polar environment. Size-exclusion chromatography proved that all peptides were monomeric while they are bound to the stationary phase. Interestingly, this study confirmed the ability to predict the retention behaviour of peptides with α-helical structures and subsequently to deduce the presence of such phenomenon in any peptide based on its retention data [24].

Several factors can affect the retention time of a peptide, including: (i) amino acid composition and the relative hydrophobicity of each amino acid residue; (ii) peptide chain length, in which longer peptides may be eluted at shorter retention time due to a stabilized secondary structure which led to some amino acids being masked from interacting with the stationary phase; and finally (iii) the sequence-dependent effect, which can be divided into nearest neighbour and conformational effects. The former is amino acid dependent but independent of the conformation. The conformational effect could alter the overall hydrophobicity of the peptide as a result of adopting certain conformational structures in comparison to the same peptide when it is present in a random coil conformation (lacking a unique conformation). The aim of this study was to demonstrate the presence of a preferred domain of binding in α-helical peptides and investigate how this domain of binding can affect the behaviour of peptides with the stationary phase. Additionally, the study was trying to locate which amino acid or part of the sequence is responsible for this preferred domain of binding [24].

Zhou et al. considered two sets of peptides with the following lengths: 7, 14, 21, 28, and 35 amino acid residues. All peptides have the same amino acid constituents, albeit with different sequences [24].

Ac-Lys-Cys-Ala-Glu-Gly-Glu-Leu-[Lys-Leu-Glu-Ala-Gly-Glu-Leu]$_n$-amide and Ac-Lys-Cys-Ala-Glu-Leu-Glu-Gly-[Lys-Leu-Glu-Ala-Leu-Glu-Gly] $_n$-amide, where n = 1–4.

The *N*-terminal was acetylated, and the *C*-terminal was amidated to eliminate any ionic interactions as a result of different charges that might develop in different pH environments. The sequence of Cys residue in the set B peptides is similar to a protein with a known α-helical coiled-coil structure, tropomyosin. However, the Cys in this study is at the *N*-terminus of the adopted model, while in the original tropomyosin, it is in the internal position 190. The same chromatographic elution conditions as per the of work of Guo et al. were adopted [9].

The set B peptides showed a larger retention time than set A peptides, with the same chain length, which is ascribed to the presence of preferred domain of binding in the B set. The difference in the retention times between the two sets have increased as the chain length increased, from 2.9 min for the 14 amino acid residues and up to 7.3 for the 35 amino acid residues [24]. Several studies showed a relationship (either linear or exponential) between the molecular weight and the retention time for the peptides which bind to the stationary phase on their monomeric form [21,32].

The retention times were compared of both peptide sets A and B with another peptide included in the study, the S peptide, which has no tendency to form α-helical structure. The S peptide is a series of five peptides with the following sequences: Ac-(Gly-Leu-Gly-Ala-Lys-Gly-Ala-Gly-Val-Gly)$_n$-amide, where n = 1–5 comprising in total 10–50 residues. Furthermore, the same positive charges were also considered in the comparison study (+1 to +5). Such a comparison would help in estimating the importance of α-helical conformation on adsorption and thus the elution process. The same molecular weight range was considered in the S peptide and the A and B sets as follows: 826–3894 and 789–3479 Da, respectively. The predicted retention time was calculated using Equation (5), developed by Mant [21], where the correction for the chain length was included in this equation, and the retention coefficients were obtained from the work of Guo et al. [9]. Equation (5) proved to be suitable in the case of S peptides as well as the peptides of set A. This agreement was exemplified by the low average deviation between the predicted and the observed retention times of 0.5 and 1.8 min, for S and set A peptides, respectively. Furthermore, a good correlation between the predicted and the observed retention times in both sets was obtained: r = 0.99. In contrast, utilizing Equation (5) to predict the retention time in case of set B peptides, which have a preferred domain of binding, resulted in high variation between the predicted and the observed retention times, i.e., around a 6 min difference in the case B peptides. To account for this phenomenon, the authors incorporated another parameter into Equation (5). They were then able to achieve better prediction for the retention times of peptides that are prone to have a preferred domain of binding with the hydrophobic stationary phase. The new term basically considers the summation of the retention coefficients of the most hydrophobic residues. If the hydrophobic residues are evenly distributed around the α-helix structure, this term would be cancelled out [24].

$$\tau_p = \sum R_c + t_0 + t_s + PA - \left(m \sum R_c \ln N + b\right) \quad (19)$$

where Equation (5) is modified by a correction factor for the preferred domain of binding (PA) calculated as $\sum R_{c,n} - \sum R_{c,(n/N)}$. Here, N represents the total number of residues in the chain, and n is the number of residues in the preferred domain of binding. Parameter $\sum R_{c,n}$ is the sum of retention coefficients in the preferred domain of binding (reflected by the most nonpolar residue). If there is no preferred domain of binding, then $\Sigma R_{c,n} = \Sigma R_{c,(n/N)}$, and PA = 0.

To determine the preferred domain of binding, a pattern recognition exercise was conducted so the distribution of the hydrophobic amino acids could be visualized. It was observed for the set A peptides that all the hydrophobic Leu residue cannot be found in one area at one time. On the other hand, this was observed with the peptides of set B. Thus, the hydrophobicity would be higher in set B than A, and accordingly, it was also the preferred domain of binding [24].

The modified equation that takes account of the preferred domain of binding yielded decreased retention time deviations to 0.7, 0.8, and 1.8 min for S, A, and B peptides, respectively. Moreover, a good correlation between the predicted and the observed retention times was obtained (r = 0.99) with the three sets of peptides using the three columns [24].

It is worth highlighting that despite there being no preferred domain of binding in the peptides of set A, using the new equation that count for this phenomenon resulted in a small enhancement in the results compared to the original equation (from 1.8 to 0.8) min. This improvement of course simply reflects the increased number of variables of fit employed in the modified equation. The significant improvement was in the case of peptides of set B (from 6.0 to 1.8 min) [24].

It was concluded that if the hydrophobic residues are evenly distributed around the α-helical structure of the peptide, then the α-helical structure will not have an enhanced contribution to the separation process as in the case of set A peptides. On the other hand, set B showed a large difference between the observed and the predicted retention times. This difference is mainly ascribed to the difference in the distribution of the hydrophobic residues around the α-helical structure of the peptide. Again, this explains the presence of a preferred domain of binding ascribed to the amphipathic nature of the α-helical structure of the peptides of set B [24].

3.10. Denaturation

Lau et al. studied the effect of the solvents and hydrophobic surfaces used in the HPLC on the secondary and quaternary structure of selected peptides [32]. A series of peptides were synthesized to study the denaturation process during the course of the HPLC work. All peptides were separated using different columns with different alkyl loadings, pore sizes and alkyl chain lengths: Alex Ultrapore RPSC C_3, 5 µm, 4.6 × 75 mm, Whatman Prtisil CCS/C_8, 5 µm, 4.6 × 250 mm, and SynChropak C_{18}, 6.5 µm, 4.1 × 250 mm. Alex and SynChropak have a 300 Å pore size, whereas the Whatman Partisil has 60 Å pores [32]. The study aimed to investigate the denaturation phenomenon, in addition to evaluating the RP columns, and to study the relationship between the retention times and the natural logarithm (ln) of the molecular weights. In general, the separation in the RP involves a linear gradient program which starts with a high percentage of the aqueous solvent, with an increasing percentage of the organic phase being programmed until the full elution is achieved. TFA is a good choice for HPLC, as silica is more stable at low pH than at high pH. Low pH can avoid the ionization of the weakly acidic silanol groups (ionising at pH 3.5 and above), thus avoiding any interactions of those groups with the basic molecules. TFA is also considered as a good stabilizer for peptides and proteins and is used to extract proteins and peptides after cleaving them from the solid supports (resins). Nevertheless, denaturation does also occur in an acidic medium, which triggered this investigational work [32].

Stationary phases usually have high loading of the alkyl chains attached, which facilitates the binding of the peptides or proteins to these hydrophobic sites. In addition, the amount of the organic solvent being used in the separation process enhances the stability of the secondary structure of the aforementioned molecules. Hence, it is highly expected that under these conditions, the denaturation of the tertiary and quaternary structure may take place (see Figure 3). This would translate into compromising the purification process, especially if the purification was designed based on the native peptide/protein conformations.

It is worth highlighting that the extent of hydrophobicity is not the same for the folded and unfolded states of peptides and proteins. This behaviour is mainly ascribed to the fact that some amino acids are being buried as a result of the folding phenomenon. In summary, avoiding denaturation is of utmost importance for efficient LC separation tasks [32].

The authors studied five synthetic peptides with the following parent sequence: Ac-(Lys-Leu-Glu-Ala-Leu-Glu-Gly)$_n$,-Lys-amide, where n = 1–5, so the peptides were: TM-8, TM-15, TM-22, TM-29, and TM-36 in length, respectively. The selection of (TM-22, TM-29, and TM-36) peptides was on the basis that they are forming a stable two-stranded α-helical coiled coils, which are stabilized by the hydrophobic interactions among the chains. High-performance size-exclusion chromatography was used to investigate the conformational structure of the peptides, including their monomeric or dimeric forms [32].

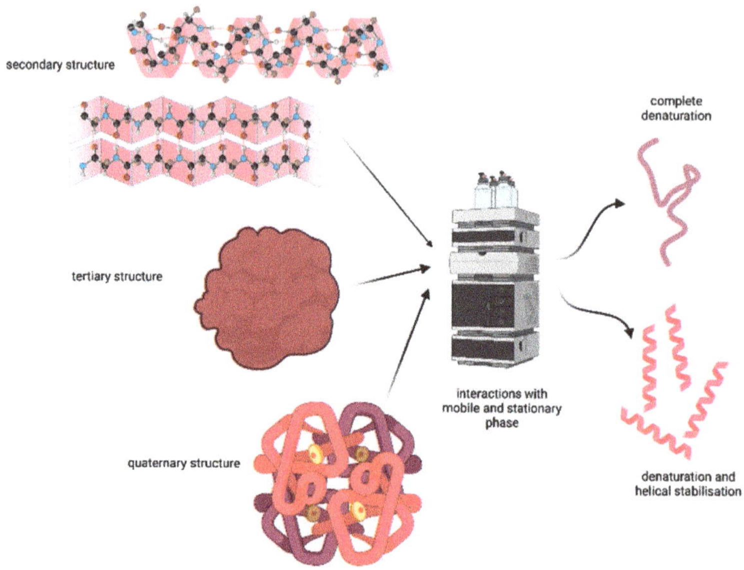

Figure 3. Denaturation of peptides and proteins during LC separation. (Created with Biorender.com.)

A mixture of the five peptides were analysed using three solvents: 0.1% aqueous TFA, which is the starting solvent for the RP-HPLC; 0.1% TFA in can, which represents the upper limit of the organic solvent usually used in the RP-HPLC; and 0.1 TFA in trifluoroethanol (TFE), which was chosen as it is does not interact with the α-helical-induced properties. Acetonitrile was considered based on the fact that it is suitable for the majority of the peptides, whereas methanol is for more hydrophilic peptides and propanol for the highly hydrophobic ones. Three columns were incorporated in the study: C_3, C_8, and C_{18}. The first three peptides (TM-8, TM-15, and TM-22) appeared as monomers, whereas the last two TM-29 and TM-36 appeared as dimers in 0.1% TFA in water [32].

A linear relationship between the natural logarithm of the monomeric molecular weights and the retention volumes was observed for the 5 peptides in the system with the organic solvent. However, for the system with the aqueous phase, the linear relationship was only obtained when using the dimeric molecular weight for the last two peptides (TM-29 and TM-36) [32].

It was observed that the tertiary and quaternary structure of TM-29 and TM-36 peptides is being disrupted in the nonpolar solvents. Obviously, the stabilizing forces for the α-helical structure are the hydrogen bonding which are highly unstable in the presence of water. Thus, the opposite is true: as the non-polarity of the medium increases, the stability of the α-helical structure will also increase [32].

There were no differences in the separation process between the different columns with respect to the alkyl chain lengths or the carbon loading. However, the best resolution was obtained with the C_{18} column. Other studies noticed that for long peptides of 30–150 residues, the most important parameter is the pore size [33]. A 300 Å pore size is superior to 100 Å, whereas 80–100 Å pores delivered poor resolution and recovery. Nonetheless, for small molecules such as the peptides in this study with 8–36 residues, the pore size had little effect on the chromatographic resolution. As for the particle size, usually, the smaller the particle size, the higher the column efficiency, and the sharper the peaks would be [32].

Plotting the natural logarithm molecular weights of the five peptides (monomeric form) versus their retention times in the RP-HPLC showed a linear relationship. As shown in another study [34] and from the size-exclusion data, these two peptides form extremely stable dimers, confirming that the hydrophobic stationary phase caused the disruption in the hydrophobic interactions among the subunits of those peptides. This resulted in them being eluted as dimers, which then dissociated upon interacting with the reversed phase media causing denaturation to occur, as indicated from the linear relationship [32].

Interestingly, denaturation was also observed even with a C_3-based column with low carbon loading. Though the organic mobile phase can cause denaturation, the hydrophobicity of the stationary phase also plays a major role in this process. In conclusion, if it is required to perform the separation and/or purification using only the native conformation, then the hydrophobicity of the stationary phase must be reduced. Furthermore, the solvent in the mobile phase must be a non-denaturing one [32].

4. Conclusions

As shown in this review, several research groups have tried to predict the retention behaviour of peptides. Although the exact behaviour cannot be accurately established, these efforts provide a useful insight into how a peptide would behave and interact with various chromatographic components. Many factors can affect the separation and the purification of peptides. Unlike the small molecules, peptides comprise different points of interaction with the separation components. For example, they can develop charges within their structure depending on the amino acid composition and the pH of mobile phase. Similarly, peptides can adopt various conformational structures which govern the extent of their adsorption. Therefore, their separation behaviour will vary according to the employed conditions. Given that the adsorption on RPLC columns is governed mainly by hydrophobic interactions, any difference in the hydrophobicity of peptide will alter their elution pattern accordingly. In fact, weighing factors are deemed necessary to improve the prediction capability of the models. In light of this, the position of amino acids within the peptide chain (*N*-, *C*-termini, or internal positions), as well as the neighbouring amino acid, influences the overall hydrophobicity. Having an a priori computational tool that considers all these factors would benefit the field for two reasons: firstly, to account for conformational and denaturation changes within the peptide structure and enhance the prediction capability, and secondly, to interpret the experimental findings and understand any unexpected or anomalous retention behaviour.

Author Contributions: All authors have read and agreed to the published version of the manuscript.

Funding: The APC was funded by Imperial College London.

Institutional Review Board Statement: Not applicable.

Informed Consent Statement: Not applicable.

Data Availability Statement: Not applicable.

Acknowledgments: The continuous support of Imperial College London is acknowledged. The authors would also like to thank Lucia Lombardi, Department of Chemical Engineering, for her contribution to designing some figures of the article.

Conflicts of Interest: The authors declare no conflict of interest.

References

1. Al Shaer, D.; Al Musaimi, O.; Albericio, F.; de la Torre, B. 2021 FDA TIDES (Peptides and Oligonucleotides) Harvest. *Pharmaceuticals* **2022**, *15*, 222. [CrossRef] [PubMed]
2. Al Musaimi, O.; Al Shaer, D.; Albericio, F.; de la Torre, B. 2020 FDA TIDES (Peptides and Oligonucleotides) Harvest. *Pharmaceuticals* **2021**, *14*, 145. [CrossRef] [PubMed]
3. Merrifield, R.B. Solid Phase Peptide Synthesis. I. The Synthesis of a Tetrapeptide. *J. Am. Chem. Soc.* **1963**, *85*, 2149–2154. [CrossRef]
4. Ferrazzano, L.; Catani, M.; Cavazzini, A.; Martelli, G.; Corbisiero, D.; Cantelmi, P.; Fantoni, T.; Mattellone, A.; De Luca, C.; Felletti, S.; et al. Sustainability in peptide chemistry: Current synthesis and purification technologies and future challenges. *Green Chem.* **2022**, *24*, 975–1020. [CrossRef]
5. Snyder, L.R.; Kirkland, J.; Dolan, J. *Introduction to Modern Liquid Chromatography*, 3rd ed.; Wiley-Interscience: New York, NY, USA, 2010.

6. Field, J.K.; Euerby, M.; Lau, J.; Thøgersen, H.; Petersson, P. Investigation into reversed phase chromatography peptide separation systems part I: Development of a protocol for column characterisation. *J. Chromatogr. A* **2019**, *1603*, 113–129. [CrossRef]
7. Al Musaimi, O.; Valenzo, O.; Williams, D. Prediction of peptides retention behavior in reversed-phase liquid chromatography based on their hydrophobicity. *J. Sep. Sci.* **2022**. [CrossRef]
8. Al Musaimi, O.; Summers, B.; Basso, A.; De la Torre, B.; Albericio, F.; Serban, S. Combining solid phase synthesis and chromatographic purification for efficient peptide manufacture. *Chim. Oggi/Chem. Today* **2020**, *38*, 18–21. Available online: https://www.teknoscienze.com/tks_article/combining-solid-phase-synthesis-and-chromatographic-purification-for-efficient-peptide-manufacture/ (accessed on 24 December 2022).
9. Guo, D.; Mant, C.; Taneja, A.; Parker, J.; Rodges, R. Prediction of peptide retention times in reversed-phase high-performance liquid chromatography I. Determination of retention coefficients of amino acid residues of model synthetic peptides. *J. Chromatogr. A* **1986**, *359*, 499–518. [CrossRef]
10. Meek, J.L.; Rossetti, Z. Factors affecting retention and resolution of peptides in high-performance liquid chromatography. *J. Chromatogr. A* **1981**, *211*, 15–28. [CrossRef]
11. Guo, D.; Mant, C.; Taneja, A.; Hodges, R. Prediction of peptide retention times in reversed-phase high-performance liquid chromatography II. Correlation of observed and predicted peptide retention times factors and influencing the retention times of peptides. *J. Chromatogr. A* **1986**, *359*, 519–532. [CrossRef]
12. Parker, J.M.; Guo, D.; Hodges, R. New hydrophilicity scale derived from high-performance liquid chromatography peptide retention data: Correlation of predicted surface residues with antigenicity and X-ray-derived accessible sites. *Biochemistry* **1986**, *25*, 5425–5432. [CrossRef]
13. Wilce, M.C.J.; Aguilar, M.; Hearn, M. High-performance liquid chromatography of amino acids, peptides and proteins: CVII. A Analysis of group retention contributions for peptides separated with a range of mobile and stationary phases by reversed-phase high-performance liquid chromatography. *J. Chromatogr. A* **1991**, *536*, 165–183. [CrossRef]
14. Horváth, C.; Melander, W.; Molnár, I. Solvophobic interactions in liquid chromatography with nonpolar stationary phases. *J. Chromatogr. A* **1976**, *125*, 129–156. [CrossRef]
15. Casal, V.; Martin-Alvarez, P.; Herraiz, T. Comparative prediction of the retention behaviour of small peptides in several reversed-phase high-performance liquid chromatography columns by using partial least squares and multiple linear regression. *Anal. Chim. Acta* **1996**, *326*, 77–84. [CrossRef]
16. Field, J.K.; Euerby, M.; Petersson, P. Investigation into reversed phase chromatography peptide separation systems part III: Establishing a column characterisation database. *J. Chromatogr. A* **2020**, *1622*, 461093. [CrossRef]
17. Wilce, M.C.; Aguilar, M.; Hearn, M. High-performance liquid chromatography of amino acids, peptides and proteins. CXXII. Application of experimentally derived retention coefficients to the prediction of peptide retention times: Studies with myohemerythrin. *J. Chromatogr.* **1993**, *632*, 11–18. [CrossRef]
18. Field, J.K.; Euerby, M.; Petersson, P. Investigation into reversed phase chromatography peptide separation systems part II: An evaluation of the robustness of a protocol for column characterisation. *J. Chromatogr. A* **2019**, *1603*, 102–112. [CrossRef]
19. Field, J.K.; Euerby, M.; Haselmann, K.; Petersson, P. Investigation into reversed-phase chromatography peptide separation systems Part IV: Characterisation of mobile phase selectivity differences. *J. Chromatogr. A* **2021**, *1641*, 461986. [CrossRef]
20. Guo, D.C.; Mant, C.; Hodges, R. Effects of ion-pairing reagents on the prediction of peptide retention in reversed-phase high-performance liquid chromatography. *J. Chromatogr.* **1987**, *386*, 205–222. [CrossRef]
21. Mant, C.T.; Burke, T.; Black, J.; Hodges, R. Effect of peptide chain length on peptide retention behaviour in reversed-phase chromatogrphy. *J. Chromatogr. A* **1988**, *458*, 193–205. [CrossRef]
22. Mant, C.T.; Zhou, N.; Hodges, R. Correlation of protein retention times in reversed-phase chromatography with polypeptide chain length and hydrophobicity. *J. Chromatogr. A* **1989**, *476*, 363–375. [CrossRef] [PubMed]
23. Chabanet, C.; Yvon, M. Prediction of peptide retention time in reversed-phase high-performance liquid chromatography. *J. Chromatogr. A* **1992**, *599*, 211–225. [CrossRef]
24. Zhou, N.E.; Mant, C.; Hodges, R. Effect of preferred binding domains on peptide retention behavior in reversed-phase chromatography: Amphipathic alpha-helices. *Pept. Res.* **1990**, *3*, 8–20. Available online: https://pubmed.ncbi.nlm.nih.gov/2134049/ (accessed on 24 December 2022).
25. Sasagawa, T.; Okuyama, T.; Teller, D. Prediction of peptide retention times in reversed-phases high-performance liquid chromatography during linear gradient elution. *J. Chromatogr. A* **1982**, *240*, 329–340. [CrossRef]
26. Sereda, T.J.; Mant, C.; Quinn, A.; Hodges, R. Effect of the alpha-amino group on peptide retention behaviour in reversed-phase chromatography. Determination of the pK(a) values of the alpha-amino group of 19 different N-terminal amino acid residues. *J. Chromatogr.* **1993**, *646*, 17–30. [CrossRef] [PubMed]
27. Creighton, T.E. *Proteins: Structures and Molecular Properties*; W.H. Freeman: New York, NY, USA, 1993.
28. Ross, J.R. Practical handbook of biochemistry and molecular biology; Edited by G D Fasman. pp 601. CRC Press, Boca Raton, Florida, USA. 1989. $00 ISBN 0-8493-3705-4. *Biochem. Educ.* **1991**, *19*, 95–96. [CrossRef]
29. Kovacs, J.M.; Mant, C.; Hodges, R. Determination of intrinsic hydrophilicity/hydrophobicity of amino acid side chains in peptides in the absence of nearest-neighbor or conformational effects. *J. Pept. Sci.* **2006**, *84*, 283–297. [CrossRef]

30. Kovacs, J.M.; Mant, C.; Kwok, S.; Osguthorpe, D.; Hodges, R. Quantitation of the nearest-neighbour effects of amino acid side-chains that restrict conformational freedom of the polypeptide chain using reversed-phase liquid chromatography of synthetic model peptides with L- and D-amino acid substitutions. *J. Chromatogr. A* **2006**, *1123*, 212–224. [CrossRef]
31. Tripet, B.; Cepeniene, D.; Kovacs, J.; Mant, C.; Krokhin, O.; Hodges, R. Requirements for prediction of peptide retention time in reversed-phase high-performance liquid chromatography: Hydrophilicity/hydrophobicity of side-chains at the N- and C-termini of peptides are dramatically affected by the end-groups and location. *J. Chromatogr. A* **2007**, *1141*, 212–225. [CrossRef]
32. Lau, S.Y.M.; Taneja, A.; Hodges, R. Effects of high-performance liquid chromatographic solvents and hydrophobic matrices on the secondary and quarternary structure of a model protein: Reversed-phase and size-exclusion high-performance liquid chromatography. *J. Chromatogr. A* **1984**, *317*, 129–140. [CrossRef]
33. Regnier, F.E. High-Performance Liquid Chromatography of Biopolymers. *Science* **1983**, *222*, 245–252. [CrossRef]
34. Lau, S.Y.; Taneja, A.; Hodges, R. Synthesis of a model protein of defined secondary and quaternary structure. Effect of chain length on the stabilization and formation of two-stranded alpha-helical coiled-coils. *J. Biol. Chem.* **1984**, *259*, 13253–13261. [CrossRef]

separations

MDPI

Article

HPLC and SFC Enantioseparation of (±)-*Trans*-β-Lactam Ureas on Immobilized Polysaccharide-Based Chiral Stationary Phases—The Introduction of Dimethyl Carbonate as an Organic Modifier in SFC

Mladenka Jurin *, Darko Kontrec and Marin Roje *

Laboratory for Chiral Technologies, Division of Organic Chemistry and Biochemistry, Ruđer Bošković Institute, Bijenička cesta 54, 10 000 Zagreb, Croatia; darko.kontrec@irb.hr
* Correspondence: mladenka.jurin@irb.hr (M.J.); marin.roje@irb.hr (M.R.)

Abstract: A series of nine racemic *trans*-β-lactam ureas were analyzed for enantiomer separation by high-performance liquid chromatography (HPLC) and supercritical fluid chromatography (SFC). The separations were performed on three immobilized polysaccharide-based chiral analytical columns (CHIRAL ART Amylose-SA, CHIRAL ART Cellulose-SB and CHIRAL ART Cellulose-SC). In HPLC mode, a normal-phase consisting of *n*-hexane/2-PrOH (90/10, *v*/*v*), a polar organic mobile phase consisting of 100% MeOH or 100% EtOH, and a non-standard mobile phase consisting of 100% dimethyl carbonate (DMC) were investigated. In SFC mode, the mobile phases CO_2/alcohol (80/20, *v*/*v*) and CO_2/DMC/alcohol (MeOH or EtOH; 70/24/6, *v*/*v*/*v* or 60/32/8, *v*/*v*/*v*) were investigated. The best achieved enantioseparation of *trans*-β-lactam ureas was obtained with an Amylose-SA column. We have shown that the green solvent dimethyl carbonate (DMC) can be efficiently used as a mobile phase in HPLC mode as well as in SFC mode along with the addition of polar organic modifiers (MeOH or EtOH).

Keywords: *trans*-β-lactam ureas; enantioseparation; chiral chromatography; high-performance liquid chromatography (HPLC); supercritical fluid chromatography (SFC); dimethyl carbonate (DMC); immobilized polysaccharide-based CSPs; green solvents

Citation: Jurin, M.; Kontrec, D.; Roje, M. HPLC and SFC Enantioseparation of (±)-*Trans*-β-Lactam Ureas on Immobilized Polysaccharide-Based Chiral Stationary Phases—The Introduction of Dimethyl Carbonate as an Organic Modifier in SFC. *Separations* **2024**, *11*, 38. https://doi.org/10.3390/separations11020038

Academic Editor: Liangliang Liu

Received: 30 December 2023
Revised: 22 January 2024
Accepted: 23 January 2024
Published: 25 January 2024

1. Introduction

Azetidin-2-ones, also known as β-lactams, are four-membered cyclic amides [1–4]. The β-lactam ring is the central structural element responsible for the antibacterial activity of β-lactam antibiotics, which are among the most commonly used types of antibiotics, such as penicillins, carbapenems, cephalosporins, nocardicins, and monobactams [5]. In addition, they show numerous other interesting pharmacological activities, such as cholesterol absorption inhibitors, human cytomegalovirus protease [6], tryptase and chymase inhibitors, thrombin inhibitors [5], and LHRH antagonists [7]. β-Lactams also have anticancer, antiviral [8], antitubercular, antifungal [6], anti-HIV, anti-inflammatory, and other biological activities [9]. β-Lactams can serve as building blocks for the synthesis of other compounds of biological and medicinal importance, such as β-amino acids, peptides, peptidomimetics, taxoids, alkaloids, and various heterocyclic molecules [7,10].

Chiral high-performance liquid chromatography (HPLC) on chiral stationary phases (CSPs) is one of the most commonly used analytical techniques for the enantioseparation of chiral β-lactams [11–20]. Pirkle et al. separated some β-lactam stereoisomers **1–7** (Figure 1) with aryl substituents at the C3 position and with alkyl or aryl substituents at the C4 position of the ring by HPLC on chiral stationary phase based on (*S*)-*N*-(3,5-dinitrobenzoyl)leucine, using *n*-hexane/2-propanol (80/20, *v*/*v*) as the mobile phase [11]. Lee et al. [12] separated the β-lactam stereoisomers with alkyl substituents in the 3-position and with aryl, furyl, or styryl substituents **8–16** (Figure 1) in the 4-position of the β-lactam ring using an (*R*)-1-{1-naphthyl)ethylamine polymer chemically bonded to spherical silica (YMC A-K03 column). In this study, *n*-hexane/dichloromethane/ethanol (70/30/2, *v*/*v*/*v*) was used as the mobile phase. Cirilli et al. [13] reported the separation of stereoisomers of a C3, C4-substituted β-lactam cholesterol absorption inhibitor (*cis*-**17** and *trans*-**17**, Figure 1) on amylose-based chiral stationary phases (Chiralpak AD-H and Chiralpak AS-H columns) in the normal-phase mode using the different binary mixtures *n*-hexane/ethanol and *n*-hexane/2-propanol as the mobile phases. Among them, amylose *tris*[(*S*)-α-methylphenylcarbamate] CSP (Chiralpak AS-H) was more effective, and the resolutions were higher than those obtained with amylose *tris*(3,5-dimethylphenylcarbamate) CSP (Chiralpak AD-H).

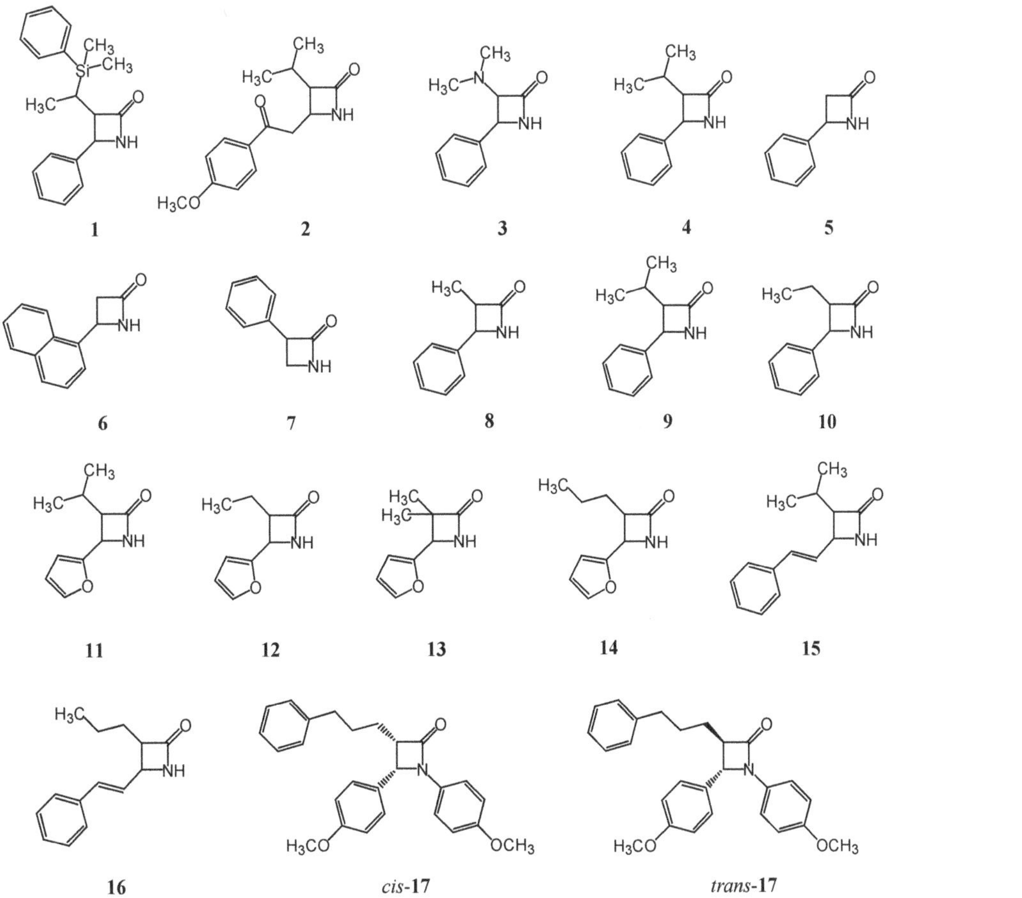

Figure 1. Chemical structures of β-lactam compounds separated by Pirkle et al. (**1–7**), Lee et al. (**8–16**), and Cirilli et al. (**17**).

Tris(phenylcarbamates) from amylose- or cellulose-based chiral stationary phases were used by Okamoto et al. [14] for the enantioseparation of various β-lactam compounds **18–35** (Figure 2). The authors used mobile phases consisting of mixtures of *n*-hexane and 2-propanol (80/20, *v*/*v* or 90/10, *v*/*v*). Most of the tested β-lactam compounds were completely resolved on the cellulose and/or amylose *tris*(phenylcarbamate) derivatives coated on the silica matrix. Pataj et al. [15] developed a direct HPLC method for the enantioseparation of nineteen racemic β-lactams **36–54** (Figure 3) on polysaccharide-based CSPs containing either amylose *tris*(3,5-dimethylphenylcarbamate) (Kromasil AmyCoat column) or cellulose *tris*(3,5-dimethylphenylcarbamate) (Kromasil CelluCoat column) as the chiral selectors. They analyzed these racemic β-lactams in normal-phase mode using mixtures of *n*-heptane with various amounts (2–10%) of polar alcoholic modifier (2-propanol). Reducing the alcohol content in the mobile phases led to better separation on the two columns tested. The amylose column Kromasil AmyCoat proved to be more suitable for the separation of the bi- and tricyclic β-lactams (all except compound **47**), whereas the 4-aryl-substituted β-lactams (all except compound **54**) were better separated on the cellulose column Kromasil CelluCoat.

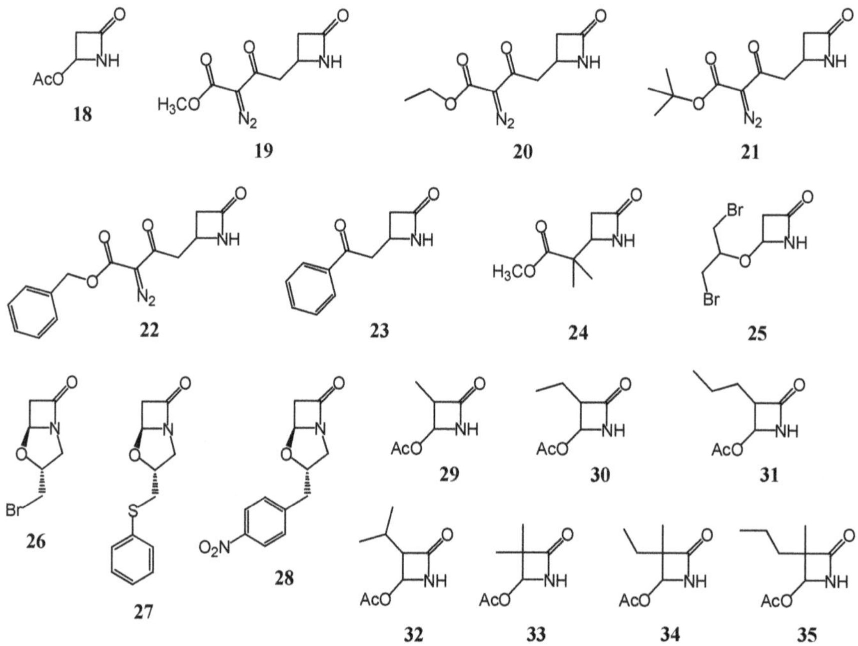

Figure 2. Chemical structures of β-lactam compounds **18–35** separated by Okamoto et al [14].

HPLC enantioseparation of the same twelve racemic bicyclic β-lactam compounds **36–47** (Figure 3) was investigated by Peter et al. [16] on two types of CSPs, one of which was cellulose *tris*(3,5-dimethylphenylcarbamate) (Chiralcel OD-RH and Chiralcel OD-H column) and the other containing a macrocyclic glycopeptide antibiotic teicoplanin (Chirobiotic T column) or teicoplanin aglycone (Chirobiotic TAG column) as a chiral selector. The authors investigated separation in different chromatographic modes. First, they investigated possibilities for HPLC separation on cellulose columns in the normal-phase mode (mixture of *n*-hexane and 2-propanol in different ration) and in the reversed-phase mode using water with different concentration of acetonitrile. In the next set of experiments, they investigated separation of compounds **36–47** on teicoplanin and teicoplanin aglycone CSPs in the polar organic mode (100% methanol or methanol/acetic acid/triethylamine, 100/0.01/0.01, *v*/*v*/*v*) and in the reversed-phase mode (water/methanol, 30/70 or 70/30, *v*/*v*; and 0.1% triethylammonium acetate pH = 4.1/methanol, 30/70 or 70/30, *v*/*v*). The result indicated that the aglycone alone was responsible for the enantioselective separation of bicyclic β-lactam compounds **36–47**. The resolution factors were higher for the aglycone CSP (Chirobiotic TAG column). Although the sugar units generally reduced the resolution of β-lactam enantiomers, they could contribute significantly to the resolution of some other compounds. The best enantioseparation of these β-lactams was obtained on the Chiralcel OD-H column in the normal-phase mode.

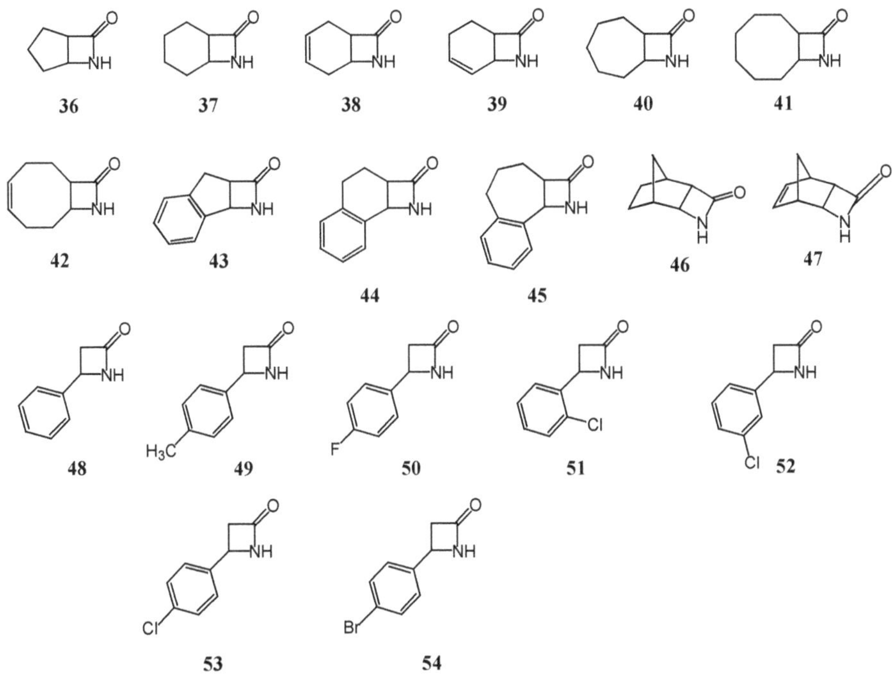

Figure 3. Chemical structures of bi- and tricyclic and 4-aryl-substituted β-lactams **36–54**.

The chiral recognition mechanisms for both polysaccharide and macrocyclic antibiotic CSP are not yet fully understood. There are interactions between enantiomers and CSP that are important for both general retention and enantioseparation. When considering retention, hydrophobic interactions (π-π interactions) are important in the reversed-phase mode, whereas hydrophilic interactions (hydrogen bonds) are important in the normal-phase mode and in the polar organic mode. However, several types of interactions can be considered in the case of enantioseparation.

Sun et al. [17] investigated the HPLC enantioseparation of the twelve β-lactam stereoisomers **36–47** (Figure 3) on three native cyclodextrin-based CSPs (α-, β-, and γ-) and on six derivatized β-cyclodextrins (acetylated, dimethylated, hydroxypropyl ether, dimethylphenyl carbamate, *S*-naphthylethyl carbamate, and *R*-naphthylethyl carbamate). On all cyclodextrin (CD) columns, the β-lactams were analyzed in the reversed-phase mode, on eight columns (except demethylated β-CD) in the polar organic mode, and on three aromatic derivatized β-cyclodextrin columns in the normal-phase mode. The dimethylphenyl carbamate β-CD proved to be the best CSP, separating eleven of twelve β-lactam compounds in the reversed-phase mode, whereas the dimethylated β-CD separated eight of twelve compounds. The other derivatized β-cyclodextrin CSPs and the native γ-cyclodextrin achieved enantioseparation for some β-lactams. The native α- and β-cyclodextrin CSPs did not separate any of the investigated β-lactams. As these β-lactams have no ionizable groups, the pH of the mobile phase has no major influence on the enantioseparation. When CD-based CSPs are used in polar organic or normal-phase media, the inner cavity is blocked by solvent molecules, which prevents the complexation of inclusions. Nevertheless, hydrophilic interactions can be enhanced in such media when solutes with hydrophilic groups bind to the polar surface of the CD. Derivatized CDs have been developed to allow additional intermolecular interactions, such as π–π interactions, hydrogen bonding, dipole–dipole interactions, and ion pairing, resulting in an improved ability for enantioseparation. In the reversed-phase mode, inclusion complexation is the dominant retentive interaction, whereas CSPs form dipolar and π-complexes in the normal-phase mode. Hydrogen bonding interactions are the most important in the polar organic mode.

Berkecz et al. [18] used macrocyclic glycopeptide antibiotic teicoplanin (Chirobiotic T column) and its aglycone (Chirobiotic TAG column) as well as a dimethylphenyl carbamate-derivatized β-cyclodextrin (Cyclobond DMP column) as chiral selectors for enantioseparation of three tricyclic chiral β-lactams **43–45** (Figure 3). These compounds contain a five- **43**, six- **44**, or seven-membered **45** aliphatic ring fused to a four-membered β-lactam ring and a benzene ring. In this study, the authors investigated the separations of compounds **36–47** in the normal-phase, polar organic, and reversed-phase modes. The size of the aliphatic ring, the nature of the CSPs, and the composition of the mobile phase influence the chiral recognition mechanism.

For the enantioseparation of the β-lactams **36–48** (Figure 3), Fodor et al. [19] used CSPs based on β-cyclodextrin (Quest-C1, Quest-C2, and Quest-C3) with HPLC in the reversed-phase mode. The Quest-C1 column, containing permethyl-β-cyclodextrin as a chiral selector, proved to be the most effective for this group of β-lactam compounds. The native β-CD (Quest C3) and its derivative hydroxypropyl-β-CD (Quest C2) CSPs showed enantioselectivity for some β-lactams. Jiang et al. used capillary electrophoresis based on β-cyclodextrin for the enantioseparation of racemic β-lactams [20].

Recently, we reported on the SFC enantioseparation of seven racemic β-lactams, **55–61** (Figure 4), on a polysaccharide-based CSPs containing either amylose *tris*(3,5-dimethylphenylcarbamate) (Chiralpak AD and Chiralcel IA columns), cellulose *tris*(3,5-dimethylphenylcarbamate) (Chirapcel OD and Chiralcel IB columns), or cellulose *tris*(4-methylphenylcarbamate) (Chirallica PST-10 column) as the chiral selectors [21]. The effect of CSP type (coated or immobilized) on the enantioseparation of *trans*-β-lactam ureas was investigated on all five columns, whereas the effect of alcoholic modifiers (methanol, ethanol, or 2-propanol), additive (isopropylamine), temperature, and backpressure were investigated on the Chirallica PST-10 column. The article demonstrated that the Chiralcel OD and Chiralpak IB columns provided better baseline separation than their amylose analogs, the Chiralpak AD and Chiralpak IA columns. The Chirallica PST-10 column separated all seven compounds tested. The effects of the three other parameters investigated—temperature, addition of isopropylamine and backpressure—showed little or no influence on the separation factor and resolution.

In our recent work, we have shown that DMC, an environmentally friendly solvent, can be efficiently used as a mobile phase in HPLC for the enantioseparation of *syn*- and *anti*-3,5-disubstituted hydantoins on the immobilized polysaccharide-based chiral stationary phases [22]. The CHIRAL ART Amylose-SA column was the most effective stationary phase, separating fourteen out of eighteen substituted hydantoins in non-standard mobile phase mode. Of the cellulose-based columns, the CHIRAL ART Cellulose-SB column proved to be more suitable for the enantioseparation of *anti*-3,5-hydantoins than the CHIRAL ART Cellulose-SC column, whereas the two cellulose columns did not exhibit enantioselectivity of *syn*-hydantoins with DMC as mobile phase.

In this study, the enantioseparation of nine chiral *trans*-β-lactam ureas, **55**, **57**, **60**, and **62–67** (Figure 4), on three different polysaccharide-type CSPs was investigated in HPLC mode with the four mobile phases *n*-hexane/2-propanol (90/10, *v*/*v*), methanol (MeOH), ethanol (EtOH), and dimethyl carbonate (DMC) and in SFC mode with the solvent mixtures CO_2/alcohol (80/20, *v*/*v*), CO_2/DMC/alcohol (70/24/6, *v*/*v*/*v*), and CO_2/DMC/alcohol (60/32/8, *v*/*v*/*v*). The aim of the present study was to introduce DMC as an organic solvent in supercritical fluid chromatography. To our knowledge, no data are available in the literature for chiral separations of racemic compounds using DMC in the SFC.

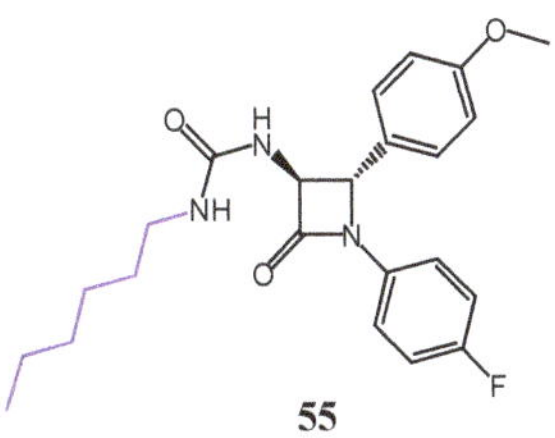

55

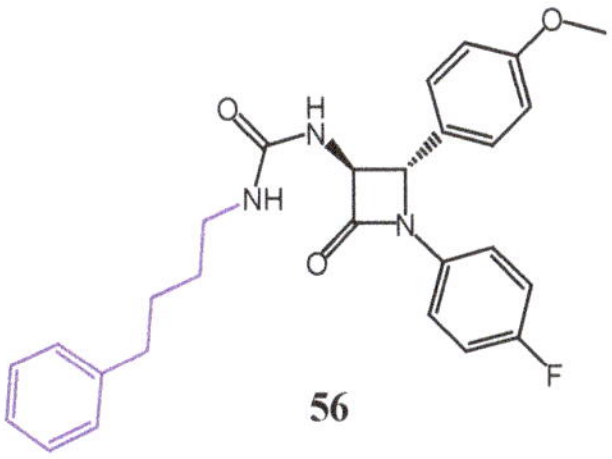

56

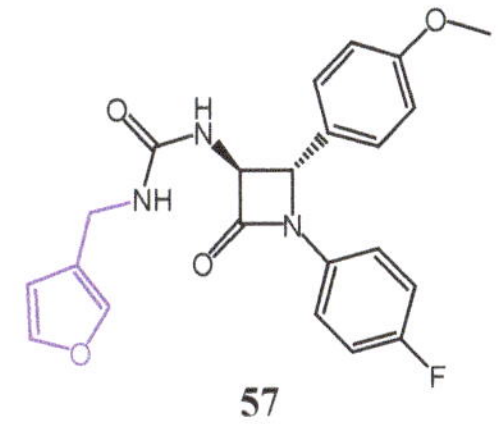

57

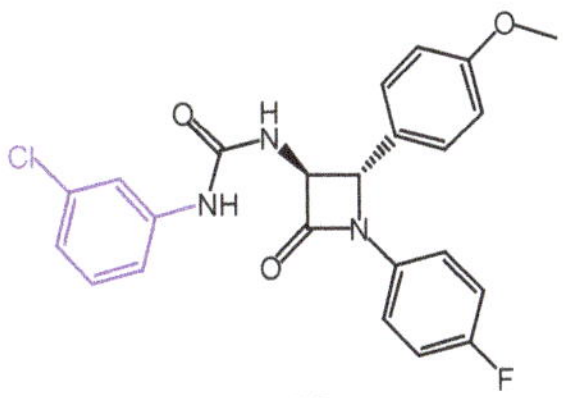

58

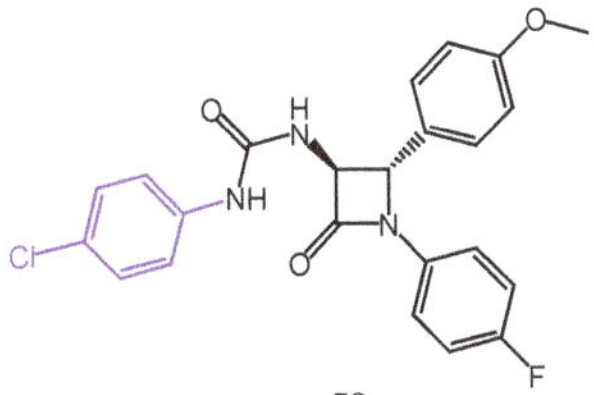

59

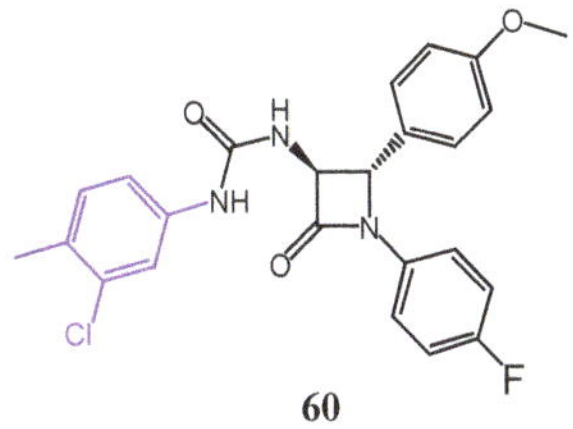

60

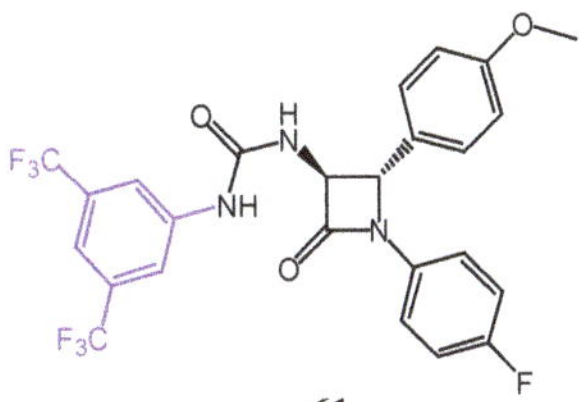

61

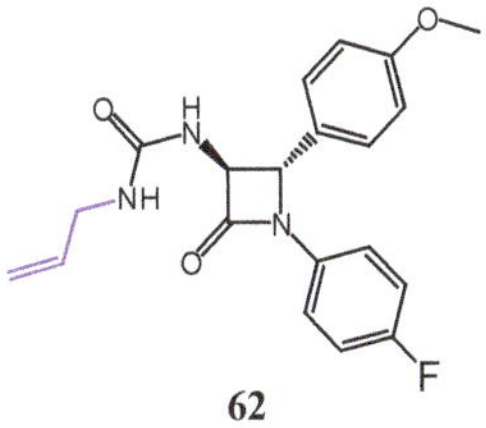

62

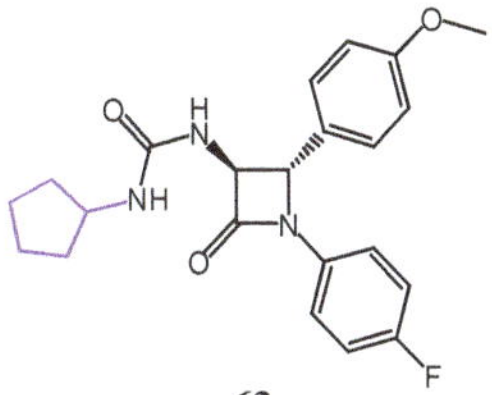

63

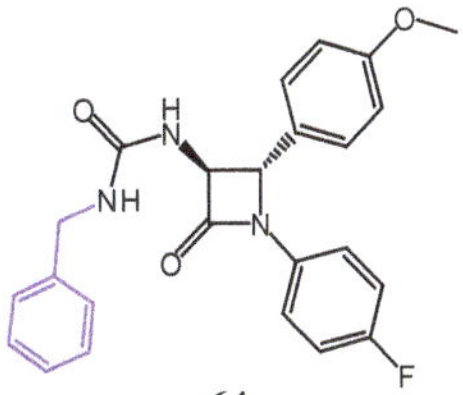

64

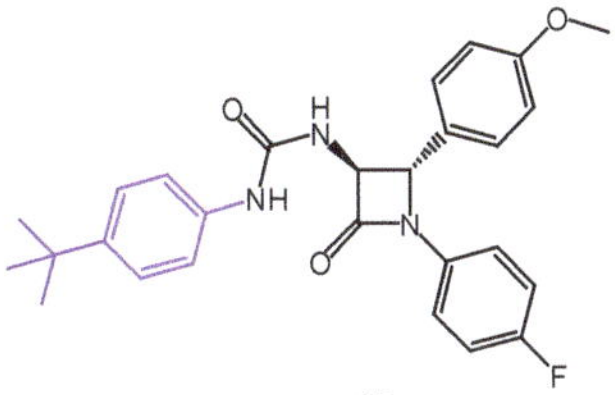

65

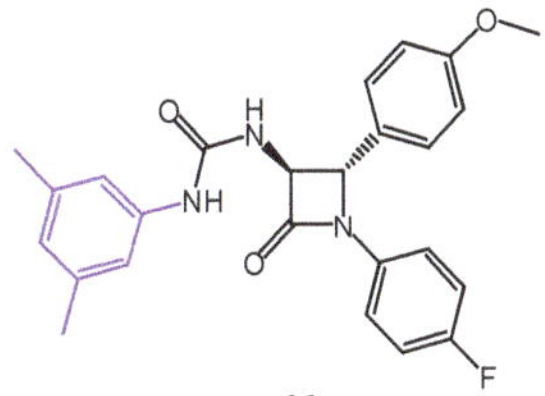

66

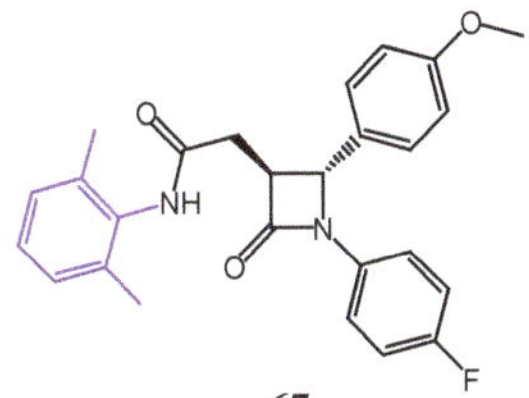

67

Figure 4. Chemical structures of (±)-*trans*-β-lactam ureas **55–67**.

2. Materials and Methods

The nine racemic β-lactam ureas were prepared in our laboratory by the addition of the corresponding isocyanate to (±)-*trans*-3-amino-β-lactam, which was prepared in three reaction steps [23].

HPLC-grade ethanol (EtOH), methanol (MeOH), 2-propanol (2-PrOH) and *n*-hexane were purchased from Honeywell (Seelze, Germany). Dimethyl carbonate (DMC) was purchased from Acros Organics (Geel, Belgium). Compressed CO_2 (class 4.5) was purchased from Messer (Zagreb, Croatia). The immobilized polysaccharide-based CSPs CHIRAL ART

Amylose-SA S-10 µm, CHIRAL ART Cellulose-SB S-10 µm, and CHIRAL ART Cellulose-SC S-10 µm were purchased in bulk from YMC (Kyoto, Japan). Empty stainless steel HPLC columns measuring 250 mm × 4.6 mm i.d. were purchased from Knauer GmbH (Berlin, Germany) and packed with the above CSPs.

HPLC analyses were performed using an Agilent 1200 Series system (Agilent Technologies GmbH, Waldbronn, Germany) consisting of a vacuum degasser, a quaternary pump, a thermostated column compartment, an autosampler, and a variable wavelength detector. The mobile phase was *n*-hexane/2-PrOH (90/10 *v*/*v*), 100% MeOH, 100% EtOH, or 100% DMC. All experiments in the normal-phase and polar and non-standard modes were performed under isocratic conditions at a flow rate of 1.0 mL min^{-1} and a column temperature of 35 °C. Detection was performed at 254 nm, and the injection volume was 20 µL. Data analysis and processing was performed using EZChrom Elite software version 3.1.7 (Agilent Technologies, Waldbronn, Germany).

The SFC analyses were performed with an Agilent 1260 Infinity II Hybrid SFC/UHPLC system (Agilent Technologies, Waldbronn, Germany). It consisted of an Infinity SFC binary pump, an Aurora A5 Fusion module, a degasser, an autosampler, a thermostated column compartment, a diode array detector, and a backpressure regulator. The system was controlled by the Open LAB CDS ChemStation Edition Rev. C01.08 software (Agilent Technologies, Waldbronn, Germany). The SFC was performed in isocratic mode at a flow rate of 4.0 mL min^{-1} and a column temperature of 35 °C in each case. The injection volume was 20 µL, and the outlet pressure was set to 15 MPa. Detection was performed at a wavelength of 254 nm using a diode array detector. The mobile phases used in the SFC consisted of liquid CO_2/alcohol (MeOH or EtOH) in a ratio of 80/20, *v*/*v* or CO_2/DMC/alcohol (MeOH or EtOH) in various ratios (70/24/6, *v*/*v*/*v* or 60/32/8, *v*/*v*/*v*). The sample solutions of the analytes were prepared by dissolving the β-lactam ureas **55**, **57**, **60** and **62–67** (Figure 4) in *n*-hexane/2-PrOH (90/10, *v*/*v*), 100% DMC, 100% MeOH, or 100% EtOH at a concentration of 0.5 mg mL^{-1} and filtered through a RC-45/25 Chromafil® Xtra 0.45 µm syringe filter (Macherey-Nagel GmbH & Co. KG, Düren, Germany).

The HPLC columns were packed with the immobilized chiral polysaccharide-based stationary phases from YMS CHIRAL ART Amylose-SA, CHIRAL ART Cellulose-SB, and CHIRAL ART Cellulose-SC. The size of the columns was 250 mm × 4.6 mm i.d., and the particle size was 10 µm. In the following text, these columns are referred to as Amylose-SA, Cellulose-SB, and Cellulose-SC. The chiral selectors in Amylose-SA, Cellulose-SB, and Cellulose-SC are amylose *tris*(3,5-dimethylphenylcarbamate), cellulose *tris*(3,5-dimethylphenylcarbamate), and cellulose *tris*(3,5-dichlorophenylcarbamate), respectively; all three are shown in Figure 5.

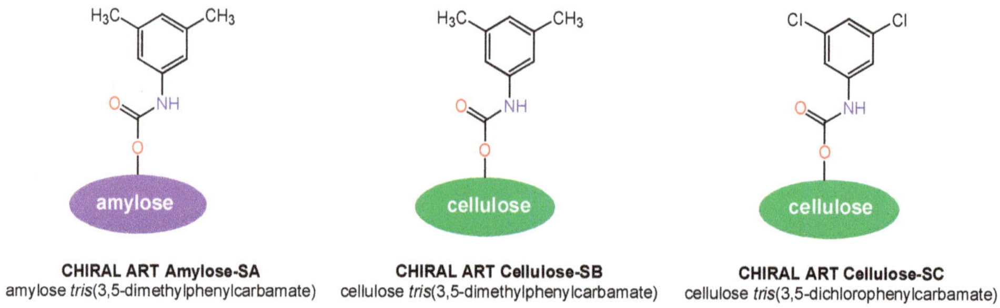

Figure 5. Chemical structure of the chiral selectors.

The retention factor (*k*) is a means of measuring the retention of an analyte on the chromatographic column [23]. It is calculated according to the following equation:

$$k = (t_R - t_0)/t_0, \quad (1)$$

where t_R and t_0 are the retention times of the analyte and the non-retained solute, respectively. The greatest gain in resolution is achieved when the k value is between 1 and 5. A high k value indicates that the sample is strongly retained and has interacted with the stationary phase for a considerable time.

The selectivity of the column is expressed by the enantioseparation factor (α). The enantioseparation factor is the ability of an HPLC method to separate two analytes from each other. It is calculated according to the following equation:

$$\alpha = k_2/k_1, \tag{2}$$

where k_1 and k_2 are the retention factors of the first and second eluted enantiomers, respectively. By definition, the selectivity is always greater than one—if α is equal to one, the two peaks are co-eluting (i.e., their retention factor values are identical). Larger selectivity values indicate better separation. As the selectivity of a separation depends on the chemistry of the analyte, the mobile phase composition, and the nature of the stationary phase, all of these factors can be altered to change or optimize the selectivity of an HPLC separation.

The resolution (R_s) indicates whether two peaks are separated from each other. It is calculated according to the following equation:

$$R_s = (t_{R2} - t_{R1})/(w_{b1} + w_{b2}), \tag{3}$$

where t_{R1} and t_{R2} are the retention times of the first and second eluted enantiomers, respectively, and w_{b1} and w_{b2} are the baseline peak widths of the first and second eluted enantiomers, respectively. According to the above definition, $R_s \geq 1.5$ means that the two peaks are baseline resolved. A higher resolution means that the peaks are well separated from each other.

In HPLC mode, the dead time (t_0), i.e., the retention time of a non-adsorbing component, was determined by injection of 1,3,5-tri-*tert*-butylbenzene, whereas in SFC mode the t_0 of the columns was determined at a first negative signal by injecting MeOH.

3. Results and Discussion

Polysaccharide-based CSPs are certainly the most dominant and widely used CSPs for the analytical and preparative separation of enantiomers in recent years due to their remarkable stability and loading capacity [24]. Polysaccharide CSPs are classified into two types, coated and immobilized, based on the chemistry of the application of the chiral selector on the chromatographic support matrix (usually silica). In the coated type, the polymeric chiral selector (amylose or cellulose derivatives) is physically coated by an adsorption process, whereas, in the immobilized type, the chiral selector is bound by a chemical process [25]. The coated polysaccharide CSPs have limited resistance to many solvents, whereas the immobilized CSPs are more robust, and can be used with non-standard solvents, such as acetone, chloroform, dichloromethane, ethyl acetate, tetrahydrofuran, etc. [26].

A screening for the enantioseparation of the nine β-lactam ureas **55**, **57**, **60**, and **62**–**67** using three different immobilized polysaccharide-type CSPs, including two cellulose-based columns, Cellulose-SB and Cellulose-SC, as well as one amylose-based column, Amylose-SA, have been performed by applying the HPLC and SFC modes. In the HPLC mode, either a standard mobile phase consisting of *n*-hexane/2-PrOH (90/10, *v*/*v*) or a polar organic mobile phase consisting of 100% alcohol (MeOH or EtOH) was used. The immobilized-type chiral columns contain a chiral selector covalently bound to the silica gel support, which enables the use of an extended range of the organic solvents [25,26], so, in this study, 100% DMC was used as a non-standard solvent in HPLC mode and in combination with an alcoholic modifier (DMC/alcohol; 80/20, *v*/*v*) in SFC mode. The effects of mobile phases: CO_2/alcohol (80/20, *v*/*v*), CO_2/DMC/alcohol (70/24/6, *v*/*v*/*v*), and CO_2/DMC/alcohol (60/32/8, *v*/*v*/*v*) on the separation were investigated in SFC mode. Chromatographic parameters such as retention factor of the first eluting enantiomer (k_1), separation factor (α), and resolution (R_s) are summarized in Table 1 for HPLC and Table 2 for SFC.

Table 1. HPLC chromatographic results of the enantioselective separations of (±)-*trans*-β-lactam ureas **55**, **57**, **60**, and **62–67** on Amylose-SA, Cellulose-SB, and Cellulose-SC columns.

Compound		Amylose-SA			Cellulose-SB			Cellulose-SC		
	Mobile Phase	k_1	α	R_s	k_1	α	R_s	k_1	α	R_s
55	A	7.99	1.00	–	9.53	1.81	3.22	21.58	1.96	3.86
	B	0.23	1.96	1.12	0.19	1.63	1.22	0.19	1.00	–
	C	0.40	2.13	2.00	0.23	1.00	–	0.24	1.00	–
	D	0.51	3.20	3.79	0.34	1.00	–	1.26	1.30	1.11
57	A	13.58	1.11	0.46	14.77	1.85	3.79	41.18	1.86	3.79
	B	0.16	2.88	1.77	0.18	1.44	0.82	0.16	1.00	–
	C	0.30	4.33	3.87	0.27	1.00	–	0.26	1.00	–
	D	0.35	2.29	2.28	0.22	1.00	–	0.75	1.16	0.53
60	A	10.78	1.34	1.58	11.33	2.36	4.59	13.52	1.38	1.02
	B	1.11	1.00	–	0.38	3.37	5.56	0.16	1.56	0.93
	C	0.90	1.00	–	0.31	2.55	3.95	0.13	1.62	0.58
	D	0.52	2.31	2.73	0.22	2.23	2.89	0.32	1.58	1.06
62	A	9.46	1.00	–	10.35	1.99	4.03	26.33	2.13	4.73
	B	0.09	2.89	1.21	0.12	1.67	0.90	0.13	1.00	–
	C	0.24	2.83	2.46	0.20	1.00	–	0.23	1.00	–
	D	0.46	2.43	2.92	0.31	1.00	–	1.12	1.10	0.44
63	A	2.41	2.90	1.26	8.61	1.66	2.68	18.54	1.70	3.02
	B	0.13	3.15	1.70	0.13	2.31	1.84	0.13	1.46	0.48
	C	0.22	3.18	2.58	0.14	1.64	1.19	0.16	1.00	–
	D	0.57	4.37	5.41	0.29	1.48	1.54	0.91	1.46	1.53
64	A	14.01	1.26	1.20	17.43	1.75	3.68	34.84	1.92	3.89
	B	0.24	2.96	2.16	0.30	1.33	0.97	0.39	1.00	–
	C	0.46	4.72	4.58	0.37	1.00	–	0.28	1.00	–
	D	0.47	2.49	2.97	0.30	1.00	–	0.91	1.07	0.29
65	A	12.36	1.00	–	9.39	1.94	3.58	16.13	1.39	1.78
	B	0.60	1.37	0.46	0.36	2.64	4.10	0.19	1.58	1.01
	C	0.96	1.73	1.88	0.30	2.10	2.94	0.18	1.56	0.76
	D	0.50	3.52	3.99	0.18	2.33	2.66	0.37	1.78	1.51
66	A	10.11	1.00	–	15.06	1.69	2.82	18.36	1.27	1.31
	B	0.51	1.98	1.50	0.34	3.68	5.72	0.38	1.21	0.56
	C	0.49	1.57	1.15	0.32	2.53	3.94	0.16	1.44	0.47
	D	0.55	4.19	4.91	0.22	2.04	2.59	0.37	1.73	1.55
67	A	10.24	2.17	3.67	6.81	1.09	0.38	15.18	1.38	1.74
	B	0.23	2.22	1.42	0.09	2.22	1.27	0.12	1.00	–
	C	0.27	7.04	3.82	0.15	1.47	0.92	0.17	1.00	–
	D	0.66	1.26	0.63	0.15	2.00	1.92	0.86	1.00	–

Note: k_1, retention factor of first eluting enantiomer; α, separation factor; R_s, resolution; "–", no separation; mobile phase, A, *n*-hexane/2-PrOH (90/10, *v*/*v*); B, MeOH; C, EtOH; D, DMC; flow rate, 1 mL min^{-1}; detection, 254 nm; column temperature, 35 °C.

Table 2. SFC chromatographic results of the enantioselective separations of (±)-*trans*-β-lactam ureas **55**, **57**, **60**, and **62–67** on Amylose-SA, Cellulose-SB, and Cellulose-SC columns.

Compound		Amylose-SA			Cellulose-SB			Cellulose-SC		
	Mobile Phase	k_1	α	R_s	k_1	α	R_s	k_1	α	R_s
55	A	3.40	1.54	1.65	2.04	1.80	4.04	2.50	1.00	–
	B	4.01	1.51	1.79	2.64	1.35	2.15	2.73	1.39	1.63
	C	6.82	1.68	2.10	3.43	1.32	1.95	4.96	1.22	1.00
	D	2.31	1.79	1.95	1.47	1.30	1.13	1.66	1.16	0.28
	E	9.85	2.05	3.01	5.97	1.00	–	7.55	1.52	2.32
	F	3.60	2.07	2.61	2.14	1.00	–	2.66	1.46	1.58

Table 2. *Cont.*

Compound	Mobile Phase	Amylose-SA k_1	Amylose-SA α	Amylose-SA R_s	Cellulose-SB k_1	Cellulose-SB α	Cellulose-SB R_s	Cellulose-SC k_1	Cellulose-SC α	Cellulose-SC R_s
57	A	3.21	1.95	3.01	2.83	1.42	2.79	3.16	1.00	–
	B	3.79	2.08	3.85	3.59	1.13	0.91	3.43	1.25	1.18
	C	4.58	1.93	2.95	3.47	1.24	1.57	4.89	1.13	0.54
	D	1.52	1.96	2.25	1.42	1.23	0.84	1.70	1.00	–
	E	6.58	1.80	2.88	5.26	1.00	–	6.90	1.42	1.97
	F	2.30	1.83	2.40	1.87	1.00	–	2.32	1.37	1.24
60	A	10.06	1.17	0.70	6.62	3.52	11.78	3.13	1.85	3.50
	B	8.92	1.25	1.21	5.65	3.23	10.71	2.87	1.66	2.59
	C	11.06	1.19	0.81	5.39	3.35	10.99	3.56	1.74	2.76
	D	3.50	1.26	0.88	1.99	3.27	7.82	1.04	1.77	1.67
	E	14.24	1.35	1.50	6.48	3.07	9.89	4.80	1.55	2.16
	F	4.63	1.40	1.51	2.13	2.90	6.99	1.42	1.58	1.47
62	A	2.17	1.57	1.70	1.66	1.62	2.92	2.20	1.00	–
	B	2.67	1.69	2.35	2.22	1.26	1.52	2.38	1.46	1.82
	C	4.20	1.60	1.91	2.75	1.27	1.48	4.21	1.27	1.22
	D	1.43	1.69	1.56	1.20	1.29	0.83	1.46	1.21	0.46
	E	6.20	1.74	2.35	4.59	1.00	–	6.09	1.62	2.62
	F	2.32	1.78	2.09	1.65	1.00	–	2.18	1.55	1.70
63	A	2.80	2.30	3.22	1.92	2.16	5.33	1.85	1.34	1.26
	B	3.11	2.55	4.14	1.38	3.09	4.24	2.51	1.00	–
	C	5.16	2.67	4.05	3.35	1.63	3.58	3.72	1.27	1.13
	D	1.76	2.77	3.34	1.42	1.62	2.25	1.21	1.28	0.57
	E	6.74	3.09	4.69	5.27	1.39	2.69	6.44	1.09	0.16
	F	2.78	3.19	4.36	1.86	1.39	1.70	2.31	1.00	–
64	A	5.28	6.65	4.02	4.17	1.40	3.05	4.15	1.00	–
	B	5.90	2.41	4.74	4.93	1.14	1.32	4.45	1.28	1.43
	C	7.50	2.23	3.87	4.94	1.17	1.30	6.24	1.13	0.64
	D	2.40	2.26	3.18	1.97	1.14	0.55	2.15	1.00	–
	E	10.37	2.09	3.93	7.38	1.00	–	8.83	1.39	1.94
	F	3.47	2.11	3.16	2.59	1.00	–	2.86	1.35	1.28
65	A	6.73	3.11	1.61	6.65	2.46	8.30	2.96	2.00	3.80
	B	7.49	1.33	1.54	6.29	2.23	7.50	3.01	1.82	3.05
	C	7.91	1.67	2.21	6.73	2.22	7.12	3.75	1.98	3.37
	D	2.70	1.73	2.07	2.63	2.21	5.48	1.14	1.98	2.15
	E	10.14	2.01	3.35	7.74	2.20	7.18	5.43	1.77	2.88
	F	3.97	2.09	3.18	2.64	2.19	5.43	1.67	1.76	1.95
66	A	5.80	1.69	2.26	4.76	3.66	11.48	3.11	1.67	2.86
	B	5.61	1.78	2.90	4.44	3.20	10.30	3.18	1.51	2.13
	C	7.58	1.87	2.74	4.10	3.28	9.85	4.07	1.64	2.52
	D	2.54	1.97	2.55	1.57	3.22	7.03	1.21	1.67	1.59
	E	9.39	2.40	4.48	5.59	2.80	8.81	5.88	1.48	2.01
	F	3.68	2.49	3.93	1.91	2.65	6.07	1.75	1.52	1.44
67	A	3.17	2.20	3.00	1.51	1.79	3.39	2.16	1.06	0.33
	B	3.54	3.84	4.94	1.66	1.55	2.56	2.43	1.00	–
	C	4.01	2.39	3.45	1.53	1.25	0.98	3.07	1.06	0.57
	D	1.41	2.42	2.78	0.73	1.08	0.30	1.08	1.00	–
	E	6.29	2.19	3.45	2.27	1.00	–	4.46	1.00	–
	F	2.31	2.26	3.13	0.87	1.00	–	1.56	1.00	–

Note: k_1, retention factor of first eluting enantiomer; α, separation factor; R_s, resolution; "–", no separation; mobile phase, A, CO_2/MeOH (80/20, *v*/*v*); B, CO_2/EtOH (80/20, *v*/*v*); C, CO_2/DMC/MeOH (70/24/6, *v*/*v*/*v*); D, CO_2/DMC/MeOH (60/32/8, *v*/*v*/*v*); E, CO_2/DMC/EtOH (70/24/6, *v*/*v*/*v*); F, CO_2/DMC/EtOH (60/32/8, *v*/*v*/*v*); flow rate, 4 mL min^{-1}; backpressure, 15 MPa; detection, 254 nm; column temperature, 35 °C.

The analyzed racemic *trans*-β-lactam ureas **55, 57, 60**, and **62–67** (Figure 4) have the same β-lactam ring with two stereogenic centers at the C3 and C4 positions of the β-lactam ring. They contain various substituents on the ureido group attached to the C3 position of the ring, such as alkyl, hexyl, cyclohexyl, furfuryl, and various substituted phenyls. The type of the substituent in the structure of the analyte and the type of the polysaccharide selector significantly influence chiral recognition through multiple interactions. It is noted that chiral recognition of racemic solutes on polysaccharide CSPs is achieved through various types of bonding within the chiral helical grooves of the chiral selector (which form the chiral pocket), in particular through H-bonding, dipole–dipole, and π–π interactions, as well as steric effects.

3.1. HPLC Enantioseparation of (±)-Trans-β-Lactam Ureas 55, 57, 60, 62–67

Baseline separations were observed for eight compounds (except **67**) on the Cellulose-SB column with *n*-hexane/2-PrOH (9:1, *v*/*v*) as mobile phase, on the Cellulose-SC column for seven compounds (except compounds **60** and **66**) with the same mobile phase (Table 1). Comparing the data obtained with Cellulose-SB and Cellulose-SC, clearly higher retentions can be observed for all analytes on CSPs with *tris*(3,5-dichlorophenylcarbamate) moiety. The higher α and R_S values (except compounds **60, 65** and **66**) generally observed on CSPs with *tris*(3,5-dichlorophenylcarbamate) indicate more pronounced chiral selector-chiral analyte interactions of the analytes investigated.

A higher separation factor was obtained for Cellulose-SB, which proved to be a better choice than Amylose-SA for the separation of most β-lactam ureas. According to the data in Table 1, structural variations can significantly affect the retention factors. For example, on the Amylose-SA column, the k_1 of the analytes **57, 60, 62**, and **64–67** are higher than those of the other analytes. The possible reason for this could be the presence of the allyl, furfuryl, or phenyl ring in the *N*-position of the ureido moiety, which could cause additional π–π interactions between the CSP and the analytes. On the other hand, analytes **55** and **63** always have the lowest k_1 values, indicating that the interaction between these compounds and the CSP is very weak, possibly due to the long alkyl or cycloalkyl substitution in the *N*-position of the ureido group. On Cellulose-SB, the k_1 values for the ureas **63** with cyclopentyl group and for **67** with 2,6-dimethylphenyl group are the lowest than the others tested, and the possible reason for this could be the rigidity of the cyclopentyl ring or the steric effect of two methyl groups in the *ortho*-position of the phenyl ring. These results indicate that the different structural features of the CSP in combination with the mobile phase *n*-hexane/2-PrOH ultimately lead to a different stereo-environment of the chiral cavities in the CSP, resulting in different chiral selectivities.

Next, dimethyl carbonate (DMC) was investigated as a mobile phase for the separation of racemic β-lactam ureas **55, 57, 60** and **62–67**. DMC has been classified as one of the most environmentally friendly solvents, in the same class as water, short-chain alcohols and ethyl acetate [27–29]. DMC also degrades rapidly in the atmosphere/environment (over 90% in 28 days) and can therefore be considered non-toxic [27,30]. However, the applications of DMC in analytical chemistry appear to be very limited. DMC has been used as a mobile phase modifier in reversed-phase liquid chromatography (RPLC) with ICP-MS detection [31] and for normal-phase liquid chromatography (NPLC) and hydrophilic interaction liquid chromatography (HILIC) [32]. So far, there are only two examples of the use of DMC in chiral HPLC chromatography, and both were carried out in our laboratory [23,33]. For chiral separation of *anti*- and *syn*-hydantoins 100% DMC was used [23], and DMC with 10% alcoholic modifier (MeOH or EtOH) was used for marinoepoxides [33].

When DMC was used, baseline separation was achieved for eight compounds (except **67**) on the Amylose-SA column and for the five compounds **60, 63**, and **65–67** on the Cellulose-SB column. For the other compounds tested, only partial resolution was observed on these two columns. For the Amylose-SA and Cellulose-SC columns, the retention factors for the most compounds were higher when pure DMC was used as mobile phase compared to MeOH or EtOH, and this was accompanied by the higher resolution values in almost all cases. Out of the nine β-lactam ureas used in this study, ureas **55**, **57**, and **62–66** achieved the higher R_s and α-values under DMC conditions on the Amylose-SA column compared to Cellulose-SB. In a few cases, such as with the compounds **55**, **63**, **65**, and **66**, the use of DMC as the mobile phase afforded superior R_s and α on the Amylose-SA. Notably, the Cellulose-SB showed significantly higher enentioselectivity values for the ureas **60** and **65–67** compared to Cellulose-SC, which contains the cellulose *tris*(3,5-dichlorophenylcarbamate) as the chiral selector. This column was able to baseline separate only three compounds, **63**, **65**, and **66**, whereas the others were partially separated or not separated at all with DMC. The nature of DMC, the structures of the analytes, and their interactions with the CSP all play a role in enantioselectivity.

Some interesting observations resulted from the comparison of the separation systems with different alcoholic modifiers. Interestingly, higher resolution values were obtained with EtOH compared to MeOH (except **65**) on Amylose-SA, indicating that EtOH may be an advantageous alternative mobile phase for MeOH. An additional example that shows the difference in using MeOH and EtOH as a pure mobile phase is that compounds **55**, **57**, **62**, and **64** on Cellulose-SB were partially separated using MeOH, but, by changing the mobile phase to EtOH, these compounds were not separated. In the case of compounds **55**, **57**, **62**, **64**, and **67**, enantioseparation was not achieved with either MeOH or EtOH when the Cellulose-SC column was used. The other tested compounds **60**, **65**, and **66** were partially separated with both alcohol modifiers, but higher R_s were obtained for these compounds when MeOH was used as mobile phase. On the Amylose-SA column, $R_s > 1.5$ were obtained for compounds **57**, **63**, **64**, and **66** with MeOH and for compounds **55**, **57**, **62–65**, and **67** with EtOH, but better resolution and separation values were obtained with 100% EtOH (all except compound **66**).

The success rate on each immobilized stationary phase is slightly different for each mobile phase. The success rates were expressed in terms of baseline-, partially, and not-separated compounds (Table 1). On Cellulose-SC, it is not possible to separate more than three of nine tested compounds (33%) with EtOH and four compounds with MeOH (44%) (Table 1). The mobile phase *n*-hexane/2-PrOH (90/10, *v*/*v*) is the most successful and separates all nine compounds, whereas eight compounds (89%) are separated with DMC. On Amylose-SA, the traditional mobile phase *n*-hexane/2-PrOH (90/10, *v*/*v*) performs slightly worse (55%) than the polar organic (89% with MeOH and EtOH) and non-standard DMC (100%) mobile phases. The Cellulose-SB column performs worse than the other two columns, providing the separation for only five racemates (55%) with 100% DMC. The mobile phases *n*-hexane/2-PrOH and MeOH are the most successful and separate all nine compounds. On this CSP, the polar organic mobile phase (EtOH) and atypical modifier (DMC) provide the same success rates (55%). For Cellulose-SB, MeOH yields slightly more baseline and partial enantioseparations than EtOH. For Cellulose-SC, the situation is the same, i.e., MeOH yields somewhat more separations. For Amylose-SA, MeOH and EtOH yield a similar number of separations, which is significantly higher than for Cellulose-SC. It is known from the literature that alcohols of different size and bulkiness can be incorporated into the CSP structures and can also cause conformational changes in the helical structure of the chiral selectors of amylose or cellulose, which result in different stereo-environments [34–36].

Some representative HPLC chromatograms are depicted in Figure 6.

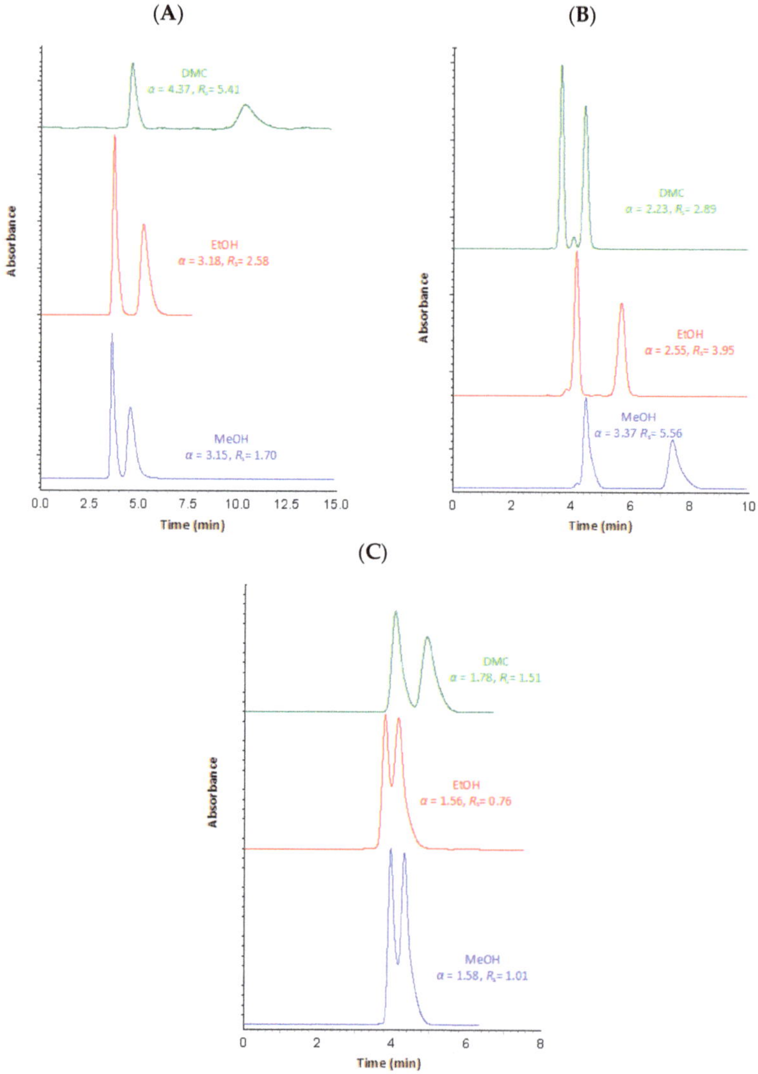

Figure 6. Effect of mobile phase composition on enantioseparation of cyclopentyl urea **63** on Amylose-SA (**A**), 3-chloro-4-methyl urea **60** on Cellulose-SB (**B**), and 4-*tert*-butylphenyl urea **65** on Celulose-SC (**C**) under HPLC conditions. Chromatographic conditions: mobile phase, MeOH (—), EtOH (—), and DMC (—); flow rate, 1 mL min^{-1}; detection, 254 nm; temperature, 35 °C.

3.2. SFC Enantioseparation of (±)-Trans-β-Lactam Ureas 55, 57, 60, and 62–67

The enantioseparation of (±)-*trans*-β-lactam ureas **55**, **57**, **60**, and **62–67** on polysaccharide CSPs in SFC mode was investigated under typical supercritical chromatography conditions, i.e., CO_2 with an alcoholic modifier, and under atypical conditions with a non-standard modifier, in this case with DMC.

The effect of the polar modifiers MeOH and EtOH in the mobile phase CO_2/alcohol on the enantioresolution for the selected (±)-*trans*-β-lactam ureas **55**, **57**, **60**, and **62–67** was investigated for all three immobilized CSPs.

The use of Amylose-SA for the enantioseparation of nine racemates showed no clear preference for the modifier. Eight compounds (except **67**) were baseline resolved with

MeOH, and the same number were baseline separated using EtOH as modifier. Herein, with EtOH as a bulk solvent component, retention values were higher in most cases on all used polysaccharide columns compared to the with the shorter alcohol MeOH. In general, better selectivity and resolution on Amylose-SA were achieved in most cases when EtOH was used as a polar modifier. The size of the alcoholic modifier seems to affect the chiral recognition mechanisms of the studied compounds of this type of polysaccharide CSP. On the Cellulose-SB column, nine analytes were baseline separated with MeOH and seven racemates with EtOH as mobile phase modifier. The Cellulose-SC column baseline separated three racemates with MeOH and five racemates with EtOH as the modifier. No separation was observed on Cellulose-SC for compounds **55**, **57**, **62**, and **64** with CO_2/MeOH (80/20, *v*/*v*) and for compounds **63** and **67** with CO_2/EtOH (80/20, *v*/*v*).

The type of alcoholic modifier can also influence retention and enantioseparation [37]. An NMR study on the effects of alcoholic modifiers on the structure and chiral selectivity of amylose-based CSP showed that the use of alcohols with different volume and concentration in the eluent can lead to differences in the chiral recognition ability of the polymer CSP as a consequence of its structural changes (e.g., crystallinity of the polymer, side chain mobility, and conformation) after incorporation of the alcohol into the CSP [34,38].

It can be concluded that both chiral stationary phases based on *tris*(3,5-dimethylphenyl carbamate) cellulose and amylose, Cellulose-SB and Amylose-SA, are suitable for the enantioseparation of the investigated β-lactam ureas **55**, **57**, **60**, and **62–67**. On the other hand, cellulose *tris*(3,5-dichlorophenylcarbamate), Cellulose-SC, showed a relatively lower separation performance. Interestingly, the chiral recognition abilities of Amylose-SA and Cellulose-SB are comparable, although they contain different polysaccharide chiral selectors.

When DMC was used as a co-solvent in the SFC, the enantiomers of compounds **55** and **62** were not eluted from the tested immobilized columns within one hour. It is important to note that, in this study, 10–40% volume percent of DMC was used under SFC conditions. Preliminary tests showed that up to a ratio of 60/40 (*v*/*v*), CO_2/DMC, the mobile phase strength was insufficient to elute these racemates. The alcohol content and type (MeOH, EtOH, and 2-PrOH) can be used to modulate retention and chiral recognition [37]. So, the alcoholic modifier (MeOH and EtOH) was added in 20% to DCM to increase elution strength for not-eluted compounds.

For the same modifier (DMC/MeOH or DMC/EtOH), as shown in Table 2, k_1 decreased with increasing DMC/alcohol volume fraction (from 24/6 to 32/8). This shows that increasing the volume fraction of the alcoholic modifier accelerates the elution rate and shortens the retention time. MeOH is a protic solvent that is also a proton donor and proton acceptor. It can form hydrogen-bonds with the β-lactam urea compounds and with the chiral stationary phase, thus competing with the compounds for hydrogen bonds. Increasing the volume fraction of MeOH (from 6% to 8%) increases this competition, reduces the interaction between the compound and the CSP, and shortens the retention time. MeOH competes with the analytes for the hydrogen bonding sites of the CSPs, whereas DMC as an aprotic solvent interacts significantly with the polymer through dipole–dipole interaction. The functional groups of the alcohols form strong H-bond complexes with the C=O and NH functional groups of the polysaccharide polymer [39]. When EtOH is used as a co-solvent instead of MeOH, the polarity of the mobile phase is reduced, the hydrogen bonding of EtOH is weaker, and the elution rate is much slower. Overall, the separation factor and resolution decreased with the increase in the volume fraction of mobile phase B, DMC/MeOH, or DMC/EtOH.

On Amylose-SA, nine separations are achieved with all four mobile phases CO_2/DMC/alcohol modifier (mobile phases C–F). For Cellulose-SB, nine separations are generated by CO_2/DMC/MeOH (mobile phases C and D) and four such separations CO_2/DMC/EtOH (mobile phases E and F), respectively, whereas Cellulose-SC generates a similar number of separations. The effects of the modifiers were rather unpredictable, as is generally the case for chiral separations. In some cases, the number of enantioseparations with all modifiers was quite similar, in other cases they differed significantly, as can be seen in Table 2. The first SFC separations with CO_2/DMC/alcoholic modifier on immobilized amylose and cellulose columns are shown, providing a new approach for the supercritical separation of the (±)-*trans*-β-lactam ureas **55, 57, 60,** and **62–67**.

The most successful composition of the mobile phase is different for each stationary phase. For Cellulose-SB, the mobile phases CO_2 with MeOH as the organic modifier perform slightly better than those with only EtOH or with alcohol modifier/atypical solvent mixture (MeOH/DMC or EtOH/DMC). For Amylose-SA and Cellulose-SC, the mobile phases with CO_2/EtOH and CO_2/EtOH/DMC provide better separation than the other mobile phases used. Overall, as described above, Amylose-SA provides the most baseline and partial enantioseparations and thus has the highest success rate under these conditions.

Some representative HPLC chromatograms are depicted in Figure 7.

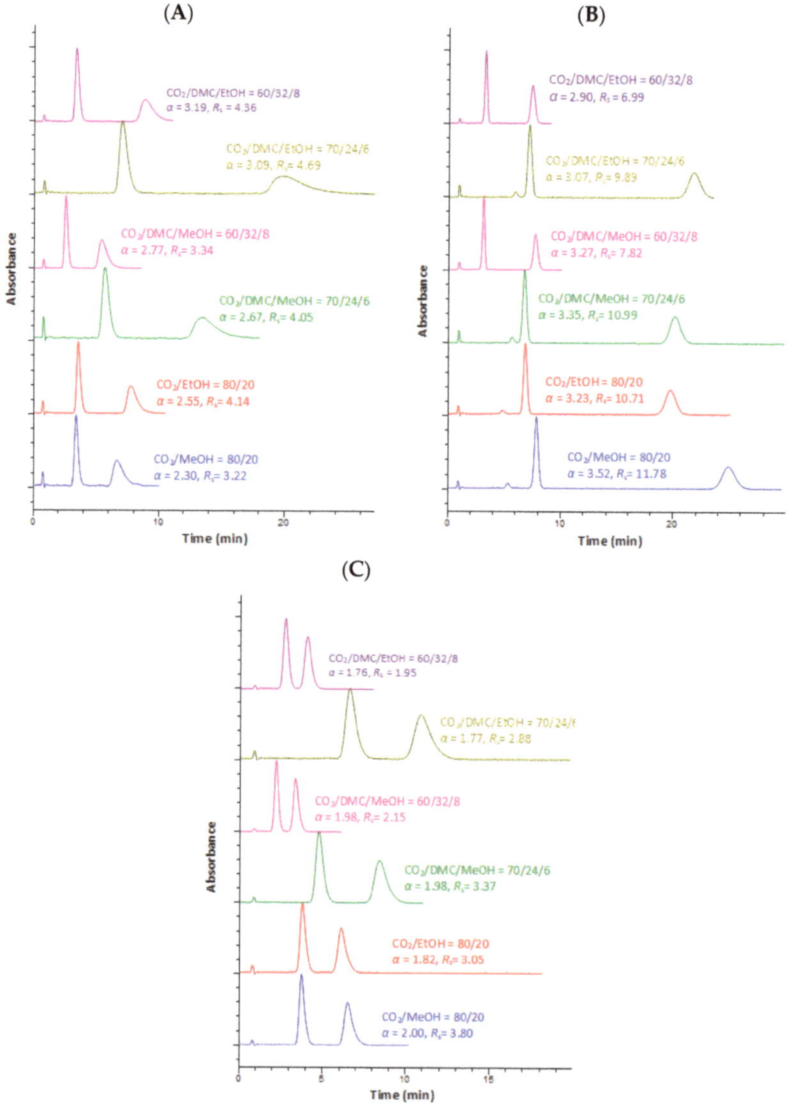

Figure 7. Effect of mobile phase composition on enantioseparation of cyclopentyl urea **63** on Amylose-SA (**A**), 3-chloro-4-methyl urea **60** on Cellulose-SB (**B**), and 4-*tert*-butylphenyl urea **65** on

Celulose-SC (**C**) under SFC conditions. Chromatographic conditions: mobile phase, CO_2/MeOH (80/20 *v*/*v*) (—), CO_2/EtOH (80/20 *v*/*v*) (—), CO_2/DMC/MeOH (70/24/6 *v*/*v*/*v*) (—), CO_2/DMC/MeOH (60/32/8 *v*/*v*/*v*) (—), CO_2/DMC/EtOH (70/24/6 *v*/*v*/*v*) (—), and CO_2/DMC/EtOH (60/32/8 *v*/*v*/*v*) (—); flow rate, 4 mL min^{-1}; detection, 254 nm; temperature, 35 °C.

4. Conclusions

In this work, the enantioseparation of nine (±)-*trans*-β-lactam ureas was performed by HPLC and SFC on three immobilized polysaccharide-based CSPs containing amylose *tris*(3,5-dimethylphenyl)carbamate (CHIRAL ART Amylose-SA), cellulose *tris*(3,5-dimethylphenyl)carbamate (CHIRAL ART Cellulose-SB), and cellulose *tris*(3,5-dichlorophenyl)carbamate (CHIRAL ART Cellulose-SC) as chiral selector.

In HPLC mode, the Cellulose-SB column is the only chiral stationary phase in this study that was able to separate all nine compounds tested with the mobile phases *n*-hexane/2-PrOH (90/10, *v*/*v*) and 100% MeOH. Of the three polysaccharide columns, Amylose-SA showed better recognition ability for the tested compounds with 100% DCM than Cellulose-SB and Cellulose-SC.

In the present study, DMC was used as an organic modifier in SFC for the first time. The elution strength of DMC was increased by adding a more polar solvent, MeOH or EtOH. Among the CSPs tested, both the Amylose-SA and Cellulose-SB were able to resolve all nine urea compounds to baseline or partially.

We have shown that the environmentally friendly solvent dimethyl carbonate (DMC) can be used efficiently as a mobile phase in HPLC mode, but it can also be used in SFC mode with the addition of the alcoholic modifiers MeOH or EtOH. The chiral stationary phase Amylose-SA provides the most baseline and partial enantioseparations and thus has the highest success rate with the mobile phase CO_2/DMC/alcohol.

Author Contributions: M.J. performed synthesis, chromatographic analyzes, literature search, and writing. D.K. and M.R. performed the study design, data analysis, revision, final approval, and handled the accountability of all aspects of the work. All authors contributed to the article and approved the submitted version. All authors have read and agreed to the published version of the manuscript.

Funding: This research was supported by the Croatian Government and the European Union through the European Regional Development Fund-the Competitiveness and Cohesion Operational Programme (KK.01.1.1.01) through the project Bioprospecting of the Adriatic Sea (KK.01.1.1.01.0002) granted to The Scientific Centre of Excellence for Marine Bioprospecting-BioProCro.

Data Availability Statement: Data are contained within the article.

Acknowledgments: We would like to thank the Croatian Government and the European Union through the European Regional Development Fund-the Competitiveness and Cohesion Operational Programme (KK.01.1.1.01) for funding this research through the project Bioprospecting of the Adriatic Sea (KK.01.1.1.01.0002) granted to The Scientific Centre of Excellence for Marine Bioprospecting-BioProCro.

Conflicts of Interest: The authors declare no conflicts of interest.

References

1. Jastrzebski, J.T.B.H.; van Koten, G. The zinc mediated condensation of amino acid esters with imines t β-lactams. *Bioorganic Med. Chem. Lett.* **1993**, *3*, 2351–2356. [CrossRef]
2. Banik, B.K.; Becker, F.F. Unprecedented stereoselectivity in the Staudinger reaction with polycyclic aromatic imines. *Tetrahedron Lett.* **2000**, *41*, 6551–6554. [CrossRef]
3. Page, M.I.; Laws, A.P. The Chemical Reactivity of β-Lactams, β-Sultams and β-Phospholactams. *Tetrahedron* **2000**, *56*, 5631–5638. [CrossRef]
4. Wang, Z.; Ni, J.; Kuninobu, Y.; Kanai, M. Copper-Catalyzed Intramolecular C(sp^3)–H and C(sp^2)–H amidation by Oxidative Cyclization. *Angew. Chem. Int. Ed.* **2014**, *53*, 3496–3499. [CrossRef]

5. Mehta, P.D.; Sengar, N.P.S.; Pathak, A.K. 2-Azetidinone—A new profile of various pharmacological activities. *Eur. J. Med. Chem.* **2010**, *45*, 5541–5560. [CrossRef]
6. Jarrahpour, A.; Ebrahimi, E.; Sinou, V.; Latour, C.; Brunel, J.M. Diastereoselective synthesis of potent antimalarial cis-β-lactam agents through a [2 + 2] cycloaddition of chiral imines with a chiral ketene. *Eur. J. Med. Chem.* **2014**, *87*, 364–371. [CrossRef]
7. Kamath, A.; Ojima, I. Advances in the chemistry of β-lactam and its medicinal applications. *Tetrahedron* **2012**, *68*, 10640–10664. [CrossRef] [PubMed]
8. Hosseyni, S.; Jarrahpour, A. Recent advances in β-lactam synthesis. *Org. Biomol. Chem.* **2018**, *16*, 6840–6852. [CrossRef] [PubMed]
9. Jarrahpour, A.; Shirvani, P.; Sinou, V.; Latour, C.; Brunel, J. Synthesis and biological evaluation of some new β-lactam-triazole hybrids. *Med. Chem. Res.* **2016**, *25*, 149–162. [CrossRef]
10. Dražić, T.; Roje, M. β-lactam rearrangements into five-membered heterocycles. *Chem. Heterocycl. Compd.* **2017**, *53*, 953–962. [CrossRef]
11. Pirkle, W.H.; Tsipouras, A.; Huyn, M.H.; Hart, D.J.; Lee, C.S. Use of chiral stationary phases for the chromatographic determination of enantiomeric purity and absolute configuration of some β-lactams. *J. Chromatogr.* **1986**, *358*, 377–384. [CrossRef]
12. Lee, C.S.; Chen, H.H. Determination of enantiomeric purity and absolute configuration of b-lactams by high-performance liquid chromatography on chiral columns. *J. Chin. Chem. Soc.* **1994**, *41*, 187–190. [CrossRef]
13. Cirilli, R.; Del Guidice, M.R.; Ferretti, R.; La Torre, F. Conformational and temperature effects on separation of stereoisomers of a C3,C4-substituted b-lactamic cholesterol absorption inhibitor on amylosebased chiral stationary phases. *J. Chromatogr. A* **2001**, *923*, 27–36. [CrossRef]
14. Okamoto, Y.; Senoh, T.; Nakane, H.; Hatada, K. Optical resolution of betalactams by chiral HPLC on *tris*(phenylcarbamate)s of cellulose and amylose. *Chirality* **1989**, *1*, 216–222. [CrossRef]
15. Pataj, Z.; Ilisz, I.; Berkecz, R.; Forró, E.; Fülöp, F.; Péter, A. Comparison of separation performances of amylose- and cellulose-based stationary phases in the high-performance liquid chromatographic enantioseparation of stereoisomers of β-lactams. *Chirality* **2010**, *22*, 120–128. [CrossRef] [PubMed]
16. Péter, A.; Arki, A.; Forró, E.; Fülöp, F.; Armstrong, D.W. Direct high-performance liquid chromatographic enantioseparation of β-lactam stereoisomers. *Chirality* **2005**, *17*, 193–200. [CrossRef]
17. Sun, P.; Wang, C.; Armstrong, D.W.; Péter, A.; Forró, E. Separation of Enantiomers of β-Lactams by HPLC Using Cyclodextrin-Based Chiral Stationary Phases. *J. Liq. Chromatogr. Relat. Technol.* **2006**, *29*, 1847–1860. [CrossRef]
18. Berkecz, R.; Török, R.; Ilisz, I.; Forró, E.; Fülöp, F.; Armstrong, D.W.; Péter, A. LC Enantioseparation of β-Lactam and β-Amino acid Stereoisomers and Comparison of Macrocyclic Glycopeptide- and β-Cyclodextrin-Based columns. *Chromatographia* **2006**, *63*, 37–43. [CrossRef]
19. Fodor, G.; Ilisz, I.; Szemán, J.; Iványi, R.; Szente, L.; Varga, G.; Forró, E.; Fülöp, F.; Péter, A. LC Enantioseparation of β-Lactam Stereoisomers through the Use of β-Cyclodextrin-Based Chiral Stationary Phases. *Chromatographia* **2010**, *71*, 29–34. [CrossRef]
20. Jiang, C.; Armstrong, D.W.; Péter, A.; Fülöp, F. Enantiomeric separation of a series of β-lactams using capillary zone electrophoresis. *J. Liq. Chromatogr. Relat. Technol.* **2007**, *30*, 1709–1721. [CrossRef]
21. Jurin, M.; Kontrec, D.; Dražić, T.; Roje, M. Enantioseparation of (±)-trans-β-lactam Ureas by Supercritical Fluid Chromatography. *Croat. Chem. Acta* **2020**, *93*, 203–213. [CrossRef]
22. Jurin, M.; Kontrec, D.; Dražić, T.; Roje, M. Enantioseparation of syn- and anti-3,5-Disubstituted Hydantoins by HPLC and SFC on Immobilized Polysaccharides-Based Chiral Stationary Phases. *Separations* **2022**, *9*, 157. [CrossRef]
23. Allenmark, S. *Chromatographic Enantioseparation: Methods and Applications*, 2nd ed.; Ellis Horwood: New York, NY, USA, 1991; pp. 49–53.
24. Chankvetadze, B. Recent trends in preparation, investigation and application of polysaccharide-based chiral stationary phases for separation of enantiomers in high-performance liquid chromatography. *TrAC Trends Anal. Chem.* **2020**, *122*, 115709. [CrossRef]
25. Dallocchio, R.; Dessì, A.; Sechi, B.; Peluso, P. Molecular Dynamics Simulations of Amylose- and Cellulose-Based Selectors and Related Enantioseparations in Liquid Phase Chromatography. *Molecules* **2023**, *28*, 7419. [CrossRef]
26. Thunberg, L.; Hashemi, J.; Andersson, S. Comparative study of coated and immobilized polysaccharide-based chiral stationary phases and their applicability in the resolution of enantiomers. *J. Chromatogr. B* **2008**, *875*, 72–80. [CrossRef]
27. Pyo, S.H.; Park, J.H.; Chang, T.S.; Hatti-Kaul, R. Dimethyl carbonate as a green chemical. *Curr. Opin. Green Sustain. Chem.* **2017**, *5*, 61–66. [CrossRef]
28. Kim, K.H.; Lee, E.Y. Environmentally-benign dimethyl carbonate-mediated production of chemicals and biofuels from renewable bio-oil. *Energies* **2017**, *10*, 1790. [CrossRef]
29. Arico, F.; Tundo, P. Dimethyl carbonate: A modern green reagent and solvent. *Russ. Chem. Rev.* **2010**, *79*, 479–489. [CrossRef]
30. Abdalla, A.O.G.; Liu, D. Dimethyl carbonate as a promising oxygenated fuel for combustion: A review. *Energies* **2018**, *11*, 1552. [CrossRef]
31. Lajin, B.; Goessler, W. Introducing dimethyl carbonate as a new eluent in HPLC-ICPMS: Stronger elution with less carbon. *J. Anal. At. Spectrom.* **2021**, *36*, 1272–1279. [CrossRef]
32. Boes, P.D.; Elleman, S.R.; Danielson, N.D. Dimethyl Carbonate as a Mobile-Phase Modifier for Normal-Phase and Hydrophilic Interaction Liquid Chromatography. *Separations* **2023**, *10*, 70. [CrossRef]
33. Buljan, A.; Roje, M. Application of Green Chiral Chromatography in Enantioseparation of Newly Synthesized Racemic Marinoepoxides. *Mar. Drugs* **2022**, *20*, 530. [CrossRef] [PubMed]

34. Wang, T.; Wenslow, R.M., Jr. Effects of alcohol mobile-phase modifiers on the structure and chiral selectivity of amylose *tris*(3, 5-dimethylphenylcarbamate) chiral stationary phase. *J. Chromatogr. A* **2003**, *1015*, 99–110. [CrossRef] [PubMed]
35. Köteles, I.; Foroughbakhshfasaei, M.; Dobó, M.; Ádám, M.; Boldizsár, I.; Szabó, Z.-I.; Tóth, G. Determination of the Enantiomeric Purity of Solriamfetol by High-Performance Liquid Chromatography in Polar Organic Mode Using Polysaccharide-Type Chiral Stationary Phases. *Chromatographia* **2020**, *83*, 909–913. [CrossRef]
36. Lämmerhofer, M. Chiral recognition by enantioselective liquid chromatography: Mechanisms and modern chiral stationary phases. *J. Chromatogr. A* **2010**, *1217*, 814–856. [CrossRef] [PubMed]
37. Zhang, X.; Li, Z.; Shen, B.; Chen, J.; Xu, X. Enantioseparation of three non-steroidal Anti-inflammatory agents on chiral stationary phase by HPLC. *J. Anal. Sci.* **2012**, *2*, 18–23. [CrossRef]
38. Kasat, R.B.; Zvinevich, Y.; Hillhouse, H.W.; Thomson, K.T.; Wang, N.H.; Franses, E.I. Direct probing of sorbent-solvent interactions for amylose *tris*(3,5-dimethylphenylcarbamate) using infrared spectroscopy, X-ray diffraction, solid-state NMR, and DFT modeling. *J. Phys. Chem. B* **2006**, *110*, 14114–14122. [CrossRef]
39. Peluso, P.; Mamane, V.; Cossu, S.; Dallocchio, R. Noncovalent interactions in highperformance liquid chromatography enantioseparations on polysaccharide-based chiral selectors. *J. Chrom. A* **2020**, *19*, 461202. [CrossRef]

separations MDPI

Article

Real-Time Authentication of Camellia Oil by Rapid Evaporative Ionization Mass Spectrometry

Jun Xiang [1,2], Qi Liu [1], Huihua Jing [2] and Xiaoqing Chen [1,*]

1 Chemistry and Chemical Engineering, Central South University, Changsha 410083, China; 212302054@csu.edu.cn (J.X.)
2 Hunan Provincial Institute of Product and Goods Quality Inspection, Changsha 410007, China
* Correspondence: xqchen@csu.edu.cn

Abstract: Camellia oil is a high-value product with rich nutrients. Recently, the adulteration of camellia oil has become an increasingly concerning issue related to human health. In this study, electric soldering iron coupled with rapid evaporative ionization mass spectrometry (REIMS) was employed for the identification and analysis of camellia oil without any sample preparation. REIMS technology coupled with chemometrics was applied to develop an analysis model for the authentication of camellia oil adulterated with soybean oil, peanut oil, rapeseed oil, sunflower oil, and corn oil (5–40%, *v*/*v*). The results showed that different types of vegetable oils could be classified using principal component analysis-linear discriminant analysis (PCA-LDA) with a correct classification of 93.8% in leave-20%-out cross-validation and 100% correctly identified in real-time recognition. The established prediction models were found to be particularly sensitive when the camellia oil samples were adulterated with 5–40% of other oils, indicating that REIMS could be a powerful tool for the authentication and adulteration analysis of camellia oil, particularly for cases where the adulteration levels are relatively high. In conclusion, the results provide valuable insights into the potential of REIMS for the rapid, accurate, and real-time authentication and adulteration analysis of camellia oil.

Keywords: electric soldering iron; rapid evaporative ionization mass spectrometry; camellia oil; adulteration

Citation: Xiang, J.; Liu, Q.; Jing, H.; Chen, X. Real-Time Authentication of Camellia Oil by Rapid Evaporative Ionization Mass Spectrometry. *Separations* **2024**, *11*, 68. https://doi.org/10.3390/separations11030068

Academic Editor: Petr Bednar

Received: 18 January 2024
Revised: 18 February 2024
Accepted: 19 February 2024
Published: 24 February 2024

1. Introduction

Camellia oil, a versatile vegetable oil, is an edible product derived from the mature seeds of the *Camellia oleifera*, a plant that is widely cultivated in southern China, specifically in regions such as Guangxi, Yunnan, and Fujian provinces, where the climate is conducive for the growth of the plant [1]. Camellia oil boasts a high content of unsaturated fatty acids and natural tea polyphenols. Camellia oil's fatty acid composition closely resembles that of olive oil, which is commonly considered an optimal oil for human consumption [2]. Due to its health-promoting characteristics, camellia oil has been referred to as "oriental olive oil", as it is rich in antioxidants and phytochemicals derived from its plant. In addition to its nutritional benefits, camellia oil has been reported to possess numerous health-promoting properties, including antioxidant, cardiovascular, and anti-tumor properties, and it is commonly used in the healthcare field to treat liver blood deficiency, expel parasites, benefit the intestines and stomach, and improve eyesight [3,4]. As a result of these benefits, camellia oil has been supported and endorsed by the Chinese government and the International Food and Agriculture Organization as a pure, natural edible vegetable oil [5]. As a result of increasing demand and change in consumer attitudes, the demand for camellia oil is also on the rise, which has led to a subsequent increase in the market price of camellia oil. However, compared to other commonly used edible oils such as rapeseed oil, soybean oil, and peanut oil, camellia oil is generally more expensive. This higher price point is largely due to the high cost of cultivation and production of camellia seeds. This has led some illegal traders to adulterate camellia oil with cheaper oils in order to maximize profits [6].

The addition of low-quality oils to camellia oil can lead to a significant reduction in its nutritional value and overall quality. The adulterated camellia oil can subsequently pose a significant risk to public health due to the unforeseen and potentially harmful effects of consuming these products. Thus, it is crucial to develop reliable and accurate methods for the identification and detection of adulteration in camellia seed oil, in order to protect the interests of legitimate producers and consumers alike, ensuring that the oil consumers are obtaining is of the highest quality possible.

Over these past few decades, numerous techniques have been developed and utilized for the authentication of edible oils, including Fourier Transform Infrared Spectroscopy (FT-IR) [5], infrared spectroscopy [7,8], Raman spectroscopy [9], fluorescence spectroscopy [10], gas chromatography-mass spectrometry [11], ultra-performance liquid chromatography quadrupole time-of-flight mass spectrometry [12], nuclear magnetic resonance [13], electronic nose [14], and electronic tongue [15], which are all highly sensitive and specific methods for the authentication of edible oils. Furthermore, many of these techniques were also utilized in the analysis and authentication of camellia oil. As reported by Dou and coworkers, the liquid chromatography coupled with tandem mass spectrometry was applied in the evaluation of the amounts of several key compounds, which include isoflavones, sinapic acid, and resveratrol [6]. Their results revealed that soybean oil contained four isoflavones concurrently, whereas peanut oil had only resveratrol, and rapeseed oil had significantly more sinapic acid than others. This method enables the detection of these compounds as chemical markers for the adulteration of camellia oil. In Wang's research, fatty acid, phytosterols, and squalene were analyzed by GC and GC-MS fingerprints in conjunction with chemometrics that were employed to identify the adulteration of camellia oil with multiple vegetable oils [11]. As a result, in PCA, Camellia oil stood out with its elevated levels of triterpene alcohols, enabling it to be distinguished from other vegetable oils. By utilizing partial least squares-discriminant analysis, adulterated camellia oil, particularly when the adulterated ratio exceeded 30%, could be accurately classified with a discrimination accuracy exceeding 92.31% of the total. On the other hand, the use of electronic noses and electronic tongues, which consist of chemical sensors with distinct selectivity for data acquisition, coupled with appropriate algorithms for signal processing, also played an important role in authentication analysis [16]. In a study conducted by Wang, they used an e-nose combined with chemometrics (PCA, LDA, ANN) for the discrimination of camellia oil adulterated with maize oil [14]. The results demonstrated that the LDA model performed better than the PCA model in distinguishing the adulterated oil. However, the quantification model still requires improvement as the ANN model showed limitations in accurately determining the percentage of maize oil in camellia oil. Further research is needed to enhance the quantification capabilities of these models. In Shan's study, four pattern recognition methods were assessed to authenticate pure camellia oil using near infrared spectroscopy [17]. The NIR spectral data was initially analyzed using unsupervised methods such as PCA and HCA. Two supervised classification techniques, DA and RBFNN, were employed to develop calibration models and predict unknown samples. The solutions achieved a correct classification rate of 98.3%. However, if the established methods require large and expensive devices or complex preprocessing, the entire detection process is time-consuming, complex, and costly, making them difficult to apply in actual situations.

Rapid evaporative ionization mass spectrometry (REIMS) is a recent form of ambient mass spectrometry, that allows in situ, real-time analysis of tissue samples [18]. Combined with mass spectrometry and a surgical diathermy device called the intelligent knife (iKnife) [19], it could achieve handheld sampling and direct analysis of the sample without preparation, generating information-rich aerosols that are then transferred via an ion transfer pathway into the mass spectrometer for accurate molecular mass analysis in a matter of seconds [20,21]. Based on REIMS, the analysis of samples does not require any sample pretreatment or liquid chromatography separation and typically takes only a few seconds to complete data collection and analysis with a high accuracy rate [22]. Meanwhile, many kinds of molecular species could be detected with the combined mass spectrometry detector.

Currently, REIMS has been successfully applied in clinical tissue resection, tumor diagnosis, microbial identification, food authentication, and other fields, providing a novel research method and approach for lipidomics research [23–26]. In terms of the identification of the species of origin or quality for food products, it is of utmost importance to have the right tools and equipment to ensure accurate and reliable results, thus providing consumers with reliable information on what they are consuming. In this respect, REIMS was used for the identification and discrimination of shrimp, animal species, *Pelodiscus sinensis*, beer, and fresh meat, etc. [25,27–30]. As a valid direct sampling analysis technique, REIMS provides numerous advantages when utilized as an instrument for authenticity testing. One notable benefit is its ability to drastically reduce analysis time, with results being obtained within seconds or minutes [31]. When coupled with high-resolution mass spectrometry, REIMS becomes even more powerful, enabling the generation of untargeted mass spectral profiles through in situ sampling [32]. Moreover, Black's research showed that REIMS could be utilized to detect different fish species [33]. However, the lipid compositions of those fish samples had little difference and the following analysis was less effective, which showed that the analyzation of data was also important. These spectral profiles of authentic samples could then be submitted to exercise and validate chemometric models, employing multivariate statistical analysis algorithms. By comparing the spectral profiles of unknown samples with the validated model through similarity scoring, real-time classification can be achieved. This approach allows for the identification of unknown samples based on their similarity to the established profiles [28].

The primary focus of this research was to explore the potential of REIMS in combination with contact heating using a soldering iron. This combination was specifically aimed at authenticating camellia oil, distinguishing it from other vegetable oils, and detecting adulteration composition and content. The goal was to utilize the capabilities of REIMS to provide reliable and efficient methods for ensuring the authenticity and quality of camellia oil. Orthogonal partial least squares discriminant analysis was performed to develop qualitative and quantitative models for the identification of camellia oil adulterated with other edible oils. This technique affords a basis for establishing the requirement of real-time and high-throughput recognition of camellia oil and expanding the application fields of REIMS.

2. Materials and Methods

2.1. Reagents and Materials

Methanol in mass spectrometry grade, isopropanol, and sodium formate in chromatography grade were bought from Merck (Darmstadt, Germany). Leucine-enkephalin was purchased from Waters (Waters Corporation, Wilmslow, UK). Six types of pure plant oils, including pressed camellia oil, soybean oil, rapeseed oil, sunflower seed oil, peanut oil, and corn oil were collected from markets. Through the communication of the manufacturer and the test by the quality supervision department, these samples were all qualified pure oil products. To simulate binary mixtures for discrimination, different proportions (2%, 5%, 10%, 20%, 30%, and 40%, *w*/*w*) of soybean oil, rapeseed oil, sunflower seed oil, peanut oil, and corn oil were mixed into camellia seed oil as adulterants. The water was obtained by a Milli-Q Reference ultra-pure water generator (Millipore, Milwaukee, WI, USA).

2.2. Sample Preparation

Approximately 20 g of oil samples were poured into a sample cup. The electric soldering iron probe (ESII probe WSD151, Weller, Berlin, Germany) was used in this study, which has a similar function to the iknife [34,35]. It was heated to 500 °C and touched the sample for 0.5 s. The generated gaseous aerosols were directly collected, passed through a high-efficiency particulate air (HEPA) filter, and subjected to the mass spectrometer for data acquisition [36,37]. Each sample was repeatedly tested 5 times, and the instruments were cleaned after 10–15 times of detection by clean solvent to avoid contamination.

2.3. Analytical Conditions

The resulting aerosol was driven by a Venturi pump with nitrogen gas (2 bar) and introduced into the Xevo G2-XS Q-ToF mass spectrometer (Waters Corporation, Wilmslow, UK) through a PTFE tube. A mixture of isopropanol and leucine-enkephalin was used as an auxiliary solvent and injected into the sample chamber via a syringe pump at a flow rate of 0.1 mL/min, which was used for cleaning impurities, enhancing signal intensity, and the mass calibration. The parameters were set as follows: the scan frequency of the mass spectrometer was 1 s, the scanning range was 50–1200 m/z, the cone voltage was 10 V, the heating voltage was 30 V, the collision energy was 20–80 eV, and all data were collected in the negative ionization mode.

2.4. Optimization of REIMS Parameters

For better data collection, the heating temperatures of the electric soldering iron probe were set from 150 °C to 650 °C. Eventually, 500 °C was chosen as the optimal heating temperature [28]. The data was collected in negative ion mode, and it was found that the signal in negative ion mode was superior to that in positive ion mode. Therefore, the negative ion mode was adopted.

2.5. Data Processing

The raw data was obtained using the instrument control software package MassLynx version 4.1 (Waters Corporation, Wilmslow, UK). The construction, real-time identification, and validation of the chemical stoichiometric models of the samples were performed using the multivariate statistical software package Live ID^{TM} (version 1.2.1172.811, Waters Corporation, Wilmslow, UK). The Live ID software was used to select the total ion current (TIC) threshold value, normalize the data results, perform grouping analysis, and establish the principal component analysis (PCA) and linear discriminant analysis (LDA) models based on those samples.

3. Results and Discussion

3.1. Mass Spectra Identification and Comparison

In order to study the chemical compositions and characteristics of different oil types, the experiments collected six representative oil samples including camellia oil, sunflower seed oil, rapeseed oil, peanut oil, corn oil, and soybean oil for further investigation. The REIMS mass spectra of six different types of oil samples were successfully generated in this study. To ensure accurate and stable data, the ion peak at m/z 554.2615 generated from the internal standard leucine-enkephalin was utilized for lock-mass correction [26]. The scanning requirements in this study set a threshold of the TIC to be no less than 1×10^7, guaranteeing sufficient data for analysis.

Figure 1 illustrates the TIC chromatograms and mass spectra of the six representative oil samples. The scan range was set from 50 m/z to 1200 m/z, which is sufficient to excite and detect various compounds present in the oil samples, including phenolic acids, amino acids, fatty acids, triglycerides, phospholipids, and other components. To ensure the accuracy and reliability of the test results, we adjusted the instrument parameters and ran multiple tests to ensure the repeatability of the results. The TIC chromatograms indicated distinct signal patterns for each oil category. Among them, camellia oil and sunflower seed oil exhibit relatively high signals, followed by rapeseed oil and peanut oil, while corn oil and soybean oil display comparatively weaker signals. Importantly, all signals exceed 2×10^7, indicating that the electric soldering iron as our ionization method was able to produce satisfactory signal acquisition. In the comprehensive mass spectrum analysis using the REIMS technique, many small molecules could be detected. These included free fatty acids, monoglycerides, diglycerides, triacylglycerides, and others [34]. The former detected masses were between m/z 50 and 1200. However, for the purposes of building a superior statistical model, the mass range was reduced from 50–1200 m/z to 50–1000 m/z, since most ions are distributed within the mass range of 50–950 m/z. The reduction was performed to ensure that the most relevant signals for each class of molecule were captured and analyzed, which would result in a more accurate statistical model. During the ionization process, fatty acids in the m/z 100–500 range and glycerophospholipids in the m/z 600–900 range were found to be dominant [38,39]. This observation can be attributed to the high abundance and low desolvation enthalpy of fatty acids and phospholipids in the oil samples. To identify potential biomarkers, the main fragment ions were selected and identified by the software. As compiled in Table 1, these fragment ions serve as candidate biomarkers for further analysis and characterization of the oil samples.

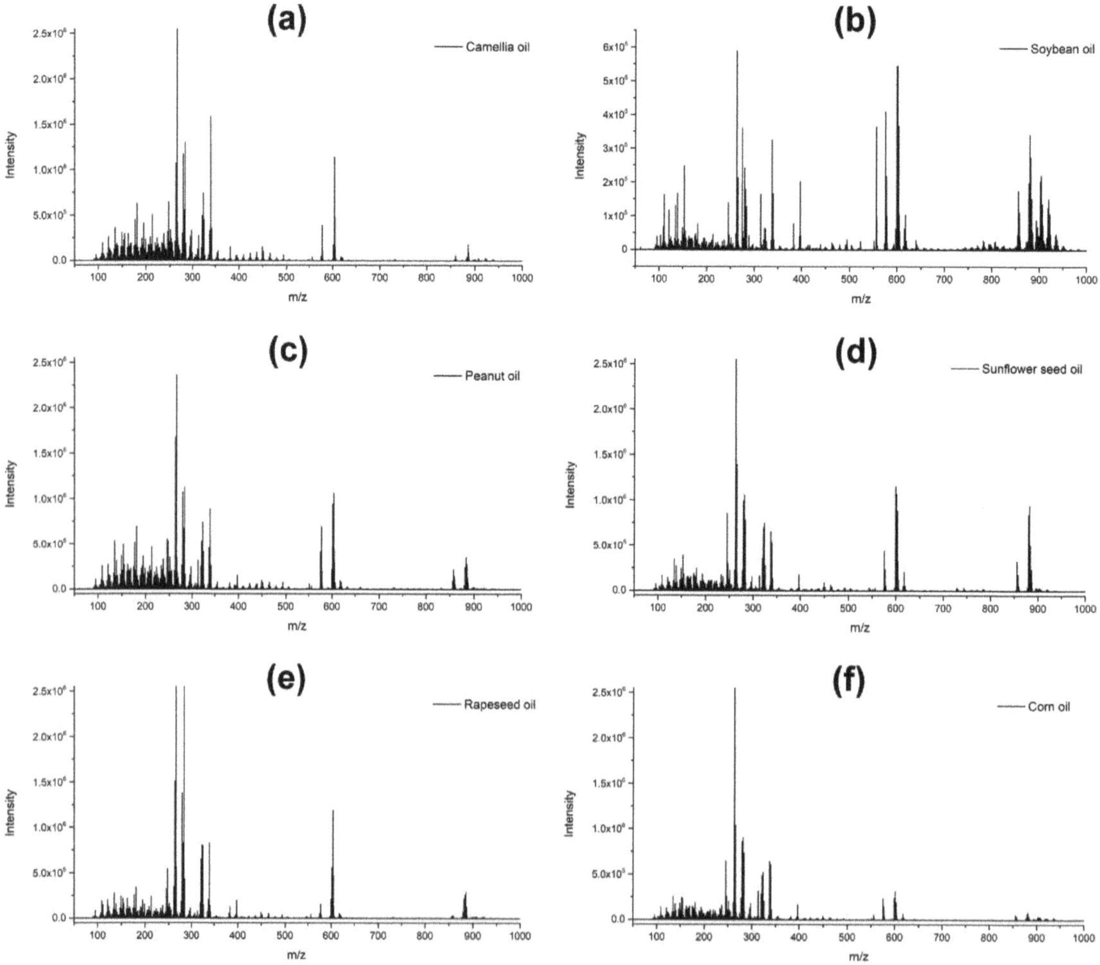

Figure 1. The mass spectrums of (**a**) camellia oil, (**b**) soybean oil, (**c**) peanut oil, (**d**) sunflower seed oil, (**e**) rapeseed oil, and (**f**) corn oil.

Table 1. Putative attribution of the corresponding spectrometric peak ranged from m/z 250 to 900 in samples.

Significant Ion (m/z)	Tentative Assignment	Main Class	Formula
271.2259	Juniperic acid	Fatty acids and conjugates	$C_{16}H_{32}O_3$
275.1991	Parinaric acid	Fatty acids and conjugates	$C_{18}H_{28}O_2$
277.2156	Pinolenic acid	Fatty acids and conjugates	$C_{18}H_{30}O_2$
305.2462	Sciadonic acid	Fatty acids and conjugates	$C_{20}H_{34}O_2$
307.2616	20:2 (7Z,14Z)	Fatty acids and conjugates	$C_{20}H_{36}O_2$
321.2767	21:2 (5Z,14Z)	Fatty acids and conjugates	$C_{21}H_{38}O_2$
337.309	22:1 (7Z)	Fatty acids and conjugates	$C_{22}H_{42}O_2$
339.3248	Behenic acid	Fatty acids and conjugates	$C_{22}H_{44}O_2$
367.3553	Lignoceric acid	Fatty acids and conjugates	$C_{24}H_{48}O_2$
395.3858	Cerotic acid	Fatty acids and conjugates	$C_{26}H_{52}O_2$
727.4859	PG (P-16:0/18:3)	Phosphatidylglycerols	$C_{40}H_{73}O_9P$
729.5021	PG (O-16:0/18:3)	Phosphatidylglycerols	$C_{40}H_{75}O_9P$
753.501	PG (P-18:0/18:4)	Phosphatidylglycerols	$C_{42}H_{75}O_9P$
755.5164	PG (O-18:0/18:4)	Phosphatidylglycerols	$C_{42}H_{77}O_9P$
853.7217	TG (15:0/18:4/19:0)	Triacylglycerols	$C_{55}H_{98}O_6$
881.752	TG (18:1/18:1/18:2)	Triacylglycerols	$C_{57}H_{102}O_6$
883.7704	TG (18:1/18:1/18:1)	Triacylglycerols	$C_{57}H{=}O_6$
887.7182	TG (15:1/18:4/22:3)	Triacylglycerols	$C_{58}H_{96}O_6$
889.7326	TG (18:3/18:3/19:1)	Triacylglycerols	$C_{58}H_{98}O_6$

3.2. Model Building and Validation

To accurately develop the chemometric model, the raw spectrometric data were meticulously acquired from authenticated samples of six different types of oil, including camellia oil (the batch number of samples, n = 28, four samples per batch), soybean oil (n = 19, five samples per batch), rapeseed oil (n = 21, four samples per batch), sunflower seed oil (n = 13, five samples per batch), peanut oil (n = 23, four samples per batch), and corn oil (n = 26, four samples per batch) for a total of 552 pieces of data. These comprehensive analyses include comprehensive data dimensionality reduction and precise pre-grouping of all data points, followed by advanced cluster analysis [40], aimed at revealing subtle spectral variations and underlying structures. PCA-LDA analysis was strategically utilized to provide a detailed representation of the classification of the different types of oils. The insightful PCA-LDA analysis facilitated clear discrimination and clustering between the different oil types, with samples of the same type grouping closely together, as displayed in Figure 2, providing a visual representation of the classification results [41]. Some parameters of the PCA-LDA model were set as follows: PCA components: 25, linear discriminants: 4, outlier setting: by standard deviation, and mass range: 50–1000 m/z. After meticulous model construction, its accuracy was critically evaluated using a robust and trusted leave-20%-out cross-validation approach, resulting in an impressive 93.8% correct classification rate, as demonstrated in Table 2, reinforcing the accuracy and reliability of the developed chemometric model [42]. Compared to other common methods of discriminating between various types of oil samples, the REIMS method possesses distinct advantages as it does not require any prior sample preparation, providing an efficient solution for the investigation of oil samples. Unlike other analytical techniques, such as GC-MS and LC-MS, REIMS is a direct analysis technique that requires only a small sample volume, making it ideal for complex and trace analysis. The method does not require the tedious and time-consuming process of pretreatment that is often necessary with other analytical techniques. The acquisition and analysis of data can be completed within a matter of seconds, allowing for a rapid and straightforward data analysis, thereby eliminating the lengthy analysis time of traditional liquid chromatography mass spectrometry technology. In addition, REIMS is capable of generating comprehensive molecular-scale information, enabling the identification and quantification of various differential compounds or metabolite ions. The ability to analyze the distribution of these ions within the oil sample can aid in the differentiation of oil types and provide valuable insights into the biological or chemical composition of the sample [25].

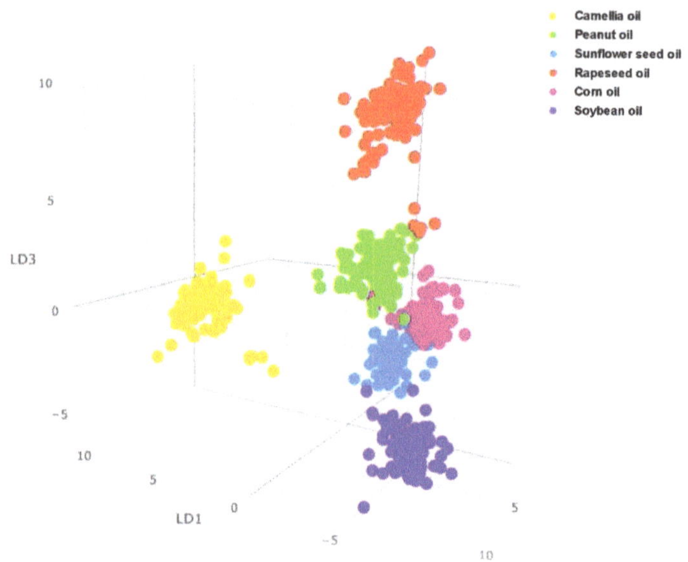

Figure 2. Three-dimensional visualization of the principal component analysis (PCA)/linear discriminant analysis (LDA) model.

Table 2. Results of the leave-20%-out.

	Camellia Oil	Peanut Oil	Sunflower Seed Oil	Rapeseed Oil	Corn Oil	Soybean Oil	Outlier	Total
Camellia oil	137	0	0	0	0	0	4	141
Peanut oil	0	110	0	0	0	0	5	115
Sunflower seed oil	0	0	57	0	0	0	8	65
Rapeseed oil	0	0	0	92	0	0	8	100
Corn oil	0	0	0	0	124	0	6	130
Soybean oil	0	0	0	0	0	86	9	95

In this study, by using the "live-recognition" function, real-time identification of oil types for each sampling was acquired almost instantly. The LiveID™ software compared the spectral data to the database and calculated a similarity score, allowing for the real-time classification of "unknown" samples [43]. In addition, some samples were analyzed live using the prototype recognition software, which had not been previously used for generating chemometric models. As a result, all of the 24 samples were correctly classified. To provide a visual representation, Figure S1 (in Supplementary File) illustrates the software interface of LiveID, with the left list recording historic results and the right circle indicating the sample identity [29]. These results demonstrate that the established experimental method and model can effectively differentiate various plant oils and provide real-time identification. The transferability of the oil detection model demonstrated here represents a significant and powerful starting point for the potential establishment of the novel technique in the food application field. In the context of the food industry, the ability to build a model in a dedicated analytical laboratory, and then utilize this model in another laboratory for authenticity evaluation and protection of pure oil products, represents a significant advance in the effort to combat food fraud and counterfeiting. This approach could provide increased protection to consumers against counterfeit or adulterated products. By developing models that can discriminate between authentic and fraudulent products, it becomes possible to provide a more reliable means of authentication and to better safeguard the reputation of these products [43].

Following the construction of the chemometric models, the corresponding loading plots for the principal components were meticulously examined to identify the significant ions responsible for sample differentiation. In particular, Figure 3 depicts the loading plots for the first three components, contributing to 53%, 26%, and 7% of the discrimination, respectively. These plots offer a valuable insight into the key ions that play a crucial role in distinguishing the samples. In both cases, the base peaks were predominantly composed of fatty acid species, such as linoleic acid (m/z 279.23) in the negative direction [44]. The loading plots provide valuable information for further characterization and differentiation of oil samples.

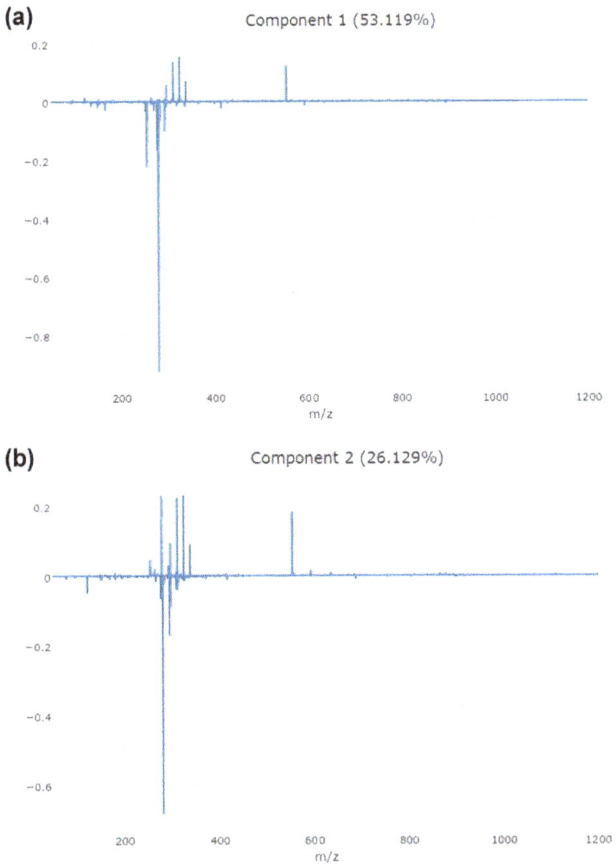

Figure 3. *Cont.*

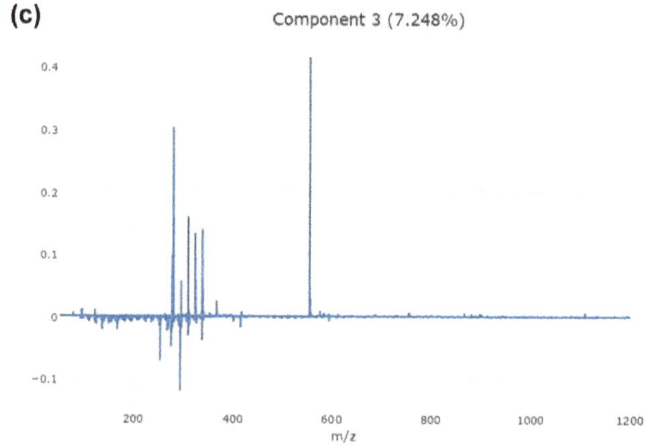

Figure 3. Loadings plot PCA/LDA model showing the m/z features responsible for the discrimination in the (**a**) first component, (**b**) second component, and (**c**) third component in the region between 500–1200 m/z.

3.3. Discrimination of Pure and Doped Camellia Oil

Discrimination between pure camellia oil and doped camellia oil is challenging due to the inherent similarities in their chemical makeups. The complex and intricate blend of components that make up plant oils makes it difficult to accurately differentiate between pure and doped camellia oil. Compared to the search of the individual and separated marker compounds, the chemometric approach based on a fingerprint presents greater promise and advance. In the present study, the capability of the established model was investigated with pure camellia oil together with adulterated camellia oil with soybean oil, rapeseed oil, sunflower seed oil, peanut oil, and corn oil with different levels (2%, 5%, 10%, 20%, 30%, 40%).

Figure 4 illustrates the 2D PCA-LDA analysis of the pure camellia oil and adulterated camellia oil samples. Within the five types of adulteration, the pure camellia oil samples plotted as blue points consistently appeared in the left part of the LD1 plot (the negative region). This phenomenon was consistent with the position of acacia honey in Cao's study [31]. In contrast, the adulterated camellia oil samples were positioned in the right part of the LD1 plot (the positive region). And the position of adulterated samples was apparently far apart from that of pure camellia oil. However, the camellia oil added with corn oil was an exception. Several addition levels (lower than 30%) were still in the left part of LD1 [31]. As the proportion of adulteration increased gradually, the sample locations shifted from negative to positive along the LD1 axis. Although the trends were less apparent for samples with lower proportions of adulteration (<10%), for example, the addition of sunflower seed oil from 2% to 20% appeared less regularly than others, the changes were more distinct in samples with higher proportions of adulteration (10%, 20%, and 40%), displaying nearly linear trends towards the 100% adulterated samples. Additionally, the direction of movement is quite different from the pure camellia oil, which could be easily distinguished by operators. Nonetheless, the adulterated camellia oil with corn oils were not moved in a linear way to corn oil. It was more like an arc type movement, which was found in other research in the assessment of fruit juice adulteration (published in Chinese). Consequently, it is evident that adulterated camellia oil can be distinguished from pure camellia oil. Even at an adulteration level as low as 2% with peanut oil, soybean oil, or rapeseed oil, the adulterated camellia oil samples were discernible from the pure samples. On the other hand, differentiating between adulteration with sunflower seed oil or corn oil was more challenging due to the presence of unclear and non-linear trends. Nonetheless, when the adulteration level exceeded 20%, these oils could still be easily distinguished.

These findings demonstrate the potential of combining REIMS with chemometric analysis in investigating camellia oil adulteration.

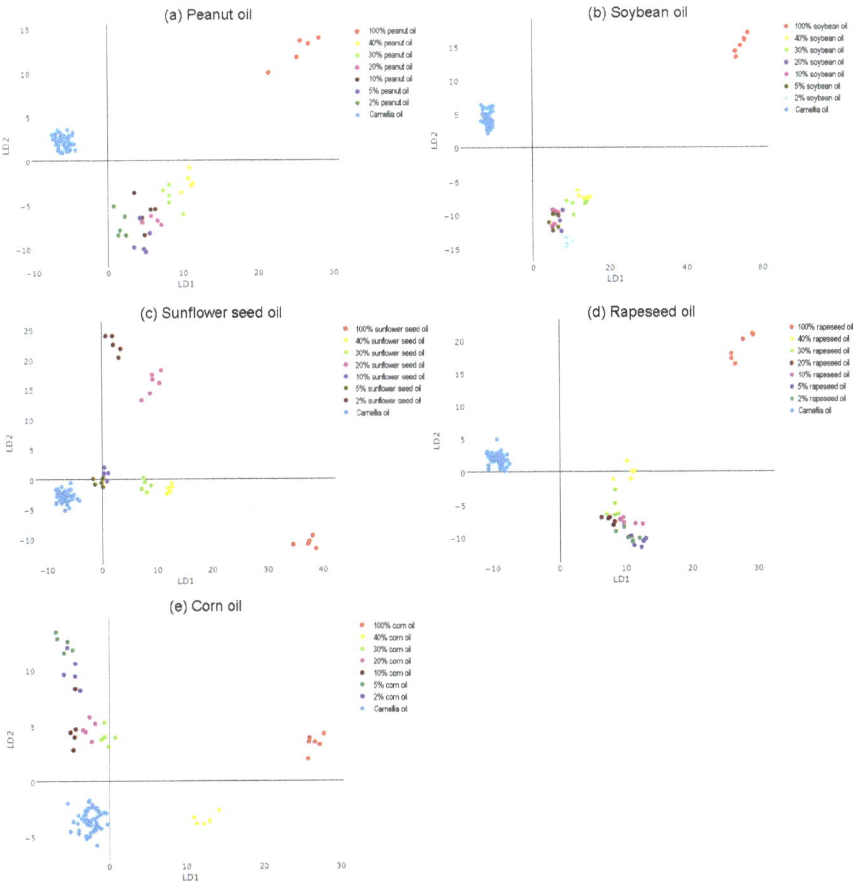

Figure 4. Two-dimension PCA-LDA analysis of camellia oil adulterated with (**a**) peanut oil, (**b**) soybean oil, (**c**) sunflower seed oil, (**d**) rapeseed oil, and (**e**) corn oil at different levels.

4. Conclusions

In this research, these detailed investigations proved that the application of REIMS, in combination with contact heating by using a soldering iron, provided an effective means for the thorough analysis of vegetable oil samples. Not only did it successfully authenticate the authenticity of camellia oil, but it also differentiated it from other vegetable oils and identified if it was adulterated with low-cost oils. The identification of these adulterations was accomplished in a rapid and accurate manner. The correct classification rate evaluated by the leave-20%-out cross-validation approach was 93.8%, and all of the 24 samples were correctly classified with the "live-recognition" function. The key to achieving this was the successful development and implementation of a PCA-LDA model. This model enabled the differentiation of camellia oil, which was adulterated with peanut oil, soybean oil, or rapeseed oil, from the pure sample. However, the adulteration trends of sunflower seed oil and corn oil were found to be more complex and required further detailed exploration using MS data mining techniques. In the following work, a comparison of the REIMS results with other analytical techniques would be beneficial. More types of vegetable oils including olive oil, flaxseed oil, and sesame oil could be added to the sample bank to make the model better. With the aid of new accessories, REIMS has the potential to become a powerful tool in the fields of food, medicine, and cosmetics, contributing significantly to these areas with its ability to provide accurate and timely results.

Supplementary Materials: The following supporting information can be downloaded at: https://www.mdpi.com/article/10.3390/separations11030068/s1, Figure S1: Real-time recognition of (a) camellia oil, (b) peanut oil, (c) sunflower seed oil, (d) rapeseed oil, (e) corn oil, and (f) soybean oil using the prototype recognition software.

Author Contributions: Conceptualization, Q.L.; data curation, H.J.; formal analysis, J.X.; methodology, Q.L.; project administration, X.C.; writing—original draft, J.X. and H.J.; writing—review and editing, X.C. All authors have read and agreed to the published version of the manuscript.

Funding: This work was supported by the Science and Technology Project of Hunan Market Supervision Administration (2023KJJH16).

Data Availability Statement: The data is unavailable due to privacy restrictions.

Conflicts of Interest: The authors declare no conflicts of interest.

References

1. Cheng, X.; Yang, T.; Wang, Y.; Zhou, B.; Yan, L.; Teng, L.; Wang, F.; Chen, L.; He, Y.; Guo, K.; et al. New Method for Effective Identification of Adulterated Camellia Oil Basing on Camellia Oleifera-Specific DNA. *Arab. J. Chem.* **2018**, *11*, 815–826. [CrossRef]
2. Su, M.H.; Shih, M.C.; Lin, K.-H. Chemical Composition of Seed Oils in Native Taiwanese Camellia Species. *Food Chem.* **2014**, *156*, 369–373. [CrossRef] [PubMed]
3. Gao, L.; Jin, L.; Liu, Q.; Zhao, K.; Lin, L.; Zheng, J.; Li, C.; Chen, B.; Shen, Y. Recent Advances in the Extraction, Composition Analysis and Bioactivity of Camellia (Camellia Oleifera Abel.) Oil. *Trends Food Sci. Technol.* **2024**, *143*, 104211. [CrossRef]
4. Yu, J.; Yan, H.; Wu, Y.; Wang, Y.; Xia, P. Quality Evaluation of the Oil of *Camellia* Spp. *Foods* **2022**, *11*, 2221. [CrossRef]
5. He, W.; Lei, T. Identification of Camellia Oil Using FT-IR Spectroscopy and Chemometrics Based on Both Isolated Unsaponifiables and Vegetable Oils. *Spectrochim. Acta A. Mol. Biomol. Spectrosc.* **2020**, *228*, 117839. [CrossRef]
6. Dou, X.; Mao, J.; Zhang, L.; Xie, H.; Chen, L.; Yu, L.; Ma, F.; Wang, X.; Zhang, Q.; Li, P. Multispecies Adulteration Detection of Camellia Oil by Chemical Markers. *Molecules* **2018**, *23*, 241. [CrossRef]
7. Yuan, J.; Wang, C.; Chen, H.; Ye, J.; Zhou, H. Identification and Detection of Adulterated Camellia Oleifera Abel. Oils by Near Infrared Transmittance Spectroscopy. *Int. J. Food Prop.* **2016**, *19*, 300–313. [CrossRef]
8. Du, Q.; Zhu, M.; Shi, T.; Luo, X.; Gan, B.; Tang, L.; Chen, Y. Adulteration Detection of Corn Oil, Rapeseed Oil and Sunflower Oil in Camellia Oil by in Situ Diffuse Reflectance near-Infrared Spectroscopy and Chemometrics. *Food Control* **2021**, *121*, 107577. [CrossRef]
9. Li, Y.; Fang, T.; Zhu, S.; Huang, F.; Chen, Z.; Wang, Y. Detection of Olive Oil Adulteration with Waste Cooking Oil via Raman Spectroscopy Combined with iPLS and SiPLS. *Spectrochim. Acta A Mol. Biomol. Spectrosc.* **2018**, *189*, 37–43. [CrossRef]
10. Wang, T.; Wu, H.-L.; Long, W.-J.; Hu, Y.; Cheng, L.; Chen, A.-Q.; Yu, R.-Q. Rapid Identification and Quantification of Cheaper Vegetable Oil Adulteration in Camellia Oil by Using Excitation-Emission Matrix Fluorescence Spectroscopy Combined with Chemometrics. *Food Chem.* **2019**, *293*, 348–357. [CrossRef]
11. Shi, T.; Wu, G.; Jin, Q.; Wang, X. Detection of Camellia Oil Adulteration Using Chemometrics Based on Fatty Acids GC Fingerprints and Phytosterols GC–MS Fingerprints. *Food Chem.* **2021**, *352*, 129422. [CrossRef]
12. Hu, Q.; Zhang, J.; Xing, R.; Yu, N.; Chen, Y. Integration of Lipidomics and Metabolomics for the Authentication of Camellia Oil by Ultra-Performance Liquid Chromatography Quadrupole Time-of-Flight Mass Spectrometry Coupled with Chemometrics. *Food Chem.* **2022**, *373*, 131534. [CrossRef]
13. Zhu, M.; Shi, T.; Chen, Y.; Luo, S.; Leng, T.; Wang, Y.; Guo, C.; Xie, M. Prediction of Fatty Acid Composition in Camellia Oil by 1H NMR Combined with PLS Regression. *Food Chem.* **2019**, *279*, 339–346. [CrossRef]
14. Hai, Z.; Wang, J. Detection of Adulteration in Camellia Seed Oil and Sesame Oil Using an Electronic Nose. *Eur. J. Lipid Sci. Technol.* **2006**, *108*, 116–124. [CrossRef]
15. Duan, D.; Huang, Y.; Zou, Y.; He, B.; Tang, R.; Yang, L.; Zhang, Z.; Su, S.; Wang, G.; Zhang, D.; et al. Discrimination of Camellia Seed Oils Extracted by Supercritical CO2 Using Electronic Tongue Technology. *Food Sci. Biotechnol.* **2021**, *30*, 1303–1312. [CrossRef]
16. Majchrzak, T.; Wojnowski, W.; Dymerski, T.; Gębicki, J.; Namieśnik, J. Electronic Noses in Classification and Quality Control of Edible Oils: A Review. *Food Chem.* **2018**, *246*, 192–201. [CrossRef]
17. Li, S.; Zhu, X.; Zhang, J.; Li, G.; Su, D.; Shan, Y. Authentication of Pure Camellia Oil by Using Near Infrared Spectroscopy and Pattern Recognition Techniques. *J. Food Sci.* **2012**, *77*, C374–C380. [CrossRef] [PubMed]
18. Guitton, Y.; Dervilly-Pinel, G.; Jandova, R.; Stead, S.; Takats, Z.; Le Bizec, B. Rapid Evaporative Ionisation Mass Spectrometry and Chemometrics for High-Throughput Screening of Growth Promoters in Meat Producing Animals. *Food Addit. Contam. Part A* **2018**, *35*, 900–910. [CrossRef] [PubMed]

19. Tzafetas, M.; Mitra, A.; Paraskevaidi, M.; Bodai, Z.; Kalliala, I.; Bowden, S.; Lathouras, K.; Rosini, F.; Szasz, M.; Savage, A.; et al. The Intelligent Knife (iKnife) and Its Intraoperative Diagnostic Advantage for the Treatment of Cervical Disease. *Proc. Natl. Acad. Sci. USA* **2020**, *117*, 7338–7346. [CrossRef] [PubMed]
20. Song, G.; Chen, K.; Wang, H.; Zhang, M.; Yu, X.; Wang, J.; Shen, Q. In Situ and Real-Time Authentication of Thunnus Species by iKnife Rapid Evaporative Ionization Mass Spectrometry Based Lipidomics without Sample Pretreatment. *Food Chem.* **2020**, *318*, 126504. [CrossRef]
21. Jones, E.A.; Simon, D.; Karancsi, T.; Balog, J.; Pringle, S.D.; Takats, Z. Matrix Assisted Rapid Evaporative Ionization Mass Spectrometry. *Anal. Chem.* **2019**, *91*, 9784–9791. [CrossRef] [PubMed]
22. Barlow, R.S.; Fitzgerald, A.G.; Hughes, J.M.; McMillan, K.E.; Moore, S.C.; Sikes, A.L.; Tobin, A.B.; Watkins, P.J. Rapid Evaporative Ionization Mass Spectrometry: A Review on Its Application to the Red Meat Industry with an Australian Context. *Metabolites* **2021**, *11*, 171. [CrossRef] [PubMed]
23. Ross, A.; Brunius, C.; Chevallier, O.; Dervilly, G.; Elliott, C.; Guitton, Y.; Prenni, J.; Savolainen, O.; Hemeryck, L.; Vidkjær, N.; et al. Making Complex Measurements of Meat Composition Fast: Application of Rapid Evaporative Ionisation Mass Spectrometry to Measuring Meat Quality and Fraud. *Meat Sci.* **2021**, *181*, 108333. [CrossRef] [PubMed]
24. Strittmatter, N.; Jones, E.; Veselkov, K.; Rebec, M.; Bundy, J.; Takats, Z. Analysis of Intact Bacteria Using Rapid Evaporative Ionisation Mass Spectrometry. *Chem. Commun.* **2013**, *49*, 6188–6190. [CrossRef]
25. He, Q.; Yang, M.; Chen, X.; Yan, X.; Li, Y.; He, M.; Liu, T.; Chen, F.; Zhang, F. Differentiation between Fresh and Frozen-Thawed Meat Using Rapid Evaporative Ionization Mass Spectrometry: The Case of Beef Muscle. *J. Agric. Food Chem.* **2021**, *69*, 5709–5724. [CrossRef]
26. Shen, Q.; Song, G.; Zhao, Q.; Wang, P.; Yang, H.; Xue, J.; Wang, H.; Cui, Y.; Wang, H. Detection of Lipidomics Characterization of Tuna Meat during Different Wet-Aging Stages Using iKnife Rapid Evaporative Ionization Mass Spectrometry. *Food Res. Int.* **2022**, *156*, 111307. [CrossRef]
27. Lu, W.; Wang, P.; Ge, L.; Chen, X.; Guo, S.; Zhao, Q.; Zhu, X.; Cui, Y.; Zhang, M.; Chen, K.; et al. Real-Time Authentication of Minced Shrimp by Rapid Evaporative Ionization Mass Spectrometry. *Food Chem.* **2022**, *383*, 132432. [CrossRef]
28. Gao, H.; Lin, J.; Jia, X.; Zhao, Y.; Wang, S.; Bai, H.; Ma, Q. Real-Time Authentication of Animal Species Origin of Leather Products Using Rapid Evaporative Ionization Mass Spectrometry and Chemometric Analysis. *Talanta* **2021**, *225*, 122069. [CrossRef]
29. Shen, Q.; Wang, J.; Li, S.; Rao, W.; Wang, Y.; Wang, H. In Situ Rapid Evaporative Ionization Mass Spectrometry Method for Real-Time Discrimination of Pelodiscus Sinensis in Different Culturing Modes without Sample Preparation. *Food Anal. Methods* **2019**, *12*, 2699–2708. [CrossRef]
30. Cardoso, V.; Sabin, G.; Hantao, L. Rapid Evaporative Ionization Mass Spectrometry (REIMS) Combined with Chemometrics for Real-Time Beer Analysis. *Anal. Methods* **2022**, *14*, 1540–1546. [CrossRef] [PubMed]
31. Wang, H.; Cao, X.; Han, T.; Pei, H.; Ren, H.; Stead, S. A Novel Methodology for Real-Time Identification of the Botanical Origins and Adulteration of Honey by Rapid Evaporative Ionization Mass Spectrometry. *Food Control* **2019**, *106*, 106753. [CrossRef]
32. Robson, K.; Birse, N.; Chevallier, O.; Elliott, C. Metabolomic Profiling to Detect Different Forms of Beef Fraud Using Rapid Evaporative Ionisation Mass Spectrometry (REIMS). *NPJ Sci. Food* **2022**, *6*, 9. [CrossRef] [PubMed]
33. Black, C.; Chevallier, O.P.; Haughey, S.A.; Balog, J.; Stead, S.; Pringle, S.D.; Riina, M.V.; Martucci, F.; Acutis, P.L.; Morris, M.; et al. A Real Time Metabolomic Profiling Approach to Detecting Fish Fraud Using Rapid Evaporative Ionisation Mass Spectrometry. *Metabolomics* **2017**, *13*, 153. [CrossRef]
34. Song, G.; Li, L.; Wang, H.; Zhang, M.; Yu, X.; Wang, J.; Shen, Q. Electric Soldering Iron Ionization Mass Spectrometry Based Lipidomics for in Situ Monitoring Fish Oil Oxidation Characteristics during Storage. *J. Agric. Food Chem.* **2020**, *68*, 2240–2248. [CrossRef] [PubMed]
35. Song, G.; Wang, H.; Zhang, M.; Zhang, Y.; Wang, H.; Yu, X.; Wang, J.; Shen, Q. Real-Time Monitoring of the Oxidation Characteristics of Antarctic Krill Oil (*Euphausia superba*) during Storage by Electric Soldering Iron Ionization Mass Spectrometry-Based Lipidomics. *J. Agric. Food Chem.* **2020**, *68*, 1457–1467. [CrossRef]
36. Strittmatter, N.; Lovrics, A.; Sessler, J.; McKenzie, J.S.; Bodai, Z.; Doria, M.L.; Kucsma, N.; Szakacs, G.; Takats, Z. Shotgun Lipidomic Profiling of the NCI60 Cell Line Panel Using Rapid Evaporative Ionization Mass Spectrometry. *Anal. Chem.* **2016**, *88*, 7507–7514. [CrossRef] [PubMed]
37. LaFranchi, B.W.; Petrucci, G.A. A Comprehensive Characterization of Photoelectron Resonance Capture Ionization Aerosol Mass Spectrometry for the Quantitative and Qualitative Analysis of Organic Particulate Matter. *Int. J. Mass Spectrom.* **2006**, *258*, 120–133. [CrossRef]
38. Liu, T.; Wang, W.; He, M.; Chen, F.; Liu, J.; Yang, M.; Guo, W.; Zhang, F. Real-Time Traceability of Sorghum Origin by Soldering Iron-Based Rapid Evaporative Ionization Mass Spectrometry and Chemometrics. *Electrophoresis* **2022**, *43*, 1841–1849. [CrossRef]
39. Balog, J.; Kumar, S.; Alexander, J.; Golf, O.; Huang, J.; Wiggins, T.; Abbassi-Ghadi, N.; Enyedi, A.; Kacska, S.; Kinross, J.; et al. In Vivo Endoscopic Tissue Identification by Rapid Evaporative Ionization Mass Spectrometry (REIMS). *Angew. Chem. Int. Ed.* **2015**, *54*, 11059–11062. [CrossRef]
40. Lin, Y.; Wang, H.; Rao, W.; Cui, Y.; Yu, X.; Dai, Z.; Shen, Q. Rapid Evaporative Ionization Mass Spectrometry-Based Lipidomics Tracking of Grass Carp (*Ctenopharyngodon idellus*) during In Vitro Multiple-Stage Digestion. *J. Agric. Food Chem.* **2018**, *66*, 6246–6253. [CrossRef]

41. Song, G.; Zhang, M.; Zhang, Y.; Wang, H.; Li, S.; Dai, Z.; Shen, Q. In Situ Method for Real-Time Discriminating Salmon and Rainbow Trout without Sample Preparation Using iKnife and Rapid Evaporative Ionization Mass Spectrometry-Based Lipidomics. *J. Agric. Food Chem.* **2019**, *67*, 4679–4688. [CrossRef] [PubMed]
42. Wen, W.; Zhao, X.; Wang, H.; Cao, X. A Real-Time Metabolomic Profiling Approach to Identify Virgin Olive Oil Adulteration by Rapid Evaporative Ionization Mass Spectrometry. *Food Anal. Methods* **2023**, *16*, 985–996. [CrossRef]
43. Rigano, F.; Stead, S.; Mangraviti, D.; Jandova, R.; Petit, D.; Marino, N.; Mondello, L. Use of an "Intelligent Knife" (Iknife), Based on the Rapid Evaporative Ionization Mass Spectrometry Technology, for Authenticity Assessment of Pistachio Samples. *Food Anal. Methods* **2019**, *12*, 558–568. [CrossRef]
44. Cui, Y.; Wang, H.; Zhao, Q.; Zhu, X.; Wang, P.; Xue, J.; Chen, K.; Shen, Q. Real-Time Detection of Authenticity and Adulteration of Krill Phospholipids with Soybean Phospholipids Using Rapid Evaporative Ionization Mass Spectrometry: Application on Commercial Samples. *Food Control* **2021**, *121*, 107680. [CrossRef]

separations

MDPI

Article

The Simultaneous Determination of Nine Furocoumarins in *Angelica dahurica* Using UPLC Combined with the QAMS Approach and Novel Health Risk Assessment Based on the Toxic Equivalency Factor

Zhao Wang, Ke Zan, Xiao-Wen Hu, Shuai Kang, Hai-Liang Li, Tian-Tian Zuo *, Hong-Yu Jin and Shuang-Cheng Ma *

National Institutes for Food and Drug Control, No. 31 Huatuo Road, Daxing District, Beijing 102629, China; wangzhao@nifdc.org.cn (Z.W.); zanke@nifdc.org.cn (K.Z.); dreadless@126.com (X.-W.H.); kangshuai@nifdc.org.cn (S.K.); lihailiang@nifdc.org.cn (H.-L.L.); jhyu@nifdc.org.cn (H.-Y.J.)

* Correspondence: zuotiantian@nifdc.org.cn (T.-T.Z.); masc@nifdc.org.cn (S.-C.M.); Tel.: +86-010-53852464 (T.-T.Z.); +86-010-53852076 (S.C.-M.); Fax: +86-010-53852072 (S.C.-M.)

Abstract: Objective: This study aimed to provide data for the type and content of linear furocoumarins (FCs) in *Angelica dahurica* (AD) in order to assess their cumulative risks and provide a scientific basis for the rational use and quality evaluation of the medicinal AD to improve public health. Methods: A UPLC method was developed for the simultaneous determination of nine FCs initially by using imperatorin (Im) as the internal standard substance, including Im, phellopterin (Ph), isoimperatorin (Is), oxypeucedanin hydrate (Oh), byakangelicin (Bn), xanthotoxin (8-MOP), bergapten (5-MOP), byakangelicol (Bl), and oxypeucedanin (Op) in two species of *Angelica dahurica* (AD). And, the risk assessment for the total FCs in AD was explored using the hazard index combined with the toxic equivalency factor (TEF-HI) strategy for the first time. Results: The established method revealed acceptable applicability, and there were no significant differences compared with the external standard method (ESM). The quantitative results demonstrated that the total content of FCs in *Angelica dahurica* (BZ) were higher than that in *Angelica dahurica* var. *formosana* (HBZ), and there was a great difference between the Bl and Op. Moreover, the risk assessment data revealed that the risk of total FCs in AD to human health was low. Conclusions: The established UPLC method that determined nine FCs in AD using a single marker could solve the problem of difficulty in obtaining a chemical reference substance with high purity and requiring a long determination time. And, the TEF-HI risk assessment approach associated with FCs in ADs could guide the rational utilization of toxic FCs in ADs in the progress of improving public health safety. In short, the whole systematic strategy provides a scientific basis for rational quality evaluation and the healthy use of related herbal medicines.

Keywords: *Angelica dahurica*; furocoumarin; quantitative analysis of multicomponents by single marker (QAMS); risk assessment (RA); toxic equivalency factors (TEFs); hazard index (HI)

Citation: Wang, Z.; Zan, K.; Hu, X.-W.; Kang, S.; Li, H.-L.; Zuo, T.-T.; Jin, H.-Y.; Ma, S.-C. The Simultaneous Determination of Nine Furocoumarins in *Angelica dahurica* Using UPLC Combined with the QAMS Approach and Novel Health Risk Assessment Based on the Toxic Equivalency Factor. *Separations* **2023**, *10*, 508. https://doi.org/10.3390/separations10090508

Academic Editor: Petr Bednar

Received: 15 August 2023
Revised: 4 September 2023
Accepted: 5 September 2023
Published: 15 September 2023

1. Introduction

Furocoumarins (FCs) are a class of natural chemical compounds that widely exists in many herbs, including some commonly used as food, cosmetics, flavorings, or herbal medicines by humans [1]. Just as its name implies, these compounds have a skeleton formed via a coumarin unit fused to a furan ring and can be divided into linear or angular isomers depending on the position of the furan ring [2]. Although FCs are widely used for medicine and food purposes, every coin has two sides, and it is reported that linear furanocoumarins, which are predominant in natural sources such as 5-methoxypsoralen (5-MOP) and 8-methoxypsoralen (8-MOP), are phototoxic and genotoxic [3–6]. And, furanocoumarin-induced hepatotoxicity has been reported to occasionally occur in humans [7]. It was once believed that FCs could interact with proteins, lipids in membranes, ribosomes, mitochondria, and, in particular, mono- and diadducts with DNA or RNA, which are the main contributors to toxicity in humans and animals [8–10]. In addition, certain FCs have been publicized to act as greatly potent inhibitors of a quantity of cytochromes P450 mono oxygenases [11], which can be interpreted as hepatotoxicity. These phenomena are defined as co-carcinogenic conditions [12]. Therefore, the U.S. Food and Drug Administration (FDA) had proscribed the use of FCs as food additives. As some research has reported, the family of FCs comprises more than 90 individual compounds, and the individual FCs shown differ in photomutagenic potency [13]. FCs are abundantly found in herbal products, especially in herbal medicine belonging to *Angelica* L., and these various FCs occur in one-plant ingredients, inevitably raising the problem of combination effects or interactions. Therefore, it is urgently required to establish feasible and reliable methods for the combined risk assessment of FCs' components in herbal medicine.

Traditional Chinese medicine (TCM) *Angelica dahurica* (AD, called BaiZhi in Chinese) is the dried root of *Angelica dahurica* (Fisch. ex Hoffm.) Benth. et. Hook. f. or *Angelica dahurica* (Fisch. ex Hoffm.) Benth. et Hook. f. var. *formosana* (Boiss.) Shan et Yuan (Fam. Umbelliferae) [14]. It is a tall perennial herb mainly distributed in the provinces of Sichuan, Hebei, Henan, Zhejiang, Anhui, Shandong, Neimenggu, Fujian, and Taiwan of China [15]. As a typical medicinal and food plant, it has the effects of expelling wind, dehumidifying, relieving pain, and relieving itching [16]. The major chemical constituents of it comprise polysaccharide, coumarins, alkaloids, benzofurans, volatile oil, fatty acids, and amino acids, and the total content of coumarins in the active ingredient of AD is more than 60% [17–19]. Modern pharmacology has shown that these chemical constituents, especially FCs, have efficacy in treating tumors, antioxidants, etc. [20,21]. But, because of the photogenotoxicity and photocarcinogenicity of FCs, the application of AD for external use is limited in China [22]. And, in October 2007, the European Medicines Agency (EMA) suggested a risk-management strategy for herbal preparations, which is rich in FCs in the paper titled "Reflection Paper on the Risk Associated with Furocoumarins Contained in Preparations of *Angelica archangelica* L." [12]. However, there are limited reports about the relevant toxicity risk assessment of AD for oral usage, and no official proposal on the potential consumption of FCs is available in China.

Nowadays, a lot of research on the identification and isolation of FCs from AD has been reported. Components in AD have been authenticated or quantified via HPLC, GC, LC-MS, GC-MS, NMR, and high-resolution mass spectrometry, such as ion trap MS and TOF-MS [23–26]. However, there has been a deficiency report about the simultaneous determination of nine FCs in AD via a single marker combined with UPLC. And, information regarding the detailed quantitative analysis of FCs in the two pieces of AD is limited. As for the risk assessments about FCs contained in plants, although EMEA has proposed some strategies, currently, the available assessment is primarily based on the data of 5-MOP and 8-MOP, owing to the fact that toxicological figures on other individual FCs in these plants are almost lacking. But, using 5-MOP and 8-MOP as a model for the risk assessment of plants which FCs are rich in might make the assessment inaccurate. Fortunately, recent researchers have proposed the concept of the toxic equivalency factor (TEF) based on photomutagenic, photocytotoxic, and photoclastogenic properties in a V79 cells experiment, which suggested that the photogenotoxic and phototoxic properties of individual furocoumarins may diverge widely but showed relatively strict additivity once occurring as multifarious mixtures [27,28]. Moreover, in 2008, the European Food Safety Authority (EFSA) conducted a series of research in experimental animals, declared that the carcinogenic influence was not triggered via a genotoxic mechanism, and then proposed the tolerable daily intake (TDI) values (0.1 mg coumarin/kg bw/day) based on the non-observed adverse effect level (NOAEL) for liver toxicity in a 2-year dog study [29]. Thus, it is significant to determine the content of various FCs in AD and to evaluate the related health risks when using medicine in a more scientific approach.

In this study, a practical QAMS method was adopted, wherein nine FCs could be determined simultaneously by using a single internal standard reference substance. And, a new risk assessment strategy for assessing the human health risk related to the consumption of AD via TEF-HI was provided to suggest the safety usage of FCs in TCM.

2. Materials and Methods

2.1. Chemicals and Reagents

Acetonitrile (MS-grade) was obtained from Merck (Overijse, Belgium). Ultrapure water was prepared by a Milli-Q system (Millipore, Billerica, MA, USA). The rest of the chemicals (analytical grade) were obtained from Sinopharm Chemical Reagent Beijing (Beijing, China). The reference standard of imperatorin (Im) and isoimperatorin (Is) were provided by NIFDC. Standard phellopterin (Ph), oxypeucedanin hydrate (Oh), byakangelicin (Bn), bergapten (5-MOP), xanthotoxin (8-MOP), byakangelicol (Bl), and oxypeucedanin (Op) were obtained from Chengdu DeSiTe Biological Technology Co., Ltd. (Chengdu, China), all with the purity above 98%.

In total, 20 batches of AD plants including 10 batches of *Angelica dahurica* (Hoffm.) *Benth. & Hook. f. ex Franch. & Sav.* (BZ, called BaiZhi in Chinese, Species No. 1) and 10 batches of *Angelica dahurica* var. formosana (Boissieu) Yen (HBZ, called HangBaiZhi in Chinese, Species No. 2) were collected from Bozhou Anhui, Anguo Hebei, Shehong Sichuan, Suining Sichuan of China, and Hong Kong, respectively, in August 2021. They were originally authenticated by Dr. Kang Shuai of the National Institutes for Food and Drug Control (NIFDC, Beijing, China). All specimens had been stored in the Chinese Herb medicine museum of NIFDC.

2.2. Instrumentation and Chromatographic Condition

An ultra-high-performance liquid chromatographic (UHPLC) system (Shimadzu LC_30AD, Shimadzu, Tokyo, Japan) equipped with a DAD detector was used. An ultrasonic bath (300 W) was also used in the process.

Chromatography was performed with a Waters ACQUITY UPLC® CSH™ C_{18} Column (2.1 mm × 100 mm, 1.7 µm) at 25 °C with a flow rate of 0.3 $mL \cdot min^{-1}$, and 254 nm was chosen as the detection wavelength. Mobile phases A and B consisted of water and acetonitrile. A gradient elution (0–10 min, 30–40% B; 10–17 min, 40–80% B; 17–20 min, 80% B) was adopted, and the injection volume was 2 µL.

2.3. Preparation of Standard Solutions

Nine standard FC substances were dissolved in methanol, respectively, and the stock solutions were made to be about 1 mg/mL. Then, the mixed reference stock solution was prepared comprising all standards including Oh, Bn, 8-MOP, 5-MOP, Bl, Op, Im, Ph, and Is at the concentration of 158.7, 49.6, 57.3, 53.6, 90.15, 258.75, 102.0, 60.1, and 102.5 mg/L, respectively. The working solutions were kept at −20 °C and were newly prepared each day via the appropriate dilution of the stock with methanol.

2.4. Preparation of BZ Samples

Before preparation, cut the 20 batches of BZ samples into small pieces and powder them through a No. 2 sieve, respectively. Then, accurately weigh 0.5 g of powdered sample into a 50 mL conical flask, add 20 mL of 70% methanol, and sonicate (300 W) the mixture for 20 min. Filter and transfer the filtrate to a 50 mL volumetric flask. Repeat the steps once. Then, wash the residue, filter paper, and the funnel with the appropriate volume of 70% methanol and combine the solutions into the same volumetric flask. Increase the volume of the volumetric flask up to the mark with 70% methanol. Filter the solution through a 0.22 µm nylon filter before analysis.

2.5. Method Validation

The established UPLC method was validated through complete methodology, including specificity, accuracy rate (using recovery rate), precision, reproducibility, linearity (or calibration curves), covers, limits of detection (LOD), limits of quantification (LOQ), stability, and durability.

Among these, Radix Ophiopogonis (one commonly used TCM) was chosen as a blank matrix to perform the specificity, LOD, and LOQ tests by adding an appropriate amount of the relevant reference. The LOD and LOQ values for the nine FCs were defined as signal-to-noise (S/N) ratios of 3 and 10, respectively. The mixed reference stock solution was diluted to six concentration levels, and calibration curves (Y = aX + b) were established where the X- and Y-axes represented the peak area and the concentration of every furocoumarin, respectively. The correlation coefficient (r) was calculated to inspect the linearity of every calibration curve. The precision was evaluated using six consecutive injections of the same mixed standard solutions. The repeatability was measured by analyzing six independently prepared sample solutions of the same batch. The stability was assessed by injecting the same prepared sample solution at different times (0, 2, 4, 8, 12, 24 h). The relative standard deviation (RSD %) was analyzed at the same time to estimate the results of precision, repeatability, and stability. Finally, the recovery test was performed to verify the accuracy of the method. The mixed reference stock solution of nine FCs at certain levels were precisely added to 0.25 g of the above sample power. Then, six parallel solutions were created according to the above sample preparation requirements, calculating the average recovery rate (n = 6) and analyzing the RSD of every component.

2.6. Theory of QAMS

The principle of the QAMS method is that the content of the component is proportional to the detector response under particular conditions, and different effective components of TCM exist on intrinsic functions and proportional relations, which make the calculation of relative correction factors (RCFs) feasible [14]. Based on the RCFs, the content of each analyst can be determined independently according to an internal standard substance. In this case, the calculation of RCFs is crucial in this study. First, Im was chosen as the internal standard substance, and then the RCFs for the rest of the FCs were calculated using the following multipoint method (Equation (1)). Thereafter, the content of the other FCs was determined according to Equation (2) [30].

The equation for RCFs was calculated as follows,

$$f_{s/i} = f_i / f_s = A_i \times C_s / (A_s \times C_i) \tag{1}$$

The quantification of the other investigated FC components is calculated as follows,

$$C_i = (A_i \times C_s) / (f_{s/i} \times A_s) \tag{2}$$

where A_s and A_i represent the peak areas of the internal standard (Im) and other FCs in standard solutions or in the AD samples, respectively. $f_{s/i}$ represents the average of the RCFs between Im and each of the other FC components. C_s and C_i represent the concentrations of Im and the other investigated FC components in standard solutions or in the AD samples.

Additionally, for the location of target chromatographic peaks, the relative retention time (RRT) was introduced to position the chromatographic peaks of the other investigated FC components according to Equation (3) [31],

$$t_{i/s} = t_i / t_s \tag{3}$$

where $t_{i/s}$ is the RRT of the analysts and t_i and t_s are the retention times of the other investigated FC components and Im, respectively.

2.7. Quantification of AD Samples

Both the newly established QAMS and ESM methods were used to determine and analyze the samples. Each sample was determined 3 times, and then the mean was calculated. The accuracy of the two methods was assessed using standardized mean difference (SMD), which was calculated according to Equation (4) [32],

$$SMD = [(C_{ESM} - C_{QAMS})/C_{ESM}] \times 100\% \quad (4)$$

where C_{ESM} and C_{QAMS} represent the mean content of the FCs determined by QAMS and ESM, respectively.

2.8. Healthy Risk Assessment

2.8.1. Assessment of Daily Intake of AD Based on the EQF

To examine the health risk of FCs in AD, a real-life exposure situation was considered by estimating the EDI (μg/kg bw/day) based on the EQF as in Equation (5) [33],

$$EDI = EF \times Ed \times IR \times C \times EQF/W \times AT \quad (5)$$

where EDI is the estimated daily intake of FCs in AD; EF is the abbreviation of exposure frequency, which is 90 days/year [33]; and Ed represents exposure time, which according to our previous study is 20 years [33]; and IR is short for the daily intake rate of AD (g/day). Based on the guidelines of ChP [14], the mean IR is 6.5 g/day. C represents the content of the FCs in AD (mg/g) and EQF is the equivalency factor. It was suggested that the value for 5-MOP should be set at 1.00, then the value for Oh, Bn, 8-MOP, Bl, Op, Im, Ph, and Is would be 0.16, 0.01, 0.20, 0.01, 0.076, 0.15, 0.01, and 0.15, respectively. Then, W symbolizes the average of body weight, which is 70 kg based on the date given in adults by the EFSA and AT represents the mean duration of exposure to AD, which is usually 365 days/year × 70 years.

2.8.2. Risk Characterization Based on Health-Based Guidance Values

Risk characterization is commonly expressed by calculating risk metrics including hazard index (HI), Margins of Exposure (MoE), or Reference Point Index (RPI). For hazardous substances with health-based guideline values, an HI strategy is recommended. As the EFSA had allocated a TDI for coumarins via diet, the HI value can be calculated according to Equation (6) [33].

$$HI = \left(\sum_{i=1}^{n} EDI_i \times 10\right)/TDI \quad (6)$$

In Equation (6), HI represents the hazard index. $\sum_{i=1}^{n} EDI_i$ is the calculated daily intake of the nine FCs in AD and 10 is the safety factor, which indicates that the daily intake of FCs from TCM and its products is not greater than 10% of the daily exposure (including diet). TDI is the tolerable daily intake value (mg/kg bw/day). The total TDI of nine FCs in each sample was calculated in this study. According to the research of EFSA, the TDI of coumarins is 0.1 mg/kg bw/day in this study. Once the HI value is >1, the exposure risks of the analyst should not be ignored.

3. Results and Discussion

3.1. Extraction Procedures and UPLC Condition Optimization

To extract coumarins from AD more completely, the extraction solvents (100%, 70%, 50%, and 30% ethanol), extraction methods (ultrasound and reflux), extraction times (20, 30, and 40 min), and number of extractions were investigated. The extraction efficiency showed that adding 20 mL of 70% methanol, sonication (300 W) for 20 min, and repeating the extraction once could extract the coumarins exhaustively.

First, for chromatographic conditions, the chromatographic columns of ACQUITY UPLC® BEH Amide (2.1 mm × 100 mm, 1.7 μm), ACQUITY UPLC® CSH™ C_{18} (2.1 mm × 100 mm, 1.7 μm), CORTECS® UPLC® C_{18} (2.1 mm × 100 mm, 1.6 μm), ACQUITY UPLC® BEH SHIELD (2.1 × 50 mm, 1.7 μm), Thermo Syncronis aQ (2.1 mm × 100 mm, 1.7 μm), and Phenomenex OOD-4475-AN Kinetex C_{18} 100A (2.1 mm × 100 mm, 1.7 μm) were investigated for studying the separation of complex components in AD. ACQUITY

UPLC® CSH™ C_{18}, CORTECS® UPLC® C_{18}, and Thermo Syncronis aQ were selected for the establishment of the QAMS method. A CSH column had the best separation effect and was selected for its stability and excellent reproducibility for both high- and low-pH compounds. The methodology and durability of this experiment were studied with a Shimadzu UPLC instrument (LC-30AD), an ultrahigh performance liquid chromatograph.

Second, the DAD was used to scan the combined working solution at full wavelength to obtain better sensitivity, and the maximum absorption wavelength for each of the nine FCs was obtained. It is shown that, at 254 nm, the nine FC components exhibited relatively good absorbance.

Third, different mobile phases, for example, 0.1% formic acidacetonitrile, wateracetonitrile, and watermethanol with different proportional gradient elutions, were compared. The baseline condition showed that the mobile phase system of acetonitrile was better than that of methanol. The adoption of a 0.1% formic acid solution had no obvious effect on the CSH column; therefore, wateracetonitrile was used as the mobile phase. At the same time, the design of several chromatographic gradient methods, flow rates (0.2, 0.3, and 0.5 mL/min), and column temperatures (25 °C, 30 °C, and 35 °C) were optimized. Finally, we confirmed an acetonitrile water system for the mobile phase at 25 °C with a flow rate of 0.3 mL/min for analysis. Under the optimized conditions, the nine FCs were all separated from the baseline within 20 min. The UPLC chromatograms of the nine mixed standard solutions and the representative AD samples are shown in Figure 1.

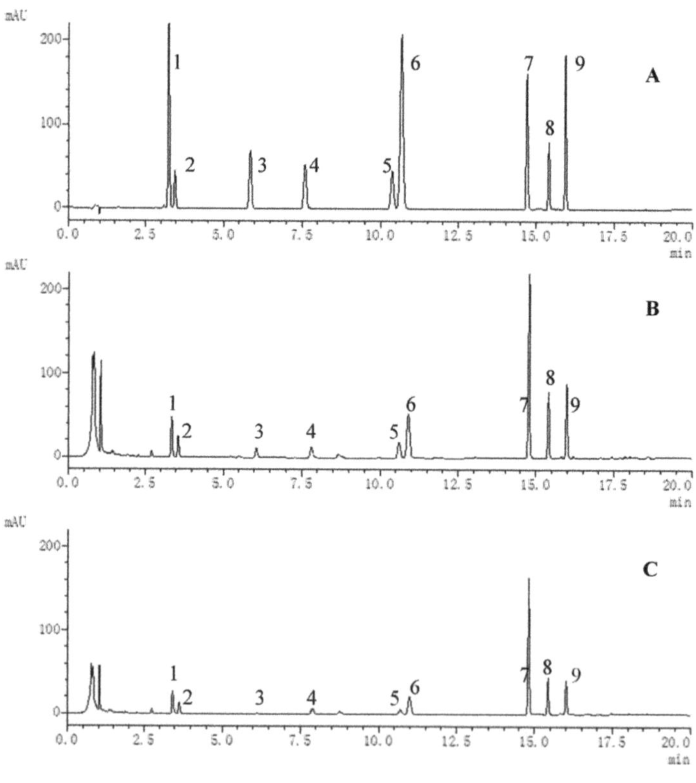

Figure 1. Representative UPLC chromatograms of the standards (**A**), the AD sample (BZ, Species No. 1) (**B**), and the AD sample (HBZ, Species No. 2) (**C**). Peak identification: oxypeucedanin hydrate (Oh, 1), byakangelicin (Bn, 2), xanthotoxin (8-methoxypsoralen, 8-MOP, 3), bergapten (5-methoxypsoralen, 5-MOP, 4), byakangelicol (Bl, 5), oxypeucedanin (Op, 6), imperatorin (Im, 7), phellopterin (Ph, 8), and isoimperatorin (Is, 9).

3.2. Method Validation

The established UPLC protocol was evaluated for linearity, the limit of detection (LOD), the limit of quantification (LOQ), precision, stability, and accuracy. The correlation coefficients of the nine FCs were >0.9999 within the respective detection test ranges. The LOD and LOQ values of each of the FCs varied from 0.01–0.02 mg/mL to 0.04–0.07 mg/mL (Table 1). The RSD values of precision, repeatability, and stability for all nine FCs were <2.75% (Table 2). The overall recovery rate (n = 6) ranged from 90.07% to 107.94%, with RSD values of <3.00% (Table 2). These results indicated that the proposed method was precise, accurate, and efficient for simultaneously qualifying the nine FCs.

Table 1. Linearity (regression equation, linear range, and correlation coefficient) and sensitivity (LOD and LOQ) of the method determination of the eight FCs.

Analytes	Calibration Curves	r	/μg·mL^{-1}	/μg·mL^{-1}	/μg·mL^{-1}
1	Y = 24489800X + 4561.67	0.9999	1.270–158.7	0.01	0.05
2	Y = 16185800X + 850.89	0.9999	0.3970–49.60	0.01	0.04
3	Y = 31411100X + 1986.14	0.9999	0.4580–57.28	0.01	0.04
4	Y = 29319800X + 1010.05	0.9999	0.4290–53.65	0.01	0.04
5	Y = 16335500X + 1443.08	0.9999	0.7210–90.15	0.02	0.07
6	Y = 26801600X + 5930.63	0.9999	2.070–258.8	0.02	0.07
7	Y = 27494800X + 3435.56	0.9999	0.8160–102.0	0.01	0.04
8	Y = 28897035X + 1351.04	0.9999	0.8000–100.50	0.02	0.07
9	Y = 28331800X + 597.95	0.9999	0.8200–102.50	0.01	0.04

'1–9' represents the analytes of oxypeucedanin hydrate (1), byakangelicin (2), xanthotoxin (3), bergapten (4), byakangelicol (5), oxypeucedanin (6), imperatorin (7), phellopterin (8), and isoimperatorin (9).

Table 2. Precision, stability, repeatability, and recovery results of nine FCs.

Analytes	Precision (RSD/%, n = 6)	Stability (RSD/%, n = 6)	Repeatability (RSD/%, n = 6)	Recoveries (n = 6)	
				Mean/%	RSD/%
1	0.02	0.82	2.01	97.57	2.98
2	0.09	0.76	1.89	107.9	2.59
3	0.52	1.10	2.75	107.9	2.98
4	0.19	1.14	2.07	106.9	2.91
5	0.23	1.07	1.93	90.07	2.98
6	0.02	0.93	1.91	92.09	2.66
7	0.01	1.04	1.97	93.76	3.00
8	0.21	0.89	2.10	101.5	2.62
9	0.01	0.63	1.98	91.97	2.16

'1–9' represents the analytes of oxypeucedanin hydrate (1), byakangelicin (2), xanthotoxin (3), bergapten (4), byakangelicol (5), oxypeucedanin (6), imperatorin (7), phellopterin (8), and isoimperatorin (9).

3.3. Quantitative Analyses of Multiple Components via Single Marker

The UPLC chromatograph detected nine FCs in the two species of AD. The chemical structures of the nine FCs are shown in Figure 2. For the QAMS method, selecting a proper internal standard substance for the calculation of RCFs and RRTs of the rest of the FCs in AD is vital. In this study, Im was selected as an internal standard substance owing to its high content, good stability, relatively moderate retention value, and easy availability in AD. Therefore, it met the requirements of QAMS. The calculated RCFs of eight FCs compared with Im are shown in Table 3. The results indicated that the RSDs of these FCs were <3%, calculated from the slope at various concentrations. Moreover, the mean RCFs were 0.891, 0.585, 1.136, 1.046, 0.587, 0.968, 1.051, and 1.015.

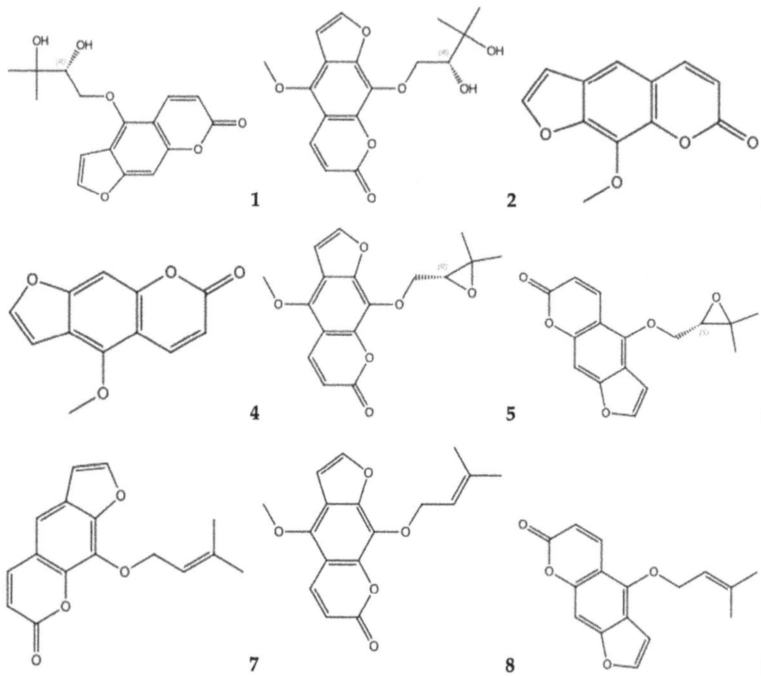

Figure 2. Chemical structures of FCs analyzed in AD: oxypeucedanin hydrate (1), byakangelicin (2), xanthotoxin (3), bergapten (4), byakangelicol (5), oxypeucedanin (6), imperatorin (7), phellopterin (8), and isoimperatorin (9).

Table 3. Relative correction factor (RCF) values of the eight FCs of the AD calculated via slope and concentration method (n = 3).

RCF	Calculated by Slope	a	b	c	d	e	f	Mean	RSD (%)
$f_{7/1}$	0.891	0.890	0.892	0.893	0.892	0.891	0.889	0.891	0.17
$f_{7/2}$	0.589	0.574	0.583	0.585	0.587	0.588	0.589	0.585	0.91
$f_{7/3}$	1.142	1.124	1.131	1.136	1.137	1.140	1.142	1.136	0.58
$f_{7/4}$	1.066	1.016	1.025	1.041	1.054	1.060	1.061	1.046	1.85
$f_{7/5}$	0.594	0.568	0.584	0.589	0.590	0.593	0.593	0.587	1.57
$f_{7/6}$	0.975	0.961	0.965	0.967	0.968	0.970	0.971	0.968	0.46
$f_{7/8}$	1. 051	1.055	1.053	1.052	1.049	1.048	1.051	1. 051	0.25
$f_{7/9}$	1.030	1.008	1.009	1.011	1.012	1.014	1.018	1.015	0.76

'f_i' represents the RCFs of oxypeucedanin hydrate (1), byakangelicin (2), xanthotoxin (3), bergapten (4), byakangelicol (5), oxypeucedanin (6), phellopterin (8), and isoimperatorin (9) compared with imperatorin (7). 'a–f' indicates six different concentration levels. Taking imperatorin (7) as an example, its concentration levels are 0.816 μg/mL, 2.04 μg/mL, 4.08 μg/mL, 10.2 μg/mL, 20.4 μg/mL, and 102 μg/mL, respectively.

Further, the system suitability and durability of QAMS were evaluated. First, different instruments and diverse columns were investigated by calculating the RSD values of each RCF. Subsequently, the Shimadzu UPLC instrument with a CSH column was selected to assess the effects of changed flow rates (0.28, 0.30, and 0.32 mL/min) and column temperatures (23 °C, 25 °C, and 27 °C) on RCFs. The results show that these potential factors, including instruments, columns, flow rates, and column temperatures, displayed no substantial influence on RCFs based on the RSDs of < 3% (Table 4), which demonstrated that Im and the other eight FCs were robust under these conditions. Finally, three chromatographic columns with two different UPLC instruments assessed the RRTs. The results showed that all RSDs were <3%, implying that the RRT method could be used to locate the chromatographic peak component of nine FCs in AD (Table 5). The RRTs were 0.224, 0.237, 0.401, 0.518, 0.703, 0.726, 1.055, and 1.085.

Table 4. Effects of different instruments, columns, column temperatures, and flow rates on RCFs of nine FCs (n = 3).

Conditions	Items	$f_{7/1}$	$f_{7/2}$	$f_{7/3}$	$f_{7/4}$	$f_{7/5}$	$f_{7/6}$	$f_{7/8}$	$f_{7/9}$
Waters ACQUITY UPLC H-Class	Column 1	0.898	0.582	1.110	1.051	0.599	0.973	1.055	1.018
	Column 2	0.890	0.541	1.125	1.082	0.547	0.967	1.049	1.042
	Column 3	0.881	0.591	1.041	1.045	0.597	0.966	1.062	1.021
Shimadzu UPLC Instrument (LC-30AD)	Column 1	0.891	0.585	1.136	1.046	0.587	0.968	1.051	1.015
	Column 2	0.878	0.523	1.121	1.103	0.571	0.978	1.048	1.038
	Column 3	0.902	0.576	1.064	1.045	0.590	0.994	1.049	1.016
	Mean	0.890	0.566	1.100	1.062	0.582	0.974	1.052	1.025
	RSD (%)	1.05	4.88	3.47	2.32	3.39	1.10	0.51	1.16
Column temperature (°C)	23	0.890	0.603	1.141	1.024	0.577	0.971	1.0518	1.018
	25	0.891	0.585	1.136	1.046	0.587	0.968	1.050	1.015
	27	0.891	0.582	1.122	1.053	0.594	0.962	1.0475	1.014
	Mean	0.891	0.590	1.133	1.041	0.586	0.967	1.050	1.016
	RSD (%)	0.07	1.92	0.87	1.46	1.46	0.47	0.21	0.21
Flow rates (mL/min)	0.28	0.872	0.569	1.114	1.041	0.579	0.966	1.049	1.006
	0.30	0.891	0.585	1.136	1.046	0.587	0.968	1.050	1.015
	0.32	0.925	0.601	1.151	1.052	0.595	0.971	1.0524	1.034
	Mean	0.896	0.585	1.134	1.046	0.587	0.968	1.051	1.018
	RSD (%)	3.00	2.74	1.64	0.53	1.36	0.26	0.17	1.41

Column 1—ACQUITY UPLC® CSHTM C_{18}(2.1 mm × 100 mm, 1.7 μm). Column 2—CORTECS® UPLC® C_{18} Column (2.1 mm × 100 mm, 1.6 μm). Column 3—Thermo Syncronis aQ (2.1 mm × 100 mm, 1.7 μm).

Table 5. Relative retention values measured in different instruments and different columns (n = 3).

Instrument	Column	$t_{7/1}$	$t_{7/2}$	$t_{7/3}$	$t_{7/4}$	$t_{7/5}$	$t_{7/6}$	$t_{7/8}$	$t_{7/9}$
Waters ACQUITY UPLC H-Class	ZORBAX Eclipse XDB-C_{18}	0.225	0.239	0.401	0.509	0.701	0.730	1.049	1.087
	ZORBAX Eclipse plus C_{18}	0.221	0.235	0.397	0.517	0.706	0.725	1.051	1.084
	ACQUITY UPLC HSS, T3	0.223	0.235	0.399	0.529	0.703	0.719	1.064	1.081
Shimadzu UPLC Instrument (LC-30AD)	ZORBAX Eclipse XDB-C_{18}	0.229	0.241	0.414	0.511	0.705	0.726	1.052	1.089
	ZORBAX Eclipse plus C_{18}	0.222	0.236	0.398	0.518	0.706	0.726	1.052	1.084
	ACQUITY UPLC HSS, T3	0.223	0.234	0.398	0.522	0.698	0.728	1.063	1.083
mean		0.224	0.237	0.401	0.518	0.703	0.726	1.055	1.085
RSD/%		1.29	1.14	1.61	1.41	0.46	0.51	0.62	0.27

't_i' represents the relative retention time (RRT) of oxypeucedanin hydrate (1), byakangelicin (2), xanthotoxin (3), bergapten (4), byakangelicol (5), oxypeucedanin (6), phellopterin (8), and isoimperatorin (9) compared with imperatorin (7).

3.4. Quantification and Method Assessment

BZ and HBZ are the two original plant sources of AD in China. Both the ESM and QAMS methods were performed simultaneously to determine the nine FCs in 20 AD samples. Subsequently, the SMD was calculated to assess the differences between the two methods. The results are shown in Supplementary Table S1. The absolute SMD values ranged from 0.02 % to 1.32 %, demonstrating that the accuracy of the content results obtained using QAMS could be accepted. Hence, it is feasible to determine the nine FCs in AD using QAMS simultaneously.

The results show that all samples of AD contain nine FCs, and the mean content of Oh, Bn, 8-MOP, 5-MOP, Bl, Op, Im, Ph, and Is are 0.47, 0.37, 0.08, 0.19, 0.37, 1.11, 1.95, 0.57, and 0.66 mg/g, respectively. As shown in Figure 3, the total content of the nine FCs varied from 4.38 to 9.37 mg/g, with those in BZ being relatively higher than in HBZ. Moreover, the most abundant ingredient of the nine quantified FCs in the two types of samples is Im. The content of Bl and Op were found to be substantially different in both samples.

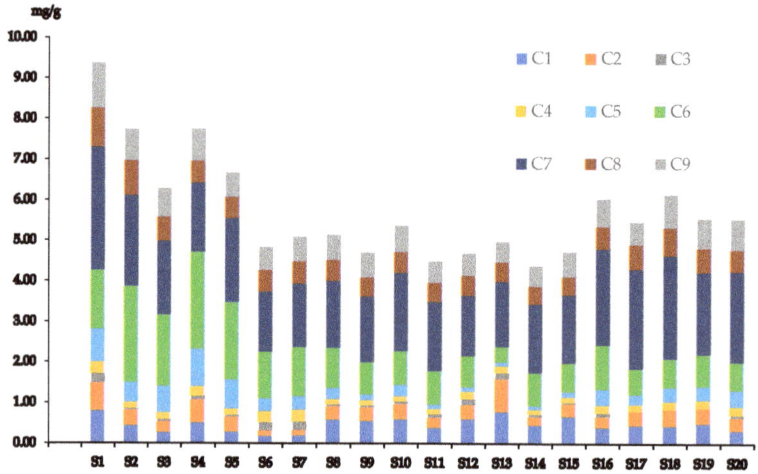

Figure 3. Comparison of FCs in 20 batches of samples. C_i represents the average content of oxypeucedanin hydrate (1), byakangelicin (2), xanthotoxin (3), bergapten (4), byakangelicol (5), oxypeucedanin (6), imperatorin (7), phellopterin (8), and isoimperatorin (9).

3.5. Risk Assessment

In 1986, the IARC3 working group concluded that 8-MOP combined with ultraviolet radiation was carcinogenic to humans, based on sufficient evidence from both animal and human data [34]. The safety evaluation of furocoumarin via diet has been a concern for a long time. However, because of the toxicological data on all individual FCs being almost absent, this assessment mainly used the data on 5-MOP and 8-MOP, which are both used as the basis for risk assessment. To the best of our knowledge, furocoumarin is never found in plants as a single compound but always as complex mixtures of many individual furocoumarins. An overvalued scenario may emerge if a risk assessment is performed using only one type of furocoumarin, such as 8-MOP, as a basis for calculating the total toxic equivalent. However, diverse results are obtained when genuine mixtures of FCs in given herbal preparations are studied. Recently, the Raquet and Schrenk Group proposed an equivalency factor for phototoxicity and genotoxicity of furocoumarin mixtures. This provides the possibility for a toxicity estimation of different FCs and can be considered as the TEF. Herein, we calculated the EDI values based on the TEF. 5-MOP was selected as a basis or called a reference compound for the total risk assessment of FCs in AD owing to its strong photomutagenicity, as testified in previous studies and through its abundance in medicinal plants and food [12]. The TEF values of Oh, Bn, 8-MOP, Bl, Op, Im, Ph, and Is calculated based on research in the literature were 0.16, 0.01, 0.20, 0.01, 0.076, 0.15, 0.01, and 0.15, respectively [28]. The results showed that the EDI values of the total FCs in AD calculated without the TEF (0.029–0.061 mg/kg bw/day) were very close to the TDI values (0.1 mg/kg bw/day). However, after correction with the TEF, the EDI ranged from 0.004 to 0.008 mg/kg bw/day, revealing a reduced order of magnitude. According to the established HI-TEF strategy, the HI values of all the samples were <1, indicating that AD posed a low risk, which is consistent with the clinical and animal toxicity research [35,36]; the details are presented in Table 6.

Table 6. Estimated daily intake (EDI) and hazard index (HI) in 20 batches of AD samples.

Batch No.	Source	Species	EDI_1	EDI_2	EDI_3	EDI_4	EDI_5	EDI_6	EDI_7	EDI_8	EDI_9	ΣEDI Based on TEF	ΣEDI without TEF	HI Values Based on TEF	HI Values without TEF
B1	Bozhou, Anhui	1	0.00084	0.00005	0.00029	0.00194	0.00005	0.00072	0.00298	0.00006	0.00109	0.0080	0.0613	0.80	6.13
B2	Anguo, Hebei	1	0.00045	0.00003	0.00005	0.00105	0.00003	0.00118	0.00220	0.00006	0.00075	0.0058	0.0506	0.58	5.06
B3	Anguo, Hebei	1	0.00028	0.00002	0.00006	0.00116	0.00004	0.00087	0.00179	0.00004	0.00069	0.0050	0.0411	0.50	4.11
B4	Anguo, Hebei	1	0.00053	0.00004	0.00011	0.00157	0.00006	0.00118	0.00168	0.00004	0.00077	0.0060	0.0507	0.60	5.07
B5	Anguo, Hebei	1	0.00030	0.00002	0.00005	0.00103	0.00005	0.00095	0.00203	0.00003	0.00059	0.0050	0.0436	0.50	4.36
B6	Jiaozuo, Hebei	1	0.00020	0.00001	0.00025	0.00185	0.00002	0.00056	0.00147	0.00003	0.00055	0.0049	0.0316	0.49	3.16
B7	Hong Kong	1	0.00021	0.00001	0.00027	0.00194	0.00002	0.00060	0.00153	0.00004	0.00059	0.0052	0.0333	0.52	3.33
B8	Hong Kong	1	0.00062	0.00002	0.00005	0.00085	0.00002	0.00049	0.00161	0.00003	0.00060	0.0043	0.0336	0.43	3.36
B9	Hong Kong	1	0.00058	0.00002	0.00006	0.00083	0.00001	0.00039	0.00160	0.00003	0.00059	0.0041	0.0308	0.41	3.08
B10	Hong Kong	1	0.00063	0.00002	0.00007	0.00086	0.00002	0.00041	0.00189	0.00003	0.00062	0.0046	0.0351	0.46	3.51
B11	Shehong, Sichuan	2	0.00041	0.00002	0.00010	0.00092	0.00001	0.00041	0.00167	0.00003	0.00050	0.0041	0.0293	0.41	2.93
B12	Suining, Sichuan	2	0.00063	0.00002	0.00018	0.00125	0.00001	0.00038	0.00147	0.00003	0.00053	0.0045	0.0306	0.45	3.06
B13	Suining, Sichuan	2	0.00082	0.00005	0.00019	0.00112	0.00001	0.00019	0.00157	0.00003	0.00048	0.0045	0.0325	0.45	3.25
B14	Suining, Sichuan	2	0.00048	0.00001	0.00007	0.00081	0.00001	0.00040	0.00166	0.00003	0.00048	0.0040	0.0286	0.40	2.86
B15	Suining, Sichuan	2	0.00070	0.00002	0.00004	0.00083	0.00001	0.00035	0.00164	0.00003	0.00059	0.0042	0.0309	0.42	3.09
B16	Suining, Sichuan	2	0.00041	0.00002	0.00010	0.00130	0.00003	0.00054	0.00232	0.00004	0.00067	0.0054	0.0395	0.54	3.95
B17	Hong Kong	2	0.00048	0.00002	0.00000	0.00108	0.00002	0.00031	0.00241	0.00004	0.00053	0.0049	0.0357	0.49	3.57
B18	Hong Kong	2	0.00046	0.00003	0.00000	0.00129	0.00002	0.00035	0.00250	0.00004	0.00079	0.0055	0.0402	0.55	4.02
B19	Hong Kong	2	0.00052	0.00003	0.00000	0.00130	0.00002	0.00039	0.00199	0.00004	0.00070	0.0050	0.0363	0.50	3.63
B20	Hong Kong	2	0.00034	0.00002	0.00007	0.00140	0.00003	0.00034	0.00219	0.00004	0.00073	0.0052	0.0361	0.52	3.61

'B1–B20' represents the batch numbers of AD samples. '1,2' represents the species numbers of AD samples. 1—*Angelica dahurica* (Hoffm.) Benth. & Hook. f. ex Franch. & Sav. 2—Angelica dahurica var. formosana (Boissieu) Yen.

The assessment in this study is quite conservative when compared with real clinical applications. AD as a part of TCM may rarely be used for a whole lifetime. In most cases, the duration of TCM usage will not exceed 2 years. In this exposure scenario, the EDI values of the total FCs in the 20 batches of AD samples are not more than 0.00023 mg/kg bw/day, which indicates that the risk of taking AD is extremely low. In addition, for the FCs with a TEF value of <0.01, we calculated under an assumption of a TEF equal to 0.01. Conversely, some uncertainties may cause underestimation in the risk calculation. For example, owing to the limited detection technology, some FCs of a low dose cannot yet be evaluated. In this study, we assumed that the toxicity of FCs in AD was the dose addition (DA) type. DA indicates that every single toxicant in the mixture contributes to the whole toxicity in proportion to its toxic unit, for example, its concentration or its potency. This implies that mixture toxicity will rise with the addition of individual agents. However, this is not the case. The joint toxicity of different FCs may be considerably larger or smaller than the toxicity (if any) associated with the NOAEL of each FC on its own [37,38]. Although there are numerous gaps and uncertainties in knowledge about the FCs from TCM, the established HI-TEF approach offers one possible strategy for the risk assessment of FCs originating from TCM.

4. Conclusions

In the present study, a fast, simple, credible, and sensitive evaluation method was developed for the accurate determination of nine FCs using single marks in 20 min based on UPLC. This method was successfully used for the content determination of AD samples in 20 batches. Thus, the problem of preparing a bulk of expense reference standards was effectively solved. Moreover, risk assessment methods for FCs from TCM in terms of toxicity were established for the first time, using AD as an example, based on the HI-TEF strategy. The cumulative risk assessment results revealed that the potential health risks of AD medicines were acceptable. Overall, the comprehensive application of the QAMS method alongside the HI-TEF risk assessment approach was meaningful, which could provide a significant reference for the rational utilization and quality control of FCs in TCM.

Supplementary Materials: The following supporting information can be downloaded at: https://www.mdpi.com/article/10.3390/separations10090508/s1, Table S1: Contents of the 9 FCs in AD determined by ESM and QAMS methods (mg/g, *n* = 3).

Author Contributions: Z.W.: conceptualization, formal analysis, data acquisition, examination, writing—original draft. K.Z.: methodology, validation. X.-W.H.: investigation. S.K.: collection and identification of samples. H.-L.L.: investigation. H.-Y.J.: investigation, data curation. T.-T.Z.: writing—review and editing. S.-C.M.: writing—review and editing. All authors have read and agreed to the published version of the manuscript.

Funding: This research was funded by the Traditional Chinese Medicine Modernization Project, under grant numbers 2022YFC3501505.

Data Availability Statement: The datasets used and/or analyzed during the current study are available from the corresponding author on reasonable request. Others come from previously reported studies and datasets, which have been cited.

Conflicts of Interest: The authors declare no conflict of interest.

References

1. Melough, M.M.; Cho, E.; Chun, O.K. Furocoumarins: A review of biochemical activities, dietary sources and intake, and potential health risks. *Food Chem. Toxicol.* **2018**, *113*, 99–107. [CrossRef] [PubMed]
2. Santana, L.; Uriarte, E.; Roleira, F.; Milhazes, N.; Borges, F. Furocoumarins in medicinal chemistry. Synthesis, natural occurrence and biological activity. *Curr. Med. Chem.* **2004**, *11*, 3239–3261. [CrossRef] [PubMed]
3. Scott, B.R.; Pathak, M.A.; Mohn, G.R. Molecular and genetic basis of furocoumarin reactions. *Mutat. Res.* **1976**, *39*, 29–74. [CrossRef] [PubMed]
4. Sayre, R.M.; Dowdy, J.C. The increase in melanoma: Are dietary furocoumarins responsible? *Med. Hypotheses* **2008**, *70*, 855–859. [CrossRef] [PubMed]

5. Wu, S.W.; Cho, E.; Feskanich, D.; Li, W.Q.; Sun, Q.; Han, J.L.; Qureshi, A.A. Citrus consumption and risk of basal cell carcinoma and squamous cell carcinoma of the skin. *Carcinogenesis* **2015**, *36*, 1162–1168. [CrossRef] [PubMed]
6. Wu, S.W.; Han, J.L.; Feskanich, D.; Cho, E.; Stampfer, M.J.; Willett, W.C.; Qureshi, A.A. Citrus Consumption and Risk of Cutaneous Malignant Melanoma. *J. Clin. Oncol.* **2015**, 33, 2500–2508. [CrossRef]
7. Abraham, K.; Wöhrlin, F.; Lindtner, O.; Heinemeyer, G.; Lampen, A. Toxicology and risk assessment of coumarin: Focus on human data. *Mol. Nutr. Food Res.* **2010**, *54*, 228–239. [CrossRef]
8. Müller, L.; Kasper, P.; Kersten, B.; Zhang, J. Photochemical genotoxicity and photochemical carcinogenesis—Two sides of a coin? *Toxicol. Lett.* **1998**, *102–103*, 383–387. [CrossRef]
9. Kitamura, N.; Kohtai, S.; Nakagaki, R. Molecular aspects of furocoumarin reactions: Photophysics, photochemistry, photobiology, and structural analysis. *J. Photochem. Photobiol. C Photochem. Rev.* **2005**, *6*, 168–185. [CrossRef]
10. Bordin, F. Photochemical and photobiological properties of furocoumarins and homologues drugs. *Int. J. Photoenergy* **1999**, *1*, 761394. [CrossRef]
11. Guo, L.Q.; Yamazoe, Y. Inhibition of cytochrome P450 by furanocoumarins in grapefruit juice and herbal medicines. *Acta Pharmacol. Sin.* **2004**, *25*, 129–136. [PubMed]
12. European Medicines Agency. Reflection Paper on the Risk Associated with Furocoumarins Contained in Preparations of *Angelica archangelica* L. EMEA/HMPC/317913/2006 [EB/OL]. 2007. Available online: https://www.ema.europa.eu/en/documents/scientific-guideline/final-reflection-paper-risks-associated-furocoumarins-contained-preparations-angelica-archangelica-l_en.pdf (accessed on 9 January 2021).
13. Raquet, N.; Schrenk, D. Relative photomutagenicity of furocoumarins and limettin in the hypoxanthine phosphoribosyl transferase assay in V79 cells. *Chem. Res. Toxicol.* **2009**, *22*, 1639–1647. [CrossRef] [PubMed]
14. Chinese Pharmacopeia Commission. *Pharmacopoeia of People's Republic of China*; Chemical Medical Science Press: Beijing, China, 2020.
15. Wang, Q.Q.; Li, Y.N.; Wang, S.G.; Xiang, Z.D.; Dong, W.C.; Li, X.Y.; Wei, Y.M.; Gao, P.; Dai, L. A review of the historical records, chemistry, pharmacology, pharmacokinetics and edibility of *Angelica dahurica*. *Arab. J. Chem.* **2023**, *16*, 1878–5352. [CrossRef]
16. Zhu, C.; Wang, M.; Guo, J.; Su, S.L.; Yu, G.; Yang, Y.; Zhou, Y.; Tang, Z. *Angelica dahurica* Extracts Attenuate CFA Induced Inflammatory Pain via TRPV1 in Mice. *Evid. Based Complement. Altern. Med.* **2022**, *2022*, 4684830. [CrossRef] [PubMed]
17. Xie, Y.; Zhao, W.; Zhou, T.; Fan, G.; Wu, Y. An efficient strategy based on MAE, HPLC-DAD-ESI-MS/MS and 2D-prep-HPLC-DAD for the rapid extraction, separation, identification and purification of five active coumarin components from Radix Angelicae Dahuricae. *Phytochem. Anal.* **2010**, *21*, 473–482. [CrossRef] [PubMed]
18. Zhao, H.; Feng, Y.L.; Wang, M.; Wang, J.J.; Liu, T.; Yu, J. The *Angelica dahurica*: A Review of Traditional Uses, Phytochemistry and Pharmacology. *Front. Pharmacol.* **2022**, *13*, 896637. [CrossRef] [PubMed]
19. Zhou, B.; Liu, P.; Chen, J.; Xu, Y.; Chen, D.W.; Duan, J.A. Analysis and evaluation of coumarins and polysaccharides in *Angelica dahurica* from different habitats. *Acta. Univ. Med.* **2015**, *31*, 68–73.
20. Zhou, Y.; Na, L. Research progress on medicinal and food homology of *Angelica dahurica*. *Asian-Pac. Tradit. Med.* **2022**, *18*, 213–217.
21. Pang, X.Y.; Jing, Y.S.; Zheng, Y.G.; Wu, L.F. Chemical constituents of *Angelica dahurica* and its therapeutic effect on nervous system diseases. *Chin. J. Pharmacol. Toxicol.* **2021**, *35*, 690–691.
22. Liu, S.; Huang, X.; Lin, Y.; Lin, H.; Wang, Q.; Liu, Z. Skin phototoxicity study of radix Angelicae dahuricae water extract using rats by oral administration. *J. Hunan Agric. Univ. Nat. Sci.* **2018**, *44*, 424–429.
23. Kang, J.; Zhou, L.; Sun, J.; Han, J.; Guo, D.A. Chromatographic fingerprint analysis and characterization of furocoumarins in the roots of *Angelica dahurica* by HPLC/DAD/ESI-MSn technique. *J. Pharm. Biomed. Anal.* **2008**, *47*, 778–785. [CrossRef]
24. Zhang, H.; Gong, C.G.; Lv, L.; Xu, Y.J.; Zhao, L.; Zhu, Z.Y.; Chai YFZhang, G.Q. Rapid separation and identification of furocoumarins in *Angelica dahurica* by high-performance liquid chromatography with diode-array detection, time-of-flight mass spectrometry and quadrupole ion trap mass spectrometry. *Rapid Commun. Mass Spectrom.* **2009**, *23*, 2167–2175. [CrossRef] [PubMed]
25. Yang, W.Q.; Zhu, Z.X.; Song, Y.L.; Qi, B.W.; Wang, J.; Su, C.; Tu, P.F.; Shi, S.P. Dimeric furanocoumarins from the roots of *Angelica dahurica*. *Nat. Prod. Res.* **2017**, *31*, 870–877. [CrossRef] [PubMed]
26. Shu, P.; Li, J.P.; Fei, Y.Y.; Zhu, H.Q.; Yu, M.Z.; Liu, A.Q.; Niu, H.Y.; Zou, S.M.; Wei, X.L.; Ju, Z.Y.; et al. Isolation, structure elucidation, tyrosinase inhibitory, and antioxidant evaluation of the constituents from *Angelica dahurica* roots. *J. Nat. Med.* **2020**, *74*, 456–462. [CrossRef] [PubMed]
27. Lohr, C.; Raquet, N.; Schrenk, D. Application of the concept of relative photomutagenic potencies to selected furocoumarins in V79 cells. *Toxicol. Vitr.* **2010**, *24*, 558–566. [CrossRef] [PubMed]
28. Raquet, N.; Schrenk, D. Application of the equivalency factor concept to the phototoxicity and -genotoxicity of furocoumarin mixtures. *Food Chem. Toxicol.* **2014**, *68*, 257–266. [CrossRef] [PubMed]
29. EFSA. Scientific opinion of the panel on food additives, flavourings, processing aids and materials in contact with food on a request from the European Commission on Coumarin in flavourings and other food ingredients with flavouring properties. *EFSA J.* **2008**, *6*, 793–808.
30. Huang, J.; Yin, L.; Dong, L.; Quan, H.F.; Chen, R.; Hua, S.Y.; Ma, J.H.; Guo, D.Y.; Fu, X.Y. Quality evaluation for Radix Astragali based on fingerprint, indicative components selection and QAMS. *Biomed. Chromatogr.* **2018**, *32*, 4343. [CrossRef] [PubMed]

31. Peng, Y.; Dong, M.; Zou, J.; Liu, Z. Analysis of the HPLC fingerprint and QAMS for Sanhuang gypsum soup. *J. Anal. Methods Chem.* **2018**, *2018*, 5890973. [CrossRef]
32. Dong, Y.; Guo, Q.; Liu, J.; Ma, X. Simultaneous determination of seven phenylethanoid glycosides in Cistanches Herba by a single marker using a new calculation of relative correction factor. *J. Sep. Sci.* **2018**, *41*, 1913–1922. [CrossRef]
33. Zuo, T.T.; Jin, H.Y.; Zhang, L.; Liu, Y.L.; Nie, J.; Chen, B.L.; Fang, C.F.; Xue, J.; Bi, X.Y.; Li, Z.; et al. Innovative health risk assessment of heavy metals in Chinese herbal medicines based on extensive data. *Pharmacol. Res.* **2020**, *159*, 104987–104999. [CrossRef]
34. International Agency for Research on Cancer (IARC). *5-Methoxypsoralen [A]//IARC Monographs on the Evaluation of Carcinogenic Risks to Humans*; World Health Organization: Lyon, France, 1986.
35. Zhao, W.H.; Cao, Y.X.; Liu, J.; Gao, Y.W.; He, J.Y.; Cao, L.Q. Research on chronic toxicology of Compound Radix Angelicae Dahuricae capsule. *J. Shanxi Med. Univ.* **2006**, *37*, 160–163.
36. Zheng, G.Y.; Ma, Y.Y.; Lu, X.L.; Hai, M.Z.; Guo, P.L. Effect of sulfuring on the acute toxicity of the angelica extract. *Pharm. Clin. Chin. Mater. Medica* **2012**, *3*, 43–50.
37. Kortenkamp, A.; Evans, R.; Faust, M.; Kalberlah, F.; Scholze, M.; Wolz, S.U. Investigation of the state of the science on combined actions of chemicals in food through dissimilar modes of action and proposal for science-based approach for performing related cumulative risk assessment. *EFSA Support. Publ.* **2012**, *9*, 1–300. [CrossRef]
38. European Food Safety Authority. Opinion of the Scientific Panel on food additives, flavourings, processing aids and materials in contact with food (AFC) related to Flavouring Group Evaluation 6 (FGE.06): Straight-and branched-chain aliphatic unsatured primary alcohols, aldehydes, carboxylic acids, and esters from chemical groups 1 and 4. *EFSA J.* **2004**, *2*, 108.

separations

MDPI

Article

Determining the Hydrophobicity Index of Protected Amino Acids and Common Protecting Groups

Varshitha Gavva [1,†], Othman Al Musaimi [1,*,†], Colin Bent [2] and Daryl R. Williams [1]

[1] Department of Chemical Engineering, Imperial College London, London SW7 AZ, UK; d.r.williams@imperial.ac.uk (D.R.W.)
[2] Eli Lilly and Company, Indianapolis, IN 46285, USA
* Correspondence: o.al-musaimi@imperial.ac.uk
† These authors contributed equally to this work.

Abstract: Peptides are in great demand in the pharmaceutical arena and a majority of these peptides contain 20 or more amino acids. They are infrequently synthesised using the fragment condensation approach. A key limitation in adopting this approach more commonly is that protected peptide fragments with high purity are often required prior to the final condensation steps. It is hypothesized that understanding the hydrophobic nature of the protected amino acids will assist with designing optimal fragment purification processes when needed. Whilst a myriad of hydrophobicity indices are reported in the literature for unprotected amino acids, the literature lacks any data regarding the protected amino acids which form the key precursor for the fragment condensation task. In this current study, hydrophobicity indices for protected amino acids with common α-amino and sidechain protecting groups were experimentally determined. Different positions for each amino acid within the peptide chain were considered, namely at the *C*-terminal and *N*-terminal as well as internal positions. These data give deep insights on the hydrophobicity of each amino acid with respect to its position in the peptide chain. The data acquired in this research facilitated the prediction of the retention time of protected peptide fragments with an uncertainty of less than ±1.5%.

Keywords: protected peptides; separation; hydrophobicity index; retention time prediction; purification; protecting groups; cyclisation; fragment condensation

Citation: Gavva, V.; Al Musaimi, O.; Bent, C.; Williams, D.R. Determining the Hydrophobicity Index of Protected Amino Acids and Common Protecting Groups. *Separations* **2023**, *10*, 456. https://doi.org/10.3390/separations10080456

Academic Editor: Aleksandra Mišan

Received: 31 July 2023
Revised: 18 August 2023
Accepted: 18 August 2023
Published: 20 August 2023

1. Introduction

Macromolecular chains usually consisting of 10 to 50 amino acids which are linked together by amide bonds are known as peptides [1]. The number, type and order of the amino acids within each peptide differentiate them from each other, and they are generally classified as medium-sized molecules [2]. A human body consists of more than 7000 peptides that occur naturally, and these peptides play a vital role by helping in the physiobiological process within the living organism [1]. Peptides functions include but are not limited to antibodies, transducing signals, helping in cellular growth and reproduction, and in the communication between cells [1,3]. Due to the technological advancements witnessed in the synthetic and biotherapeutic formulation strategies, drug delivery and purification techniques of peptides, they are now of great interest for the pharmaceutical industry [1,4,5]. Peptides have high target affinity and specificity, and can degrade easily or stay in the targeted location for a long time depending on the intended use. All these positive attributes make peptides one of the most important drugs in the market today, with a total of 26 approvals between 2016 and 2022 [6]. Research projects are now being undertaken to further study and exploit peptides in a range of applications, as they can exhibit excellent performance not only in the field of medical therapeutics, but also in tissue engineering, cosmetics, biosensing, and antibacterial agents, among other fields [7,8].

In 2006, there were more than 40 FDA-approved peptides on the market and 400 in clinical trials. Currently, there are 120 FDA-approved peptides drugs on the market, 500 in preclinical trials and 140 in clinical trials. These data clearly show that the demand for peptides on the market is increasing and growing fast [8–12]. The peptide market is expected to grow from USD 29 Bn to USD 51 Bn between 2020 and 2026 [8].

Some of the peptides on the market are synthesised using solution phase synthesis, especially if they are short peptides and easy to synthesise [1,2,13,14]. However, the Nobel laureate Merrifield revolutionised the field of peptide chemistry by introducing solid-phase peptide synthesis (SPPS) methodology [15]. SPPS is the method of choice for synthesising peptides both in research and industrial settings. This methodology helped in synthesising longer and more complicated peptides following the stepwise approach [15]. However, the efficiency starts to diminish with peptides of more than 30 amino acids, and sometimes 20 amino acids depending on the type of amino acids [16,17]. The fragment condensation method proved to be a better alternative to the stepwise approach for these longer peptides, as the desired sequence is divided into multiple fragments which are synthesised using the SPPS stepwise approach, and later these fragments are merged into the final product. T20 is one of these drugs which is currently on the market and is synthesised using this approach [18]. A blood–brain barrier peptide called TD2.2 was also successfully synthesised following the fragment condensation approach [19]. Protected peptide fragments are also important for synthesising other peptide families, such as cyclic peptides [20].

It is vital to maintain the purity standard of peptides, as it is set by good manufacturing practice (GMP), and it is also important to study their applications further. HPLC is the most popular technique for this purpose due to its ability to separate the structurally related impurities from the target peptide [8,21,22]. Significant improvements in the chromatographic techniques have been witnessed for both analytical and manufacturing scales of purification [22,23].

Among the physicochemical properties of peptides, hydrophobicity is the major property that governs the chromatographic separation of peptides [22–24]. Various studies have established the hydrophobic indices of unprotected amino acids [22], which allow the predicting of the retention times of peptides with a known amino acid sequence, hence saving experimental time, costs and laboratory efforts [22]. However, there are no data in the literature with respect to the hydrophobicity of protected amino acids which in turn could be used in predicting the retention time of protected peptide fragments, the main precursors in the fragment condensation methodology.

In this work, we aim to establish a hydrophobicity index for the protected amino acids with common α-amino and sidechain protecting groups. This approach will be helpful in predicting the separation and purification processes of protected peptide fragments, which are key precursors in the fragment condensation reactions. Hence, to fully understand their hydrophobicity, 20 natural amino acids were considered in a model peptide, as well as various positions within the peptide sequence, including acid and peptide amide, and acetylated forms were investigated using reversed-phase liquid chromatography (RPLC).

2. Materials and Methods

2.1. Materials

PuroSynth CTC (1.0 mmol/g) and Merck Sieber amide (0.75 mmol/g) resins with manufacturer's specifications were used. Other solvents and reagents that were used in the process include diethyl ether, *N,N*-dimethylformamide (DMF), trifluoroacetic acid (TFA), tri-isopropylsilane (TIS), dichloromethane (CH_2Cl_2), chloroform, *N,N*-diisopropylethylamine (DIEA), piperdine, acetonitrile (CH_3CN), *N,N*′-diisopropylcarbodiimide (DIC) and methanol. These reagents, amino acids, and organic solvents were used without further purification unless otherwise stated. Analytical HPLC was performed on a Shimadzu LC20 system using Lab solution software for data processing with a Symmetry Luna C_{18} (3.6 µm, 4.6 × 150 mm) column. Another HPLC and columns were used to verify the model: a Shimadzu LC40 system using Lab solution software for data processing using Symmetry Luna C_8 (5 µm, 4.6 × 150 mm). The mobile phase flow rate was 1.0 mL/min and UV detection was at 220 nm. Mobile phase A was 0.1% TFA in H_2O, and mobile phase B was 0.1% TFA in CH_3CN.

2.2. *Peptide Synthesis*

2.2.1. Incorporation Procedure

CTC resin

The first amino acids were incorporated onto CTC resin using dry CH_2Cl_2. CTC resin was swelled in CH_2Cl_2 for 10−20 min. The Fmoc-amino acids (2 equiv) were dissolved in a minimum amount of the CH_2Cl_2 (0.5 mL/100 mg resin) and sonicated for 10 min. DIEA (4 equiv) was then added to the solution, which in turn was added to the previously swelled resin and allowed to react for 1 h under mechanical shaking. After this, MeOH (80 µL/100 mg of resin) was added to endcap any unreacted chloride of the CTC resin. Finally, the resin was washed twice with CH_2Cl_2 and vacuum dried.

Sieber amide resin

The first amino acids were incorporated onto Rink amide resin using dry DMF. Rink Amide resin was swelled in DMF for 10−20 min. Fmoc was removed using 20% piperidine/DMF and the mixture was allowed to shake for 2 and 7 min. The Fmoc-amino acids (3 equiv) and OxymaPure (3 equiv) were dissolved in a minimum amount of the DMF (0.5 mL/100 mg resin) and sonicated for 10 min. DIC (3 equiv) was then added to the solution, which in turn was added to the previously swelled resin and allowed to react for 1 h under mechanical shaking. Finally, the resin was washed twice with CH_2Cl_2 and dried over vacuum.

2.2.2. Peptide Assembly

Peptides were synthesised following the standard methodology performed in our laboratory (3 equiv. of Fmoc-AA-OH, 3 equiv. of OxymaPure, 3 equiv. of DIC) in DMF and then shaken for 1 h. Fmoc was then removed (see under Rink amide resin section).

2.3. *Cleavage Protocols*

2.3.1. Protected Fragments

Peptide resin was placed in a syringe fitted with porous polyethylene filter. It was then swelled with CH_2Cl_2 for 10 min. The solvent was then filtered off, the cleavage solution (2 mL, 2% TFA (*v*/*v*) in CH_2Cl_2) was added per 100 mg of the peptide resin, and the syringe was closed with a cap and shaken for 30 min at room temperature. Finally, the filtrate was collected over CH_3CN (4.0 mL), and the cleaved resin was washed 4 times with CH_3CN (1 mL each time). The filtrates were again collected in the same tube. An aliquot (5 µL) was then injected into the HPLC system to record the retention time.

2.3.2. Unprotected Fragments

CTC resin

The final synthesised peptide was cleaved from the resin using TFA/TIS/H_2O (95:2.5:2.5) (1 mL/100 mg) under mechanical shaking for 60 min. Chilled diethyl ether was then added (5 times the cleavage solution volume), and the solution was kept in an ice bath for 30 min. The solution was then centrifuged for 5 min at 5000 rpm, and the supernatant was decanted. A new amount of the ether (5 times the cleavage solution volume) was added to repeat this step. Any remaining ether was dried under N_2. Finally, the precipitate was dissolved in CH_3CN-H_2O (1:1). A small amount of the solution was injected into the HPLC system to record the retention time.

Sieber amide resin

As the peptide amides were soluble in the ether, another workup procedure was followed. First, TFA was removed from the cleavage solution under vacuum. Next, 10% (*v*/*v*) of aqueous acetic acid was added to the dried peptidyl resin to dissolve the peptide. Then, chloroform (1 mL/100 mg) was added to the mixture and then shaken vigorously. The mixture was then allowed to settle, and the lower layer (chloroform and the released protecting groups) was discarded. This step was performed twice or thrice. Finally, the peptide in the 10% (*v*/*v*) solution was diluted in CH_3CN-H_2O (1:1). A small amount of the solution was injected into the HPLC system to record the retention time.

3. Results and Discussions

Six different peptide sequences were considered, in which the target amino acid was placed at various positions to investigate its influence on the peptide's hydrophobicity (see Figure 1). The previous position has been shown to influence the hydrophobicity of amino acids [24,25].

1. Fmoc/H-Xxx-Gly-Phe-Gly-Leu-OH	4. Ac/H-Xxx-Gly-Phe-Gly-Leu-NH_2
2. H-Phe-Gly-Xxx-Gly-Leu-OH	5. H-Phe-Gly-Xxx-Gly-Leu-NH_2
3. H-Leu-Gly-Phe-Gly-Xxx-OH	6. H-Leu-Gly-Phe-Gly-Xxx-NH_2

Figure 1. Peptide sequences considered in this study. Fmoc: Fluorenylmethyloxycarbonyl protecting group; Ac: acetyl group.

Amino acids were located at the *C*-terminal, *N*-terminal and internal positions. To be able to determine the hydrophobicity of a particular protected amino acid, a short peptide with one amino acid that had a functional side chain was considered. Here, the pentapeptide Leu-enkephalin, which is an endogenous opioid peptide neurotransmitter, was considered [26]. This peptide has five amino acids, and one of them has a functional side chain which requires protection prior to any synthesis work. This amino acid will be interchanged with the 20 natural amino acids in this study. However, the sequences have been designed in a way that the target amino acid is placed next to a Gly residue to ensure unrestricted rotation around the amide bond, and hence the hydrophobicity data will be more accurate and representative for each amino acid [27]. To understand and reflect the hydrophobic nature of each amino acid, the target amino acid was placed at different positions, as follows: (i) *N*-terminal of the sequence, (ii) internal position, and (iii) *C*-terminal. These three sequences were considered for the acid peptide, with an additional three sequences of the same peptides but with an amidated *C*-terminal. Hence, the first three sequences were synthesised using 2-chlorotrityl (2-CTC) resin. The second three sequences were synthesised using Sieber amide resin. Both resins are highly acid-labile resins, where protected peptides can be cleaved from the resin with a mild acidolysis reaction [28,29].

3.1. Amino Acids Hydrophobicity

Twelve out of the twenty natural amino acids have a functional sidechain, and hence appropriate protection is deemed necessary to avoid complications during the synthesis process. The following are the common sidechain protecting groups for the Fmoc/*t*Bu peptide synthesis strategy: Trityl (Trt), *tert*-butyloxycarbonyl (Boc), 2,2,4,6,7-pentamethyldihydrobenzofuran-5-sulfonyl (Pbf). Several amino acids share the same sidechain protecting group, in which His, Asn, Gln, and Cys have the same Trt group to protect their functional sidechain, whereas Asp, Glu, Ser, Thr, and Tyr have *t*Bu, and Trp and Lys have Boc. Pbf is the sidechain protecting Arg only. Hence, it would be interesting to establish and compare the hydrophobicity index of these protecting groups on different amino acids.

Given that Gly has only H on its sidechain, the Gly-peptide analogue was considered as a reference to estimate the hydrophobicity index of the protected amino acids. Hydrophobicity indices were determined for all sequences by subtracting the retention time of the peptide with the amino acid of interest from the retention time of the Gly-peptide of the same sequence (Figure 2) (Supplementary Table S1) (Supplementary Table S2) (Figure S1).

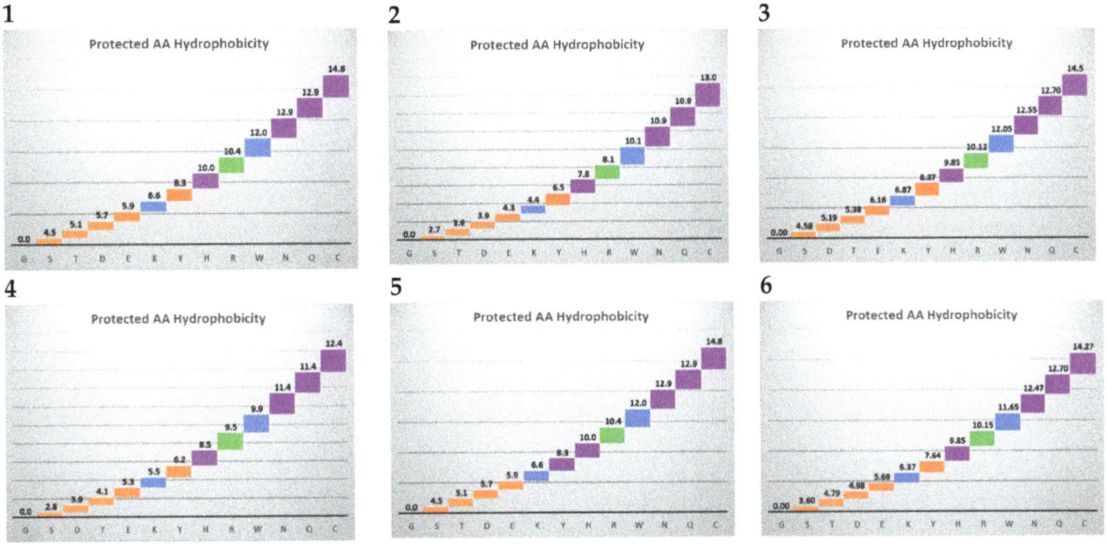

Figure 2. Hydrophobicity index for the protected amino acids in different positions of the peptide chains. The figure number represents the sequence of the peptide, as per Figure 1.

In all of the investigated peptide sequences, and regardless of the amino acid's position in the sequence, Cys(Trt) was the most hydrophobic protected amino acid and Ser(*t*Bu) was the least. The position of the amino acid does play a significant role in its hydrophobicity. In the acid peptides, amino acids showed a more hydrophobic character when placed at the *C*-terminal than when at the *N*-terminal, and the least values when they were in the internal positions (Figure 3).

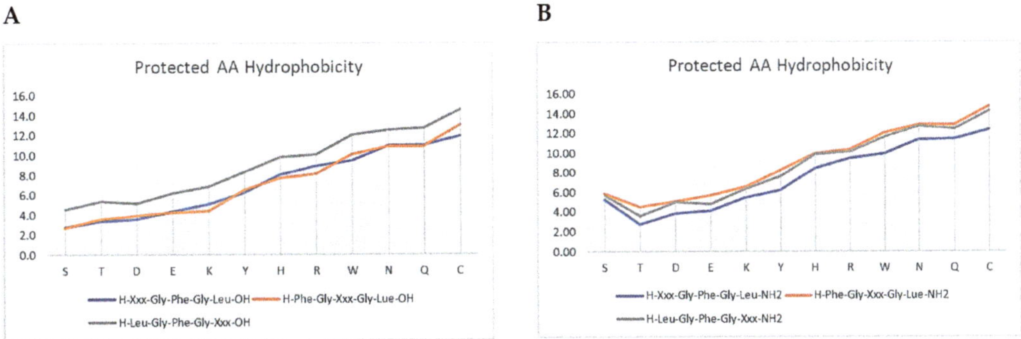

Figure 3. Hydrophobicity pattern for the protected amino acids in different positions of the peptide chains. (**A**) Acid peptide. (**B**) Peptide amide.

However, the overall hydrophobicity pattern of all the protected amino acids showed an excellent correlation. The hydrophobicity index in the first three acid peptide sequences resulted in the following r^2correlation values:

Sequence 1 versus sequence 2 $r^2 = 0.98$;
Sequence 1 versus sequence 3, $r^2 = 0.99$;
Sequence 2 versus sequence 3, $r^2 = 0.99$;
Sequence 4 versus sequence 5, $r^2 = 0.98$;
Sequence 4 versus sequence 6, $r^2 = 0.99$;
Sequence 5 versus sequence 6, $r^2 = 0.99$ (Figure S2).

The situation was a bit different in the peptide amide sequences. Even though the protected amino acids showed a higher hydrophobic character when at the *C*-terminal, they also showed a higher hydrophobic character when in the internal positions, and in some cases higher than when at the *C*-terminal. This observation is an unsurprising fact, and it is ascribed to the amidated *C*-terminal ($CONH_2$), which is hydrophobic in its nature, whereas in acid peptides there is the hydrophilic OH group instead. This phenomenon caused the protected amino acids at the *N*-terminal to show the lowest hydrophobic character in the peptide amide sequences (Figure 3).

Protected amino acids were also compared when present in acid peptides versus peptide amides, with the same position as shown below in Figure 4.

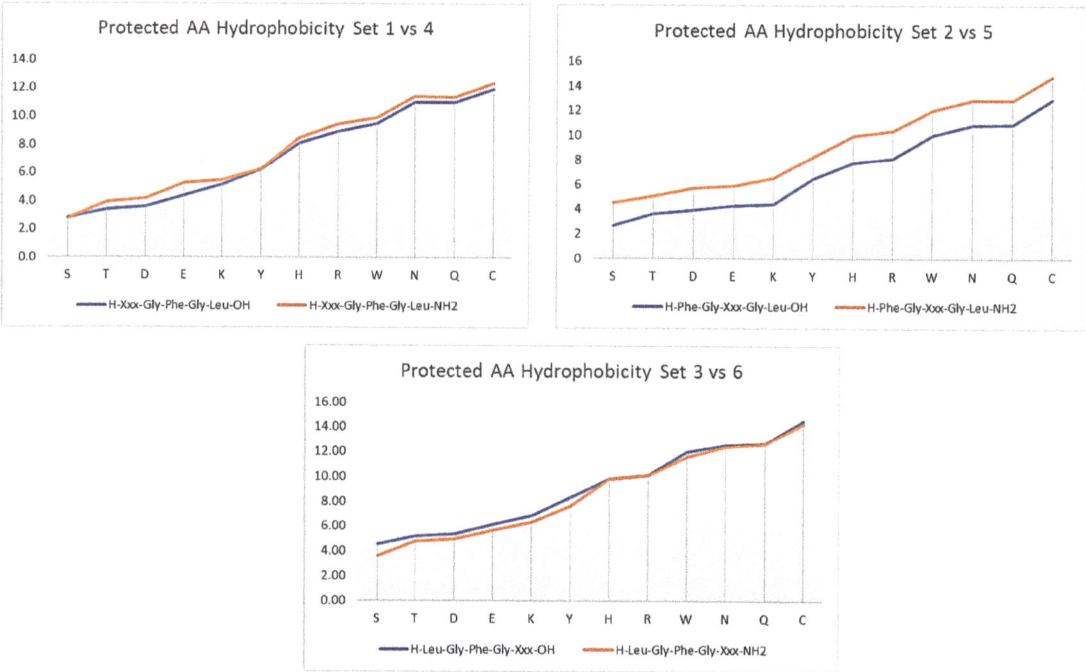

Figure 4. Hydrophobicity pattern for the protected amino acids in acid versus peptide amides.

Unsurprisingly, in all cases the peptide amide showed higher hydrophobicity than the acid form. The difference was more pronounced when the amino acid was placed in the internal positions. Excellent correlations were obtained as follows: 1 versus 4, $r^2 = 0.99$; 2 versus 5, $r^2 = 0.99$; and 3 versus 6, $r^2 = 0.98$ (Figure S3).

However, though this study focuses on protected amino acids, we also investigated the unprotected ones in order to allow the retention time prediction reported at the end of this study (Supplementary Table S2) (see Figure S1). Trp was shown to be the most hydrophobic unprotected amino acid, whereas Lys was the least. Very good correlations were also obtained in the overall hydrophobicity pattern, in which:

Sequence 1 versus sequence 2, $r^2 = 0.92$;
Sequence 1 versus sequence 3, $r^2 = 0.92$;
Sequence 2 versus sequence 3, $r^2 = 0.96$;
Sequence 4 versus sequence 5, $r^2 = 0.93$;
Sequence 4 versus sequence 6, $r^2 = 0.91$;
Sequence 5 versus sequence 6, $r^2 = 0.97$ (see Figure S2).

When comparing the same amino acid, in the same position but in the acid and peptide amide, a very good correlation was also obtained, as follows:

Sequence 1 versus sequence 4, $r^2 = 0.93$;
Sequence 2 versus sequence 5, $r^2 = 0.93$;
Sequence 3 versus sequence 6, $r^2 = 0.92$ (see Figure S3).

3.2. Protecting Groups' Hydrophobicity

Additional processes were also considered to establish the hydrophobicity index for the protecting groups themselves. The hydrophobicity indices of the twelve protected amino acids with respect to the Gly-peptide of the same sequence are shown below in Figure 5.

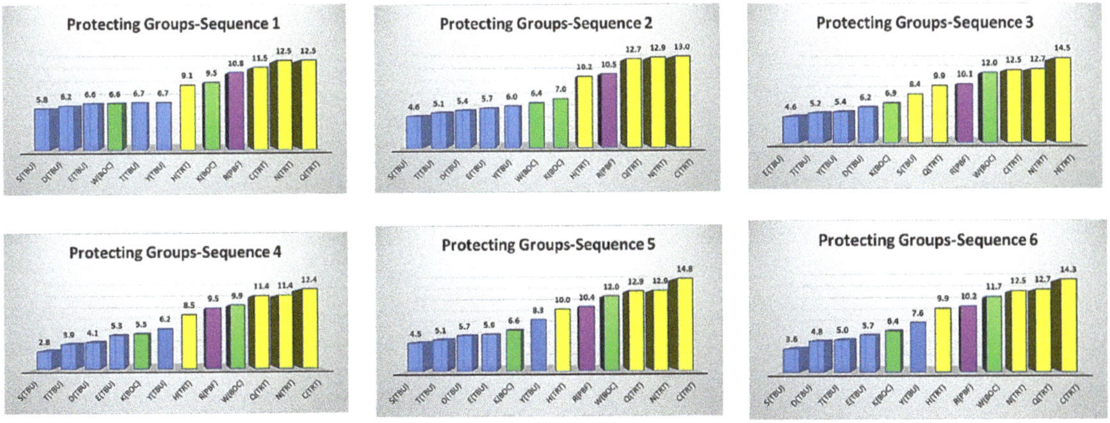

Figure 5. Hydrophobicity indices for the sidechain protecting groups.

Hydrophobicity indices were determined for the protecting groups themselves by subtracting the retention time of the protected peptide from the unprotected analogue of the same peptide. This applies to any protecting group, either at the α-amino or the sidechain of the amino acid of interest. Investigating the hydrophobicity indices of the sidechain protecting groups that can protect different amino acids showed that a large protecting group would have the same hydrophobicity index regardless of the amino acid type (Figure S4). For example, the Trt group showed a hydrophobicity index that ranged from about 10 to 14 min, with four amino acids (His(Trt), Asn(Trt), Glu(Trt) and Cys(Trt)). It is worth highlighting that Trt had the lowest hydrophobicity index of 10 min with the least hydrophobic amino acid, which was His. Other than the fact that His is hydrophilic, its ring within its structure probably experienced less interaction with the stationary phase and hence had a lower hydrophobicity value. Nevertheless, the Trt group was the most hydrophobic protecting group among all the protecting groups incorporated in this study. The next protecting group in terms of the hydrophobicity was the Pbf group, which protects the side chain of Arg only, then Boc group, and the last one was the *t*Bu group. Unlike large protecting groups, Boc and *t*Bu are small protecting groups and hence they are affected by the type of amino acid they are protecting. Their hydrophobicity ranged from 6 to 12 min and 4 to 8 min, respectively. With such small protecting groups, the influence of the amino acid and its interaction with the chromatographic components, and mainly the stationary phase, will be more pronounced than in the case of the large protecting group. All the protecting groups showed a uniform hydrophobicity pattern regardless of their position within the sequence, whether the peptide was an acid or a peptide amide one.

The protection of α-amino functionality is often carried out using Boc and/or Fmoc protecting groups. However, Fmoc is more convenient and an easier methodology that does not require special laboratory infrastructure, unlike the case for the Boc strategy [1]. Therapeutic peptides are often presented in the acetylated form to enhance their stability. Therefore, the hydrophobicity index was also determined for Fmoc amino acids, both protected and unprotected, Ac amino acids (protected and unprotected), the Fmoc protecting group and the acetyl group to understand their hydrophobicity, and such data were also helpful for the prediction task (Supplementary Tables S3 and S4) (Figure S4). For the Fmoc group, only sequence 1 was selected, while for the acetyl group sequence 4 was selected. The selection was made based on how peptide fragments are presented during the fragment condensation synthetic approach (Fmoc-fragment-OH). In the case of acetyl, as explained above, therapeutic peptides often exist as (Ac-peptide-NH_2); hence, sequence 4 was considered. There was a difference in the hydrophobicity indices of Fmoc amino acids and the Fmoc protecting group itself, where the values ranged from 9 to 27 min. The difference was even higher in the case of Ac amino acids and the acetyl group itself (0.2 to 17 min), which was again attributed to the small size of the acetyl group with respect to that of Fmoc.

3.3. *Retention Time Prediction*

Using the extensive dataset generated, the prediction of retention times for various peptide fragments was undertaken as described below. The following peptides were employed in the study: Fmoc-APPPS(*t*Bu)-OH, Fmoc-GPS(*t*Bu)S(*t*Bu)G-OH and Fmoc-E(*t*Bu)FIE(*t*Bu)-OH. The predicted retention (τ) time was calculated by summing up the hydrophobicity index of the amino acids in the peptide chains, as per Equation (1):

$$\tau = \sum R_c \tag{1}$$

where $\sum R_c$ represents the sum of the retention coefficients for all amino acid residues and the termini in the given peptide sequence.

The predictive capability in our work was satisfactory regarding the uncertainty <3.5% (Table 1).

Table 1. Prediction of peptides' retention time.

Peptide Fragment	Predicted τ (Minute)	Actual (Minute)	Difference (Minute)
Fmoc-APPPS(*t*Bu)-OH	16.9	17.5	−0.6
Fmoc-GPS(*t*Bu)S(*t*Bu)G-OH	20.5	20.3	+0.2
Fmoc-E(*t*Bu)FIE(tBu)-OH	30.7	30.9	−0.2

A Symmetry Luna C_{18} (3.6 µm, 4.6 × 150 mm) column was used, with 1.0 mL/min and UV detection was at 220 nm. The mobile phase A was 0.1% TFA in H_2O, and the mobile phase B was 0.1% TFA in CH_3CN.

It should be noted that having the Gly residue at the C-terminal will not shield the subsequent amino acid from the effect of that location. Thus, in the second peptide (Fmoc-GPS(*t*Bu)S(*t*Bu)G-OH), the hydrophobicity index of S(*t*Bu) amino acid should be taken from sequence 1 as if S(*t*Bu) is at the C-terminal position with a hydrophobicity index of 4.6 min rather than sequence 2 that represents the internal positions, provided that its hydrophobicity index in the internal position is lower (2.6 min) and influences the prediction capability with a difference of 1.7 min.

To further validate our approach, we also predicted the retention times for peptide fragments by changing the gradient profile using the same HPLC and column and repeated the procedure using another HPLC system and column. Simple correction factors must be adopted to account for these changes. Therefore, in the case of changing the gradient profile, the ratio between both gradient profiles has to be incorporated as per Equation (2), where 16.8 is the gradient profile for the gradient used to estimate the hydrophobicity index in this study (15–70% in 30 min).

$$\tau = \sum R_c \times \text{current gradient profile}/16.8 \tag{2}$$

In the case of using the new HPLC system, a standard peptide should be used as a reference, and it is recommended to consider a standard that shares common amino acids and/or protecting groups with the peptides of interest. Next, the difference between the predicted and actual retention times of the standard peptide should be subtracted as per Equation (3).

$$\tau = \sum R_c - t_s \quad (3)$$

where $\sum R_c$ represents the sum of the retention coefficients for all amino acid residues and the termini in the given peptide sequence, the current gradient profile represents the amount of organic modifier needed to elute the peptide, and t_s represents the difference between the predicted and actual retention time of the standard peptide.

The first and second peptides (Fmoc-APPPS(*t*Bu)-OH and Fmoc-GPS(*t*Bu)S(*t*Bu)G-OH) were selected to validate our approach. Our model was able to predict the retention time when changing the gradient profile. For example, using a gradient profile of 5–95% in 15 min instead of the original one, which was 15–70% in 30 min, and employing Equation (2) resulted in a good prediction accuracy. A small difference of only −0.4 min and +1.0 min was observed for the first and the second peptides, respectively. Furthermore, using another HPLC and column while keeping the same gradient profile (15–70% in 30 min) exhibited an excellent prediction capability for our model using Equation (3). While both columns had the same length and diameter, they were different in their alkyl chain length and particle size: 3.6 μm versus 5 μm and C_{18} versus C_8, respectively. After employing Equation (3) and considering H-S(*t*Bu)GFGL-OH as a standard peptide for making the t_s correction, the final prediction capability was as follows: −1.9 min and −1.2 min for the first and the second peptides, respectively.

This study illustrates the excellent versatility of our first order model to predict the RPLC elution behaviour of a protected peptides with a known composition. However, it should be highlighted that there are other factors not incorporated into this model that can alter the elution behaviour, including the structural conformations, peptide length and neighbouring group effect, among others [24]. Thus, ultimately, our model needs to be coupled with an a priori computational model which can consider all of these additional factors affecting the elution process and incorporate them into a more advanced prediction model. Nevertheless, this first order model had offered excellent predictive performance for the peptides investigated in the current study.

4. Conclusions

For the first time, the hydrophobicity indices for protected amino acids with common α-amino and sidechain protecting groups were determined for a large set of synthesised peptides. The data in this work will help in predicting the retention time of protected peptide fragments that are the main precursors in the current peptide synthesis field. These data gave deep insights into the hydrophobicity of each amino acid with respect to its position in the peptide chain. The data acquired in this research facilitated the prediction of the RPLC retention time of protected peptide fragments with an uncertainty of $<\pm 1.5\%$. Therefore, being able to predict the retention time of these protected peptide fragments will definitely save time, funds and effort during their manufacture and purification. Based on previous work reported for unprotected amino acids in the literature [22,24], we plan for this model to be enhanced by coupling it with a molecular computational tool. This enhancement will allow for more complex hydrophobic phenomena to be understood, including any potential observed deviations from the current model. We will endeavour to exploit the available models, including machine learning and computational tools, to establish a more accurate prediction model.

Supplementary Materials: Supplementary information contains the hydrophobicity indices for unprotected amino acids and for protecting groups. The following supporting information can be downloaded at: https://www.mdpi.com/article/10.33990/separations10080456/s1, Table S1: Hydrophobicity index of protected amino acids. Table S2: Hydrophobicity index for unprotected amino acids. Table S3: Hydrophobicity index for Fmoc, Fmoc-amino acid. Table S4: Hydrophobicity index for Acetyl, Ac-amino acid. Figure S1: Hydrophobicity index for the unprotected amino acids in different positions of the peptide chains. The figure number represents the sequence of the peptide as per Figure 1. Figure S2: Correlation among hydrophobicity indices for the protected and unprotected amino acids in different positions of the peptide chains. Figure S3: Correlation among hydrophobicity indices for the protected and unprotected amino acids in acid and peptide amides. Figure S4: Hydrophobicity indices for: Fmoc-amino acid (protected and unprotected), Fmoc protecting group, Ac-amino acid (protected and unprotected), and Acetyl group.

Author Contributions: Conceptualization, O.A.M. and D.R.W.; methodology, V.G. and O.A.M.; formal analysis, V.G., O.A.M. and D.R.W.; investigation, V.G. and O.A.M.; resources, O.A.M. and D.R.W.; writing—original draft preparation, V.G. and O.A.M.; writing—review and editing, V.G., O.A.M., D.R.W. and C.B.; supervision, O.A.M. and D.R.W.; project administration, O.A.M. and D.R.W.; funding acquisition, O.A.M., D.R.W. and C.B. All authors have read and agreed to the published version of the manuscript.

Funding: Funding from Eli Lilly and Company and the UK EPSRC, through the PharmaSEL-Prosperity Programme (EP/T005556/1), is gratefully acknowledged.

Institutional Review Board Statement: Not applicable.

Informed Consent Statement: Not applicable.

Data Availability Statement: Not applicable.

Conflicts of Interest: The authors declare no conflict of interest.

Abbreviations

Ac	acetyl group
Boc	*tert*-butyloxycarbonyl
CH_2Cl_2	dichloromethane
CH_3CN	acetonitrile
DIEA	*N,N*-diisopropylethylamine
DIC	*N,N′-diisopropylcarbodiimide*
DMF	*N,N*-dimethylformamide
GMP	good manufacturing practice
Pbf	2,2,4,6,7-pentamethyldihydrobenzofuran-5-sulfonyl
RPLC	reversed-phase liquid chromatography
SPPS	solid-phase peptide synthesis
*t*Bu	tert-butyl
TIS	triisopropylsilane
TFA	trifluoroacetic acid
Trt	trityl

References

1. Jaradat, D.M.M. Thirteen decades of peptide synthesis: Key developments in solid phase peptide synthesis and amide bond formation utilized in peptide ligation. *Amino Acids* **2018**, *50*, 39–68. [CrossRef] [PubMed]
2. Sachdeva, S. Peptides as 'Drugs': The Journey so Far. *Int. J. Pept. Res. Ther.* **2017**, *23*, 49–60. [CrossRef]
3. Petrou, C.; Sarigiannis, Y. Peptide synthesis: Methods, trends, and challenges. In *Peptide Applications in Biomedicine, Biotechnology and Bioengineering*; Koutsopoulos, S., Ed.; Woodhead Publishing: Sawston, UK, 2018; pp. 1–21. [CrossRef]
4. Lau, J.L.; Dunn, M.K. Therapeutic peptides: Historical perspectives, current development trends, and future directions. *Bioorg. Med. Chem.* **2018**, *26*, 2700–2707. [CrossRef] [PubMed]
5. Loffet, A. Peptides as Drugs: Is There a Market? *J. Pept. Sci.* **2002**, *8*, 1–7. [CrossRef]
6. Al Musaimi, O.; Al Shaer, D.; Albericio, F.; Torre, B.G.d.l. 2022 FDA TIDES (Peptides and Oligonucleotides) Harvest. *Pharmaceuticals* **2023**, *16*, 336. [CrossRef]
7. Al Musaimi, O.; Lombardi, L.; Williams, D.R.; Albericio, F. Strategies for Improving Peptide Stability and Delivery. *Pharmaceuticals* **2022**, *15*, 1283. [CrossRef]
8. Ferrazzano, L.; Catani, M.; Cavazzini, A.; Martelli, G.; Corbisiero, D.; Cantelmi, P.; Fantoni, T.; Mattellone, A.; De Luca, C.; Felletti, S.; et al. Sustainability in peptide chemistry: Current synthesis and purification technologies and future challenges. *Green Chem.* **2022**, *24*, 975–1020. [CrossRef]

9. Kaspar, A.A.; Reichert, J.M. Future directions for peptide therapeutics development. *Drug Discov. Today* **2013**, *18*, 807–817. [CrossRef]
10. D'Aloisio, V.; Dognini, P.; Hutcheon, G.A.; Coxon, C.R. PepTherDia: Database and structural composition analysis of approved peptide therapeutics and diagnostics. *Drug Discov.* **2021**, *26*, 1409–1419. [CrossRef]
11. Muttenthaler, M.; King, G.F.; Adams, D.J.; Alewood, P.F. Trends in peptide drug discovery. *Nat. Rev. Drug Discov.* **2021**, *20*, 309–325. [CrossRef]
12. Albericio, F. Developments in peptide and amide synthesis. *Curr. Opin. Chem. Biol.* **2004**, *8*, 211–221. [CrossRef] [PubMed]
13. Zompra, A.A.; Galanis, A.S.; Werbitzky, O.; Albericio, F. Manufacturing peptides as active pharmaceutical ingredients. *Future Med. Chem.* **2009**, *1*, 361–377. [CrossRef] [PubMed]
14. Andersson, L.; Blomberg, L.; Flegel, M.; Lepsa, L.; Nilsson, B.; Verlander, M. Large-scale synthesis of peptides. *Pept. Sci.* **2000**, *55*, 227–250. [CrossRef]
15. Merrifield, R.B. Solid Phase Peptide Synthesis. I. The Synthesis of a Tetrapeptide. *J. Am. Chem. Soc.* **1963**, *85*, 2149–2154. [CrossRef]
16. Athanassopoulos, P.; Barlos, K.; Gatos, D.; Hatzi, O.; Tzavara, C. Application of 2-Chlorotrityl Chloride in Convergent Peptide Synthesis. *Tetrahedron Lett.* **1995**, *36*, 5645–5648. [CrossRef]
17. Gongora-Benitez, M.; Tulla-Puche, J.; Albericio, F. Handles for Fmoc solid-phase synthesis of protected peptides. *ACS Comb. Sci.* **2013**, *15*, 217–228. [CrossRef]
18. Bray, B.L. Large-scale manufacture of peptide therapeutics by chemical synthesis. *Nat. Rev. Drug Discov.* **2003**, *2*, 587–593. [CrossRef]
19. Al Musaimi, O.; Morse, S.V.; Lombardi, L.; Serban, S.; Basso, A.; Williams, D.R. Successful synthesis of a glial-specific blood–brain barrier shuttle peptide following a fragment condensation approach on a solid-phase resin. *J. Pept. Sci.* **2022**, *29*, e3448. [CrossRef]
20. White, C.J.; Yudin, A.K. Contemporary strategies for peptide macrocyclization. *Nat. Chem.* **2011**, *3*, 509–524. [CrossRef]
21. Snyder, L.R.; Kirkland, J.J.; Dolan, J.W. *Introduction to Modern Liquid Chromatography*, 3rd ed.; Wiley-Interscience: New York, NY, USA, 2010. [CrossRef]
22. Al Musaimi, O.; Valenzo, O.M.M.; Williams, D.R. Prediction of peptides retention behavior in reversed-phase liquid chromatography based on their hydrophobicity. *J. Sep. Sci.* **2023**, *46*, 2200743. [CrossRef]
23. De Luca, C.; Lievore, G.; Bozza, D.; Buratti, A.; Cavazzini, A.; Ricci, A.; Macis, M.; Cabri, W.; Felletti, S.; Catani, M. Downstream Processing of Therapeutic Peptides by Means of Preparative Liquid Chromatography. *Molecules* **2021**, *26*, 4688. [CrossRef] [PubMed]
24. Musaimi, O.A.; Mercado-Valenzo, O.M.; Williams, D.R. Factors Influencing the Prediction Accuracy of Model Peptides in Reversed-Phase Liquid Chromatography. *Separations* **2023**, *10*, 81. [CrossRef]
25. Tripet, B.; Cepeniene, D.; Kovacs, J.M.; Mant, C.T.; Krokhin, O.V.; Hodges, R.S. Requirements for prediction of peptide retention time in reversed-phase high-performance liquid chromatography: Hydrophilicity/hydrophobicity of side-chains at the N- and C-termini of peptides are dramatically affected by the end-groups and location. *J. Chromatogr. A* **2007**, *1141*, 212–225. [CrossRef] [PubMed]
26. Klee, W.A.; Nirenberg, M. Mode of action of endogenous opiate peptides. *Nature* **1976**, *263*, 609–612. [CrossRef]
27. Kovacs, J.M.; Mant, C.T.; Hodges, R.S. Determination of intrinsic hydrophilicity/hydrophobicity of amino acid side chains in peptides in the absence of nearest-neighbor or conformational effects. *J. Pept. Sci.* **2006**, *84*, 283–297. [CrossRef] [PubMed]
28. Alhassan, M.; Al Musaimi, O.; Collins, J.M.; Albericio, F.; de la Torre, B.G. Cleaving protected peptides from 2-chlorotrityl chloride resin. Moving away from dichloromethane. *Green Chem.* **2020**, *22*, 2840–2845. [CrossRef]
29. Al Musaimi, O.; Gavva, V.; Williams, D.R. Greener Cleavage of Protected Peptide Fragments from Sieber Amide Resin. *ChemistryOpen* **2022**, *11*, e202200236. [CrossRef]

separations

MDPI

Article

Development and Validation for Quantitative Determination of Genotoxic Impurity in Gemfibrozil by Gas Chromatography with Mass Spectrometry

Hari Naga Prasada Reddy Chittireddy [1], J. V. Shanmukha Kumar [1,*], Anuradha Bhimireddy [1], Mohammed Rafi Shaik [2,*], Mohammad Rafe Hatshan [2], Mujeeb Khan [2], Abdulrahman Alwarthan [2] and Baji Shaik [3]

1 Department of Engineering Chemistry, College of Engineering, Koneru Lakshmaiah Education Foundation, Vaddeswaram, Guntur 522 502, Andhra Pradesh, India
2 Department of Chemistry, College of Science, King Saud University, Riyadh 11451, Saudi Arabia
3 School of Chemical Engineering, Yeungnam University, Gyeongsan 38541, Republic of Korea
* Correspondence: shanmukh_fed@kluniversity.in (J.V.S.K.); mrshaik@ksu.edu.sa (M.R.S.); Tel.: +966-11-4670439 (M.R.S.)

Citation: Chittireddy, H.N.P.R.; Kumar, J.V.S.; Bhimireddy, A.; Shaik, M.R.; Hatshan, M.R.; Khan, M.; Alwarthan, A.; Shaik, B. Development and Validation for Quantitative Determination of Genotoxic Impurity in Gemfibrozil by Gas Chromatography with Mass Spectrometry. *Separations* **2023**, *10*, 145. https://doi.org/10.3390/separations10030145

Academic Editor: Liangliang Liu

Received: 22 January 2023
Revised: 13 February 2023
Accepted: 20 February 2023
Published: 21 February 2023

Abstract: All regulatory organizations are paying close attention to the identification and measurement of genotoxic contaminants. Using conventional analytical techniques like high-performance liquid chromatography (HPLC) and gas chromatography to quantify probable genotoxic substances (PGIs) at the trace level is difficult (GC). Therefore, there is a necessity for advanced analytical techniques for the development of highly sensitive analytical procedures for the determination of trace-level PGIs in drug products and drug substances. This study's goal is to develop and evaluate an analytical technique for measuring allyl chloride, a possible genotoxic contaminant in gemfibrozil. For the detection of very low and trace levels of impurities, a gas chromatography with a triple quadrupole mass spectrometry detector (GC-MS/MS) approach was developed and validated. Using a column USP phase G27, a nonpolar and low bleed 5% diphenyl, 95% dimethylpolysiloxane, with dimensions of 30 m in length, 0.32 mm internal diameters, and 1.5 m film thickness, along with a flow rate of 2.0 mL/min and Helium (He) as a carrier gas, this method uses a thermal gradient elution program. The method was calibrated with a linearity range from 30% to 150% concentration with respect to the specification level and achieved a limit of detection (LOD) and limit of quantification (LOQ) were 0.005 ppm and 0.01 ppm, respectively, for allyl chloride. According to current ICH requirements, the method was validated, and it was discovered to be specific, exact, accurate, linear, sensitive, tough, robust, and stable. This method is suitable for determining allyl chloride in the regular analysis of Gemfibrozil.

Keywords: allyl chloride; gemfibrozil; genotoxic impurity; organochlorine; GC-MS/MS

1. Introduction

In the manufacture of pharmaceutical products for commercial use or clinical research, it is a primary responsibility of chemists, engineers, and formulators to ensure the safety of their production. The quality and purity of the raw materials utilized in the formulation, particularly the active pharmaceutical ingredient(s), are given considerable consideration while focusing on safety [1]. A drug substance will often contain a variety of low-level impurities originating from the initial ingredients, reagents, intermediates, or by-products of the synthesis or degradation processes; these must be studied and controlled to permitted parts per million (ppm) limits. A certain amount of patient risk can be tolerated when weighed against the anticipated health benefits, even though the pharmaceutical substance itself is unlikely to be fully safe. This trade-off between risk and return must be carefully considered by pharmaceutical firms and regulatory organizations on a case-by-case basis.

Impurities, however, are thought to solely cause harm and offer no benefits. Therefore, manufacturers must develop and implement their own analytical approach to get rid of them (or at least, reduce the level of these impurities to the greatest extent possible) [2].

Human cancer may be brought on by pharmaceutical genotoxic impurities (GTIs) that cause genetic mutations, chromosomal breakage, or chromosomal rearrangements [3]. Therefore, exposure to even minute quantities of these contaminants in the finished active pharmaceutical ingredient may result in significant toxicological problems [4]. Hence, chemical scientists should consider strategies to lower the synthesis and consumption of these genotoxic substances during the manufacturing process. However, completely ceasing the use of such substances or ceasing the manufacture of pollutants that are DNA reactive is not always feasible. GTIs can be crucial in drug development even though they are present in trace amounts [5] and, if properly addressed, could result in clinical holds or a delay in regulatory authorities' clearance. In order to precisely assess and regulate the quantities of GTIs in medications, analytical scientists must develop the necessary analytical procedures. For the creation of a reliable manufacturing process, as well as for assuring patient safety, adequate analytical methods are crucial. Some medications may produce GTIs through degradation during formulation or storage in addition to process impurities. For instance, hydrolytic compounds like anilines and oxidative degradation products like hydroperoxides or epoxides could be genotoxic. Additionally, excipient components may interact with API or its counter ion to create a new impurity that is potentially genotoxic, such as halogenated furanone [6]. This includes many challenges to the drug development process.

Genotoxicity is characterized as a corrosive effect on a cell's genetic material (DNA and RNA) that compromises the integrity of the cell. Mutagens include genotoxins (radiation, chemical, or physical agents). A genotoxin is a substance that exhibits genotoxicity. Genotoxins can be teratogens, which can cause birth defects, and mutagens, which can cause mutations, carcinogens, or which can cause cancer [7].

Impurities that are specified by the Food and Drug Administration (FDA), the International Council for Harmonization (ICH), and the USP can be divided into four categories: drug-related impurities, process-related impurities (PRIs), residual solvents, and heavy metals. The first category of impurities related to active pharmaceutical ingredients (APIs) is further divided into two types as a result of specific reactions, such as oxidation, dehydration, and carbon dioxide removal. The second type results from a reaction between the API and the excipients, container, or any lingering contaminants in the excipients, reagents, or solvents. Due to their link between structure and action, contaminants connected to API may pose a risk for genotoxicity, mutagenicity, and cancer [8].

Genotoxic impurities can be introduced from a variety of sources, most commonly from the starting materials used to synthesize drugs and their impurities in the form of genotoxic intermediates or process-related by-products. Additionally, genotoxic impurities are present in drug substances because of synthesis components such as solvents, catalysts, and reagents that are involved to create drugs. Furthermore, drug impurities are produced because of drug degradation during storage, exposure to light, air oxidation, or hydrolysis. The formation of chiral impurities in pharmacological substances occurs at several phases of drug production from stereoisomers of raw materials and intermediates during the synthesis of stereoselective drugs [9].

Genotoxicity statistics are useful for assessing the risks of chemicals, as well as those of food and feed, consumer goods, human and veterinary pharmaceuticals, and industrial items. Genotoxicity information is fundamental for assessing the dangers of naturally occurring environmental toxins in chemicals, food, and feed. Numerous regulatory authorities and advisory groups have suggested methods for genotoxicity testing based on this. Even at low exposure levels, genetic modifications in genetic materials, such as somatic and germ cells, have substantial negative effects on health. Several genetic illnesses are caused by mutations in proto-oncogenes, tumor suppressor genes, or DNA damage response genes by different carriers like physical and chemical. Degenerative disorders include accelerated aging, impaired immunity, cardiovascular, and neurological diseases are also caused by somatic cells with damaged DNA. The evaluation of mutagenic potential is a fundamental part of chemical risk assessment in order to prevent such negative effects of genetic damage to human health [10].

In order to assess the product and method for safety, regulatory authorities from all over the world demand regulatory data on the genotoxic potential of pharmaceutical products. Pre-clinical studies are therefore typically carried out to evaluate fundamental toxicological information of new chemical entities (NCE). Based on toxicological information, the safety and effectiveness of NCE will aid in analyzing and determining whether the drug will likely have a risk or benefit assessment during the new drug application (NDA) process. It will also aid in identifying genotoxicity risks that could result in DNA damage and its fixation [11].

Gemfibrozil (Figure 1) belongs to the class of fibric acid derivatives and has the chemical name of 5-(2,5-dimethylphenoxy)-2,2-dimethylpentanoic acid. It is mainly indicated for the treatment of hyperlipidemia [12,13]. The first synthetic route was developed by Creger et al. [14], and screened for the treatment of abnormal blood lipid levels in the year 1976 [15]. It has been marketed since 1982 due to its overwhelming ability to reduce plasma triglyceride levels [16]. Later, clinical studies proved that gemfibrozil can be used to prevent cardiovascular events by increasing HDL cholesterol and it can also be useful to control various types of signaling pathways responsible for switching of T-helper cells, inflammation, migration, cell-to-cell contact, and oxidative stress [17]. As per the FDA label, Gemfibrozil is available in the dosage form of a Tablet, which contains 600 mg Gemfibrozil free base and other excipients. Gemfibrozil has an empirical formula of $C_{15}H_{22}O_3$ and a molecular weight 250.35 g/mol. It is soluble in water and acid (0.0019%), and in dilute base (>1%). Under ordinary conditions, Gemfibrozil is stable and has a melting point at 58–61 °C.

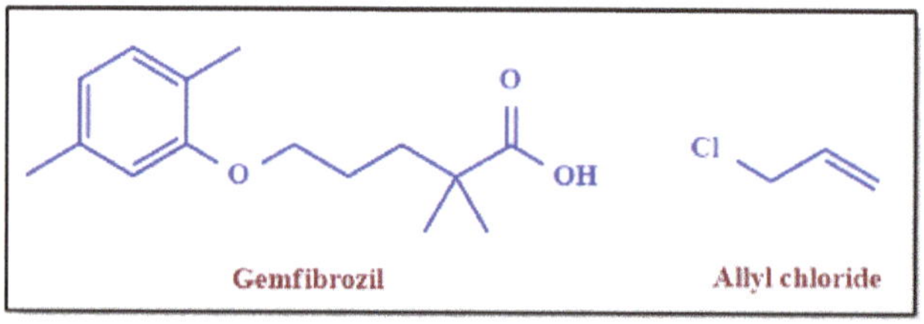

Figure 1. Structure of Gemfibrozil and Allyl chloride.

Gemfibrozil can be prepared by a variety of synthetic routes involving several purification processes to obtain high-quality drug that can fulfill all of the stringent regulatory requirements. According to ICH guidelines, the reported maximum daily dosage of gemfibrozil should not be more than 1.2 g, while impurities in the drug substance should be not exceed 0.15% [18]. Korupolu et al. have reported an efficient method for the synthesis of gemfibrozil with high purity, the process involves the O-alkylation using 2,5-dimethylphenol and isobutyl 5-chloro-2,2-dimethylpentanoate [19]. Apart from this, Gemfibrozil synthesis also involves a variety of other intermediates that can induce unwanted impurities [20]. For example, Ramachandran et al. have reported the preparation of Gemfibrozil in a multistep process involving different types of intermediates including allyl butyrate, 5-bromo-2,2-dimethylpentanoic acid methyl ester etc. [21]. This crucial intermediate of gemfibrozil is prepared by involving the reaction of isobutyl isobutyrate with allyl chloride to form allyl intermediate and then subjected to bromination with hydrogen bromide to obtain the 5-bromo-2,2-dimethylpentanoic acid isobutyl ester (Figure 2). Notably, some of the intermediates or their synthetic processes involve allylic derivatives that may remain present as impurities in the final product.

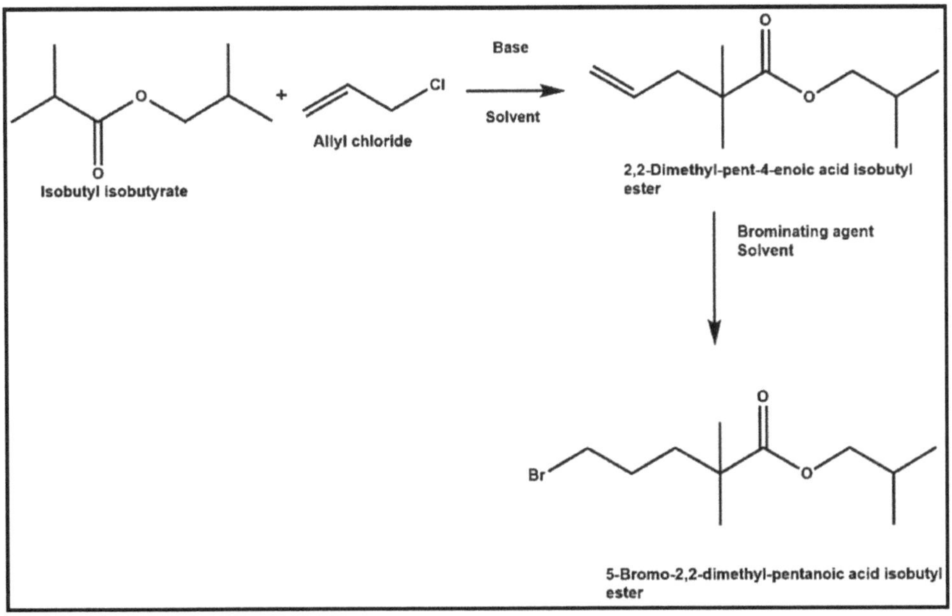

Figure 2. Reaction mechanism for the formation of 5-Bromo-2,2-dimethyl-pentanoic acid isobutyl ester.

According to the summary basis of approval by the FDA, the definite mechanism of action with Gemfibrozil was unknown. It has been demonstrated in man that it inhibits peripheral lipolysis and lowers hepatic extraction of free fatty acids, which lowers hepatic triglyceride production. Gemfibrozil mainly metabolizes through the oxidation of a ring methyl group to form a hydroxymethyl and a carboxyl metabolite [22]. Gemfibrozil is a hyperlipidemia lowering drug, which is known to adjust the level of lipid in the blood stream in patients. However, prolonged use of Gemfibrozil may cause acute liver injury, and thus it is crucial to remove the possibilities of toxic action [23]. The organ-related toxicity of Gemfibrozil can possibly be attributed to the formation of highly reactive metabolites and subsequent covalent binding of protein [24]. In the case of Gemfibrozil, it is reported to be metabolized to oxidative and glucuronide metabolites by the catalytic activity of enzyme-based catalysts in humans and animals [25]. It has been previously reported in many studies that Gemfibrozil 1-O-β-acyl glucuronide, which is one of the major metabolites, typically undergoes transacylation reactions, wherein the glucuronic acid is substituted by nucleophile, or in other cases intramolecular rearrangement, hydrolysis, and covalent binding of the α-OH aldehyde moiety to the protein occur through glycation mechanism [26]. Additionally, various sulfate metabolites of Gemfibrozil are often toxic when they are attached to the benzylic or allylic sites due to their high chemical reactivity (electrophilicity). These metabolic processes often promote the toxicity processes.

Given the importance of Gemfibrozil as adjunctive therapy to diet, there is a need to control impurities, particularly genotoxic impurities, i.e., allyl chloride (Figure 1), with robust analytical technique. Allyl chloride is generally used as an intermediate in organic chemistry to develop the drug substances. It is a chlorinated hydrocarbon that is liquid at room temperature, colorless, flammable, and volatile. The unreacted reagents sometimes cause serious health issues. In a diet, a small quantity of allyl chloride substances also causes injury to the liver and kidneys and the onset of pulmonary edema (fluid in the lungs) [27]. There is no evidence of allyl chloride causing human cancer, but in a study involving the injection of the chemical into mice's stomachs has revealed an increase in the frequency of forestomach tumors which is linked to gavage exposure to allyl chloride [28]. According to the EPA, allyl chloride is a Group C chemical that may cause cancer in people [29]. As of today, various analytical methods like HPLC, RP-HPLC, and LC determination [29–32] are known for the estimation and determination of Gemfibrozil.

However, the prior methods neither disclose the detection of allyl chloride nor the quantification thereof in Gemfibrozil. Furthermore, these methods could not be used to detect the lowest concentration of allyl chloride and are not sophisticated for trace level analysis, whereas the method developed in the current work involves the use of a more accurate analytical technique for detecting traces of allyl chloride by using mass spectroscopy. Furthermore, this method has an advantage in the minimization of solvent and time period for quantification with respect to run time. Furthermore, this method is simple, sensitive, and reproducible GC-MS/MS and validated it as per ICH guidelines [33].

2. Materials and Methods

2.1. Materials and Reagents

Allylchloride was procured from HTS Biopharma Ltd., ALEAP Industrial Area, Hyderabad (India), and GC HS grade Methanol from Fischer Chemicals (India). Gemfibrozil was gifted by Jisai Pharma Pvt Ltd. Plot No-12, Phase (4), IDA-Cherlapally, Hyderabad-500051, India.

2.2. Equipment

Mass tuning of allyl chloride impurity, method development, and validation was performed on Agilent 7890B GC system (Make: Agilent, Santa Clara, CA, USA) is connected with Agilent 7010B GC/TQ triple quadrupole equipped with electron impact ionization (EI) as MSD ion source and MRM mode. The data were collected using Mass Hunter software. Weighing of the standards and sample was done using an analytical balance (Make: Mettler Toledo; Model: ME204E, Im Langacher 44, 8606 Greifensee, Switzerland). The sample and standards were blended using a Remi vortex mixer (Make: Remi, Maharashtra, India).

2.3. Chromatographic Conditions

Chromatographic conditions of the GC-MS/MS system were optimized by using USP phase G27, a nonpolar and low bleed 5% Diphenyl, 95% Dimethylpolysiloxan, having dimensions 30 m length, 0.32 mm internal diameters,1.5 µm film thickness GC column. The oven temperature of the column was set to 40 °C as initially and held for 0 min. The temperature was raised gradually, ramping up to 250 °C at a rate of 15 °C/min, and held for 6 min. Helium was used as the carrier gas, flowing at a rate of 2.0 mL/min. The injector heater's temperature was 200 °C, and the injection volume was 2 µL.

2.4. Mass Spectrometer Conditions

MRM mode was used in GC-MS/MS system by considering precursor ion (Q1) 76 *m/z* and product ion (Q3) 41 *m/z*, MRM-1:76 amu $\rightarrow$ 41 amu were used for quantitation. The temperature of the mass source was 230 °C. The collision energy (CE) was 8 v (Table 1).

2.5. Impurity Standard and Test Sample Solution Preparation

The concentration of the allyl chloride impurity standard (0.03 ppm) was prepared in methanol (diluent). The test sample of Gemfibrozil (150 mg/mL) was prepared in the diluent. The solutions were subjected to a vortex for 5 min and mixed well. The diluent was injected as a blank.

The specification limit of allyl chloride was 0.03 ppm with respect to sample concentration. Hence, the sample concentration was optimized based on the accuracy results obtained during method development. The different sample concentrations were used for spiking of impurity and achieved recovery with the sample concentration of 150 mg/mL.

Table 1. GC-MS/MS instrument final conditions.

Instrument Setup	Details		
GC Parameters			
Chromatographic system	Agilent Technologies 7890B GC system		
GC Column	USP phase G27, 30 m length, 0.32 mm internal diameters, 1.5 µm film thickness		
Carrier Gas	Helium		
Column Mode	Constant Flow		
Gas flow	2 mL/min		
Injector (Heater)	200 °C		
Volume of injection	2 microliters		
Oven Programming	Ramping (°C/min) rate	Temperature	Hold time in minutes
	-	40 (°C)	0
	15	250 (°C)	6
Split flow	20		
Run time	20 min		
Mass spectrometry conditions			
MS system	Agilent Technologies 7010B GC/TQ		
Ion source and Detection mode	EI and MRM		
For qualification (*m/z*)	76 amu → 39 amu		
For quantification (*m/z*)	76 amu → 41 amu		
Dwell time (in milliseconds)	100		
Collision energy (CE)	8 v		
Gain Factor	20		
Detector off (MS -Off)	4 min		
Temperature of Source (°C)	230		
Temperature Transfer Line (°C)	240		
MS Quad temperature (°C)	150		

3. Results

3.1. Optimization of Mass Spectrometric Parameters

For both analytes, mass tuning was performed to determine the Q1 and Q3 values. The solubility of Gemfibrozil and allyl chloride was determined in order to identify and measure the impurity in gemfibrozil. Gemfibrozil and the allyl chloride impurity are soluble in alcohol.

By introducing a diluted solution of allyl chloride into the mass spectroscopy, mass parameters were tuned. EI was used as an ion source for establishing mass detection and finalizing Q1 and Q3 values. The MRM fragments MRM-1 is 76 amu → 41 amu and MRM-2 is 76 amu → 39 amu. Although the response rate of MRM-2 was low when compared with MRM-1, both values can be used for identification, and MRM-1 was used for quantitation of the impurity (Figure 3).

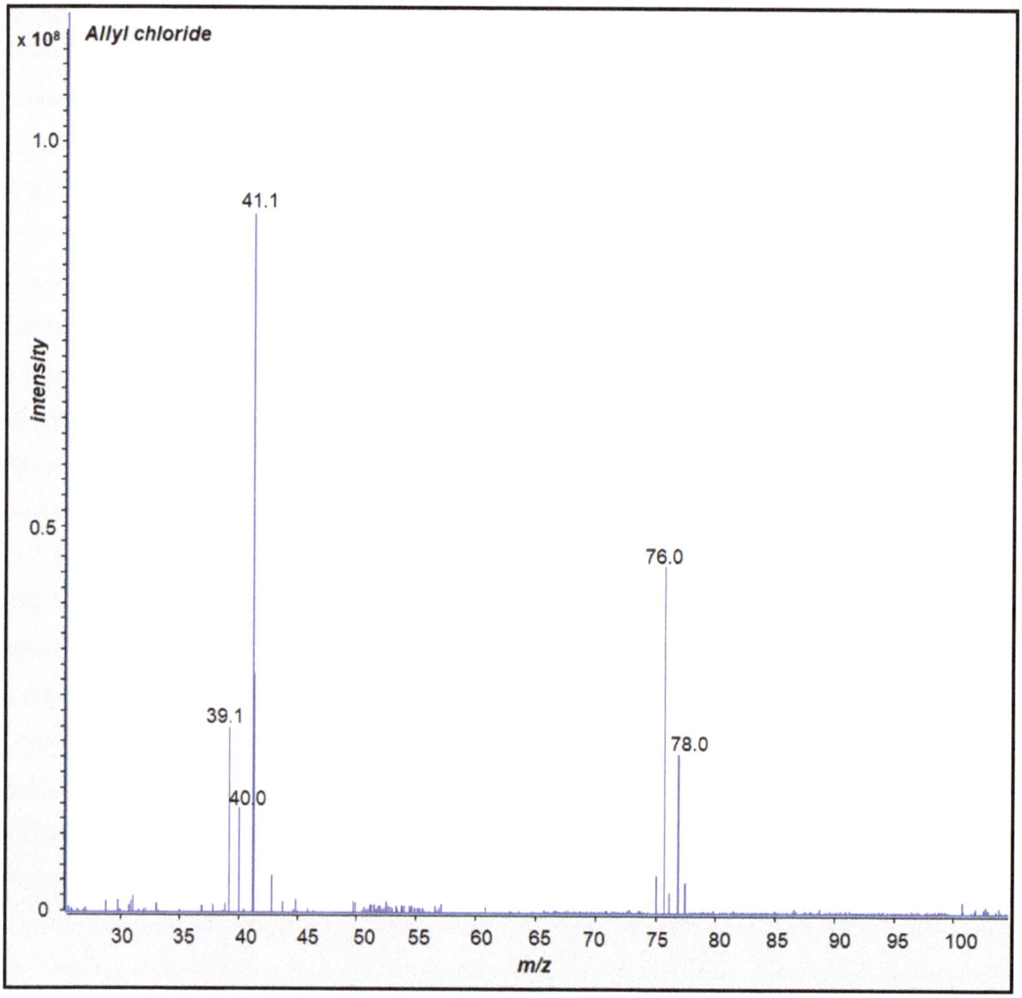

Figure 3. Mass fragmentation of allyl chloride.

3.2. Optimization of Chromatographic Conditions

Various solvents were used to establish suitable diluents. The diluent compatibility study was performed by considering low and high boiler, polar and non-polar solvents such as N-methyl-2-pyrrolidone (NMP), methanol (MeOH), dichloromethane (DCM), dimethyl sulfoxide (DMSO) and hexane. During the development, we observed that there was some solvent interference, broad peak shape, and poor response of the impurity with many diluents (the data are given in the Supplementary Materials, Figures S12–S17). However, there was no interference with the methanol, and observed that there was a very good response in ppm level concentrations, with a sharp peak and good ionization for the impurity.

The selection of columns can play a very important role during the method development. During development, different column chemistries, column lengths, diameters, and film thicknesses of columns like DB-1, DB-wax, DB-5, and DB-624 were used for the optimization. From these column chemistries, we observed that DB- 5 was more suitable with very sharp and good ionization of the impurity peak.

The final method was optimized by using helium as a carrier gas. The retention time of the impurity was about 2 min. To protect the ion source from the high concentration of gemfibrozil, applied for the detector off (MS-off) program after the impurity peak elution.

3.3. Method Validation

The validation of an analytical procedure is the process by which it is established through laboratory experiments (for the developed method) that the performance characteristics of the method meet the requirements for the intended analytical applications. The process of validation of any analytical method entails a series of studies.

The GC-MS/MS method was validated and performed as per the international conference on harmonization (ICH) guidelines in the present study. Method validation conditions are system suitability, specificity, the limit of detection, Limit of quantification, Precision for Limit of quantification, linearity, range, method precision, accuracy, robustness, intermediate precision, and solution stability.

3.4. Specificity and System Suitability

System stability and specificity were determined by injecting blank, standard, sample, and spiked sample solutions and individual impurity to check the % of RSD (relative standard deviation) and allylchloride peak retention time (RT) of all the above solutions. The obtained results were well within the limit. The standard solution and blank solution baselines were good, and there was no blank interference observed at the retention time of allyl chloride (Table 2) (Figure 4).

Table 2. Summary of method validation results.

Parameters	Acceptance Criteria	Observation
Specificity and SST	Area % RSD of allyl chloride peak should be ≤15.0.	3.5%
	RT % RSD of allyl chloride peak should be ≤5.0.	0.8%
	RT of allyl chloride peak.	2 min
	Any blank Interference	Not observed
Limit of detection	Limit of detection Concentration	0.005 ppm
	Signal-to-noise ratio should be ≥3	22
Limit of quantification	Limit of quantification Concentration	0.01 ppm
	Signal-to-noise ratio should be ≥10	45
Precision for Limit of quantification	Area % RSD of allyl chloride peak from 6 injections of LOQ solution should be ≤20.0%	4.8%
Linearity and Range	Range for allyl chloride	0.01 to 0.045 ppm
	Correlation coefficient for allyl chloride linearity solutions should be ≥0.99	0.99
	Square of the correlation coefficient for allyl chloride linearity solutions should be ≥0.99	0.99

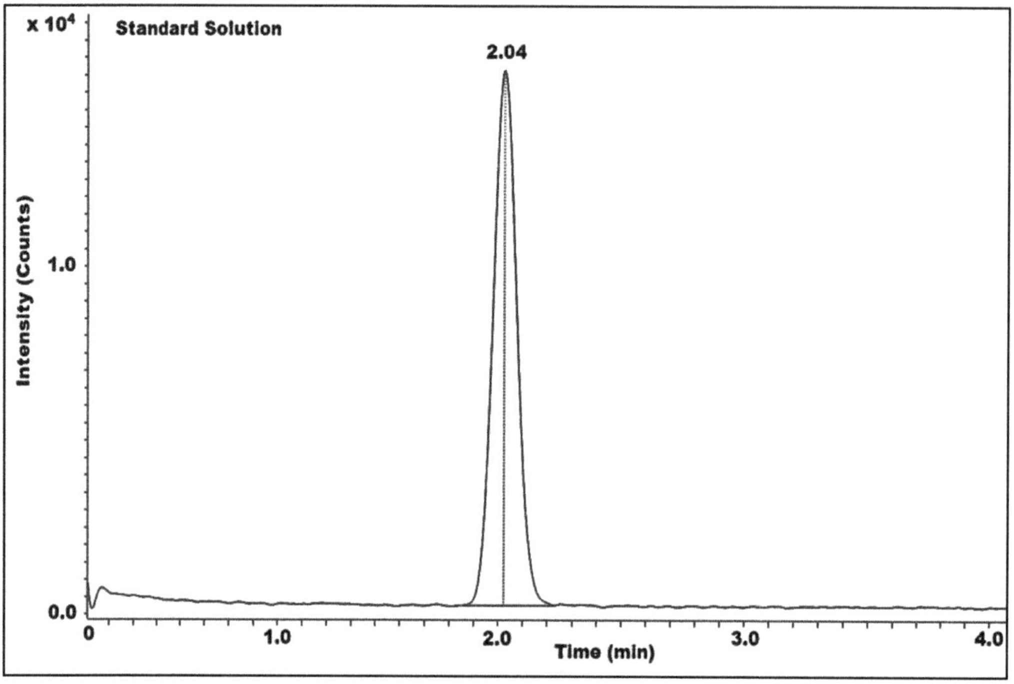

Figure 4. Allyl chloride standard solution chromatogram.

3.5. Limit of Detection, Limit of Quantification, Precision for Limit of Quantification

The smallest amount of analyte in a diluted standard solution that can be detected but not necessarily quantitated under the specified experimental circumstances is known as the limit of detection. The limit of quantification is the smallest quantity of analyte in a dilute standard solution and a sample that can be determined with acceptable accuracy and precision under the stated experimental conditions.

The limit of detection and limit of quantification were established by injecting diluted standard solutions in triplicate while measuring the impurity's concentration, resulting in a signal-to-noise ratio (s/n) of at least 3 for LOD and at least 10 for LOQ. LOD is 0.005 ppm and LOQ is 0.01 ppm, LOQ precision was tested by administering six replicate injections of the LOQ solution. The obtained s/n ratio for LOD solutions and LOQ was more than 3 and 10, respectively. For six replicate injections of LOQ precision % RSD is 4.8 (Table 2), (Figures 5–7).

3.6. Linearity

The linearity was established by injecting different known-concentration solutions of impurity with LOQ ranging up to 150% of the specified concentration level. The standard was prepared (in ppm) 0.01, 0.015, 0.0225, 0.03, 0.036, and 0.045 in the diluent, and injected at LOQ levels of 50%, 75%, 100%, 120%, and 150% levels in duplicate. The correlation coefficient (r) and square of correlation coefficient (r^2) were determined by plotting the graph of the peak area responses against concentration. Both r and r^2 were 0.99. The method was linear as a result (Table 2).

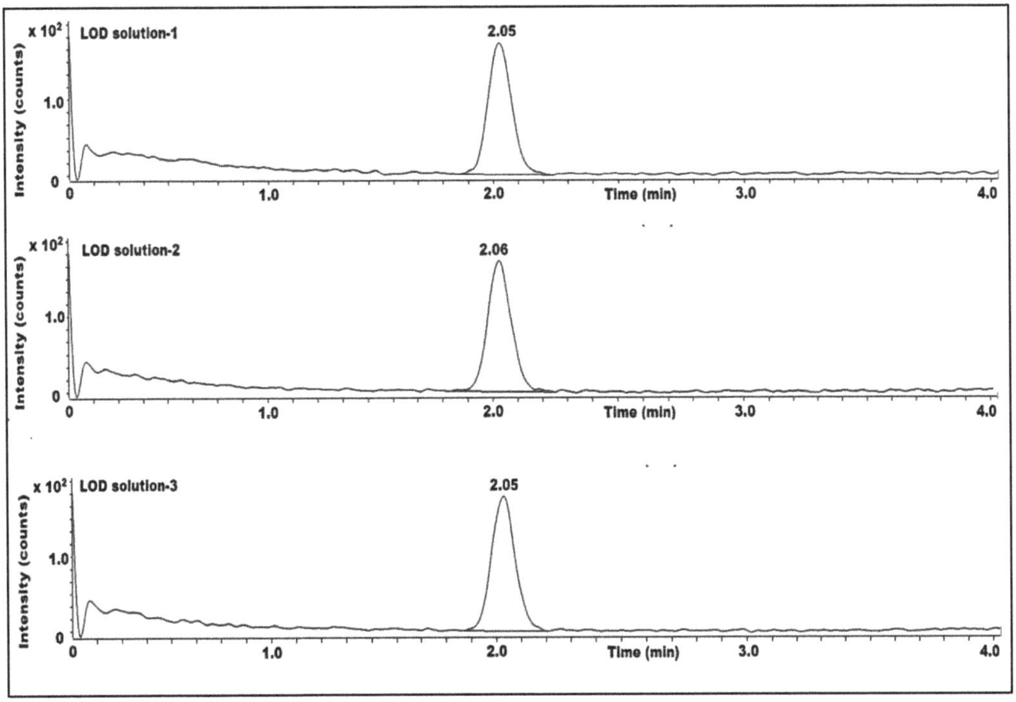

Figure 5. LOD solution chromatogram.

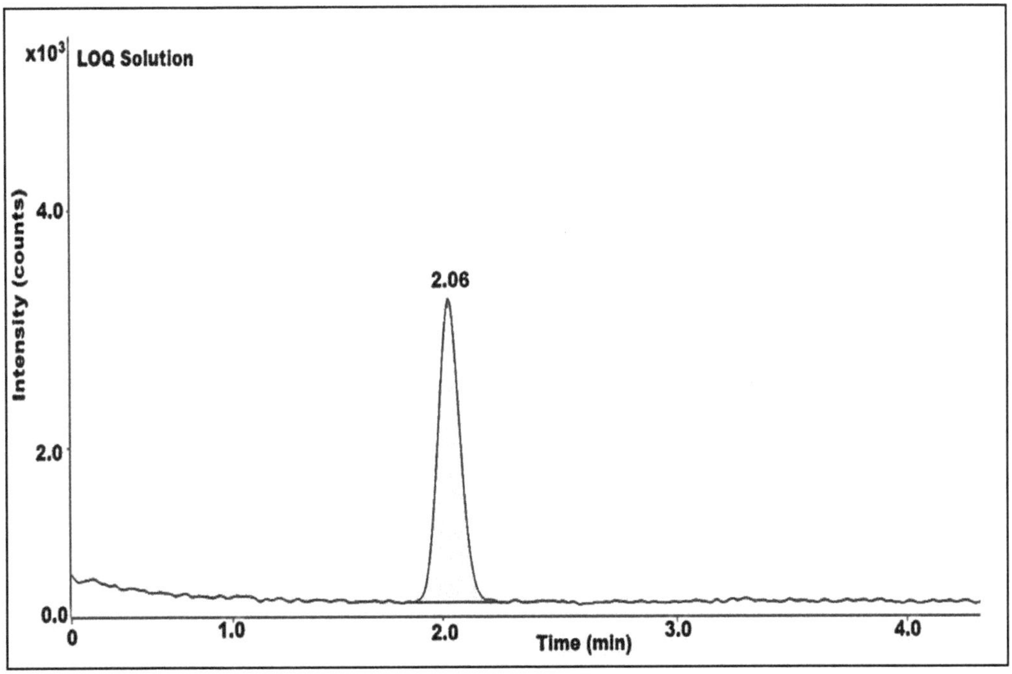

Figure 6. LOQ solution chromatogram.

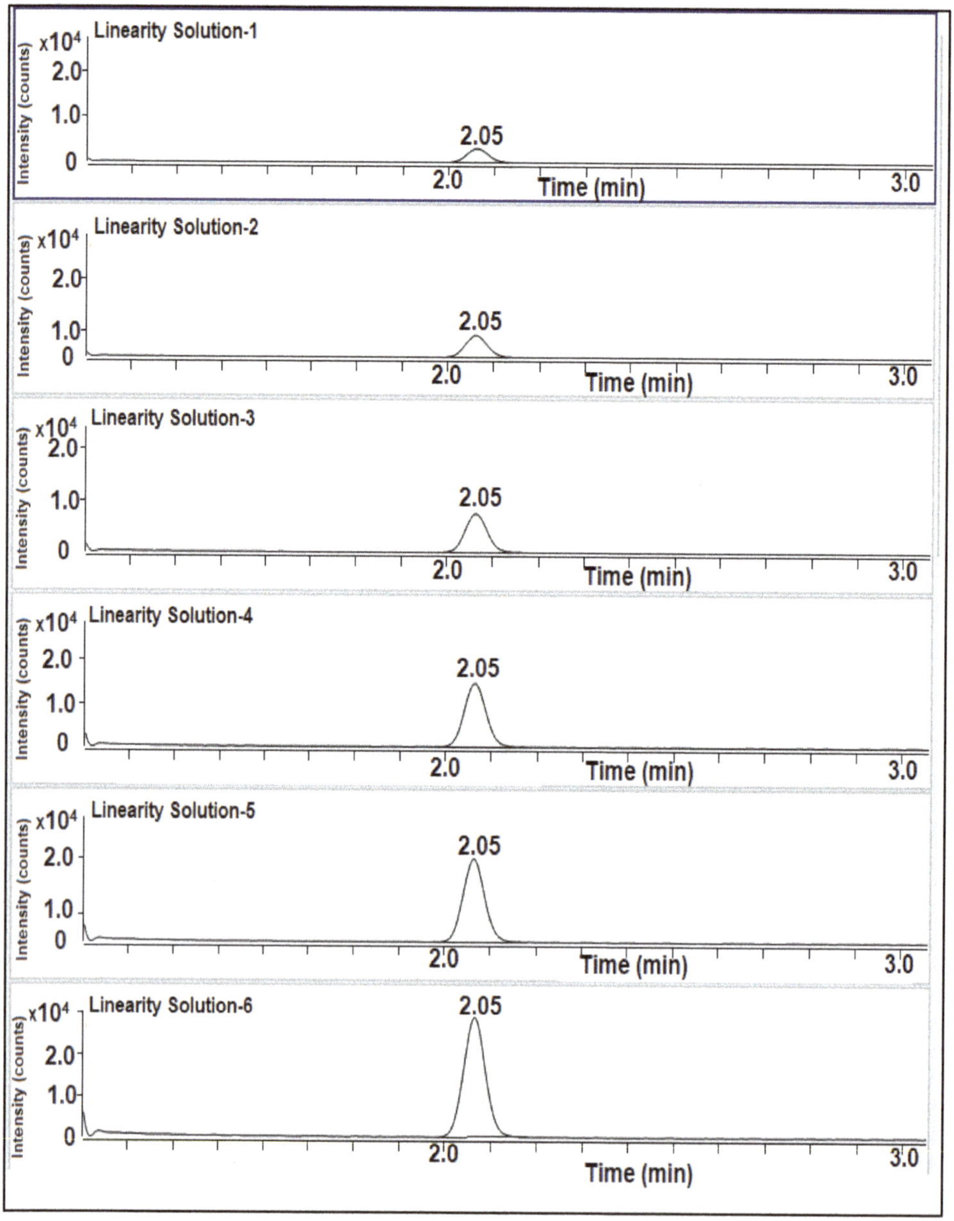

Figure 7. Linearity solution chromatogram.

3.7. Method Precision

By injecting six samples of Gemfibrozil and six spiked samples of impurity into the system at a concentration of 0.03 ppm, the method precision (MP) was established. One injection was given for each preparation. In both the sample and spike solutions, the impurity content and %RSD were calculated. As a result, the sample solutions were devoid of impurities. The %RSD for the spiked sample solutions (n = 6) was 5.5. The procedure was exact and repeatable, according to the findings (Table 3).

Table 3. Method validation results.

Parameters	Acceptance Criteria	Observation
Method precision	Area % RSD of allyl chloride peak from six preparations of spiked samples should be ≤15.0.	5.5%
Intermediate precision	Area % RSD of allyl chloride peak from six preparations of spiked samples should be ≤15.0.	4.8%
	Area % RSD of allyl chloride peak from 12 preparations of spiked samples from MP and IP should be ≤20.0.	Results with in acceptance limit.
Accuracy	The average recovery should be between 70% to 130% for LOQ spiked solution.	94.40%
	Average recovery should be between 80% to 120% for 50% spiked solution.	96.60%
	Average recovery should be between 80% to 120% for 100% spiked solution.	98.50%
	Average recovery should be between 80% to 120% for 150% spiked solution.	97.30%
Robustness	Plus flow: concentration difference and retention time of allyl chloride spiked sample.	2.1%, 1.9 min
	Minus flow: concentration difference and retention time of allyl chloride spiked sample.	1.4%, 2.1 min
	Plus oven temperature: concentration difference and retention time of allyl chloride spiked sample.	2.6%, 1.9 min
	Minus oven temperature: concentration difference and retention time of allyl chloride spiked sample.	1.7%, 2.1 min
Standard and spike solution Stability	Allyl chloride standard and spiked solutions were observed for 48 h under ambient laboratory temperature (25 ± 5 °C) and under refrigeration (2–8 °C).	Both solutions are Stable

3.8. Intermediate Precision

Repeating MP parameters with different analysts, different days, and different lot columns were used to establish the intermediate precision (IP). Gemfibrozil samples and 0.03 ppm impurity-spiked sample solutions were used to calculate the impurity's content and % RSD. As a result, the impurity is absent from the sample solutions. The %RSD for the spiked sample solutions (n = 6) was 4.8. The RSD (%) for preparations of MP and IP spiked sample at specification level less than 20.0 (n = 12). The results showed that the procedure was rugged (Table 3).

3.9. Accuracy

The accuracy was determined between LOQ and 150% level impurity concentration. Prepared triplicate solutions by spiking impurity into the sample of gemfibrozil at LOQ (0.01 ppm), 50% (0.015 ppm), 100% (0.03 ppm), and 150% (0.045 ppm). Each level was prepared in triplicate with a single injection. The impurity % recovery was calculated from the spiked sample solutions and was found to be between 80% and 120%. According to the outcomes, the method was accurate (Table 3).

3.10. Robustness

By altering the real column flow rate (plus + flow) of 2.2 mL/min and (minus − flow) of 1.8 mL/min as well as the initial column oven temperature of 42 °C (plus + oven) and 38 °C (minus − oven), robustness was determined. With standard and spike solutions, the results were compared with MP data for retention time and concentration. The impurity content % difference between MP and robustness study findings was less than 5. The technique was hence robust (Table 3).

3.11. *Standard and Spike Solution Stability*

Utilizing allyl chloride secondary intermediate stock solution and spiked samples at 100% concentration levels for up to 48 h at ambient laboratory temperature (25 ± 5 °C) and under refrigerated conditions (2–8 °C), a stability study was performed. By comparing against freshly prepared standard solutions of allyl chloride, which have been shown to be stable, the percentage recoveries of standard solutions of allyl chloride and spiked samples submitted to stability studies were calculated. The results were found to be stable (Table 3). Additionally, validation chromatograms as well as standard qualification data are provided in the Supplementary Materials (Figures S1–S16).

4. Discussion

In the pharmaceutical business, gas chromatography with electron ionization mass spectrometry is an effective analytical technique for very precise and quantitative assessments of very low levels of analytes and impurities. To determine the content of allyl chloride in gemfibrozil, an improved GC-MS/MS method was optimized. No interferences caused by other drug substances or blank were observed at the retention time of the impurity because molecular mass is more particular for each drug substance and impurity. The ability to detect allyl chloride at very low ppm levels, as opposed to those stated [29–32], is an additional benefit of this method. The developed method is straightforward and requires no additional derivatization steps. The method has the following advantages over the other methods reported in the research. A more accurate and sensitive way of detection would be to use GC-MS/MS; findings from a validation study for the suggested method showed great accuracy and precision. The limit of quantification was used to measure the sensitivity. It was established that the LOQ was 0.01 ppm. This approach is just as effective as or better than that those mentioned in other articles.

5. Conclusions

The simple, effective and reproducible GC-MS/MS method described in the present work is useful for the determination and quantification of very low traces of allyl chloride impurity in Gemfibrozil. The developed method was validated according to ICH guidelines and complied with the acceptance criteria of the analytical parameters. This method is suitable to identify the allyl chloride impurity in routine analysis of the drug substance for Gemfibrozil since it can detect the impurity at 0.005 ppm and quantify it at 0.01 ppm.

Supplementary Materials: The following supporting information can be downloaded at: https://www.mdpi.com/article/10.3390/separations10030145/s1, Figure S1: Allyl chloride standard certificate of analysis; Figure S2: Allyl chloride standard purity by GC analysis; Figure S3: Allyl chloride standard Mass data; Figure S4: Allyl chloride standard 1H-NMR data-1; Figure S5: Allyl chloride standard 1H-NMR data-2; Figure S6. Allyl chloride standard 1H-NMR data-3; Figure S7: Allyl chloride standard 1H-NMR data-4; Figure S8: Allyl chloride standard IR-Spectrum; Figure S9: MS/MS chromatogram of Blank solution; Figure S10: MS/MS chromatogram of System suitability standard solution; Figure S11: MS/MS chromatogram of spiked sample solution; Figure S12: MS/MS chromatograms of Method precision; Figure S13: MS/MS chromatograms of Intermediate precision; Figure S14: MS/MS chromatograms of 150% Accuracy; Figure S15: MS/MS chromatograms of Solution stability standard; Figure S16: MS/MS chromatograms of Solution stability spiked sample; Figure S17: MS/MS chromatograms of Allyl chloride standard poor response and improper peak in Method optimization with different diluents study.

Author Contributions: Conceptualization, H.N.P.R.C. and J.V.S.K.; methodology, H.N.P.R.C. and J.V.S.K.; formal analysis, M.R.S., M.R.H., M.K., A.A. and B.S.; investigation, H.N.P.R.C., J.V.S.K. and M.R.S.; resources, H.N.P.R.C. and J.V.S.K.; data curation, H.N.P.R.C., J.V.S.K., M.R.S. and M.R.H.; validation, A.B. and M.K.; writing—original draft preparation, H.N.P.R.C., J.V.S.K. and A.B.; writing—review and editing, H.N.P.R.C. and M.R.S.; supervision, J.V.S.K.; project administration, J.V.S.K.; funding acquisition, M.R.S. All authors have read and agreed to the published version of the manuscript.

Funding: The authors acknowledge the funding from Researchers Supporting Project number (RSPD2023R665), King Saud University, Riyadh, Saudi Arabia.

Institutional Review Board Statement: Not applicable.

Informed Consent Statement: Not applicable.

Data Availability Statement: Data are contained within the article and the Supplementary Materials.

Acknowledgments: The authors acknowledge the funding from Researchers Supporting Project number (RSPD2023R665), King Saud University, Riyadh, Saudi Arabia.

Conflicts of Interest: The authors declare no conflict of interest.

References

1. Stauffer, F.; Vanhoorne, V.; Pilcer, G.; Chavez, P.; Rome, S.; Schubert, M.; Aerts, L.; De Beer, T. Raw material variability of an active pharmaceutical ingredient and its relevance for processability in secondary continuous pharmaceutical manufacturing. *Eur. J. Pharm. Biopharm.* **2018**, *127*, 92–103. [CrossRef] [PubMed]
2. Sangshetti, J.N.; Deshpande, M.; Zaheer, Z.; Shinde, D.B.; Arote, R. Quality by design approach: Regulatory need. *Arab. J. Chem.* **2017**, *10*, S3412–S3425. [CrossRef]
3. Bolt, H.M.; Foth, H.; Hengstler, J.G.; Degen, G.H. Carcinogenicity categorization of chemicals—New aspects to be considered in a European perspective. *Toxicol. Lett.* **2004**, *151*, 29–41. [CrossRef]
4. Kondo, K.; Watanabe, A.; Iwanaga, Y.; Abe, I.; Tanaka, H.; Nagaoka, M.H.; Akiyama, H.; Maitani, T. Determination of genotoxic phenylhydrazine agaritine in mushrooms using liquid chromatography–electrospray ionization tandem mass spectrometry. *Food Addit. Contam.* **2006**, *23*, 1179–1186. [CrossRef]
5. Leblanc, B.; Charuel, C.; Ku, W.; Ogilvie, R. Acceptability of low levels of genotoxic impurities in new drug substances. *Int. J. Pharm. Med.* **2004**, *18*, 215–220. [CrossRef]
6. Froetschl, R. Presentation at the 2009 PhRMA API Workshop. *New Burn, NC, USA*, 2009; 18–21.
7. Savale, S.K. Genotoxicity of drugs: Introduction, prediction and evaluation. *Asian J. Biomater. Res.* **2018**, *4*, 1–29.
8. Liu, K.-T.; Chen, C.-H. Determination of impurities in pharmaceuticals: Why and how? In *Quality Management and Quality Control-New Trends and Developments*; IntechOpen: London, UK, 2019; pp. 1–17.
9. Gosar, A.; Sayyed, H.; Shaikh, T. Genotoxic impurities and its risk assessment in drug compounds. *Drug Des. Intellect Prop. Int. J.* **2018**, *2*, 227–232.
10. Committee, E.S. Scientific Opinion on genotoxicity testing strategies applicable to food and feed safety assessment. *EFSA J.* **2011**, *9*, 2379.
11. Shah, S.U. Importance of Genotoxicity & S2A guidelines for genotoxicity testing for pharmaceuticals. *IOSR J. Pharm. Biol. Sci.* **2012**, *1*, 43–54.
12. Honorato, J.; Masso, R.; Purroy, A. The Use of Gemfibrozil in the Treatment of Primary Hyperlipoproteinæmia. Preliminary Report. *Proc. R. Soc. Med.* **1976**, *69*, 78–79. [CrossRef]
13. Todd, P.A.; Ward, A. Gemfibrozil. *Drugs* **1988**, *36*, 314–339. [CrossRef] [PubMed]
14. Creger, P.L.; Neuklis, W.; Arbor, A. Aryloxypentane Compounds. U.S. Patent 3,707,566, 26 December 1972.
15. Creger, P.; Moersch, G.; Neuklis, W. Structure/activity relationship of gemfibrozil (CI-719) and related compounds. *Proc. R. Soc. Med.* **1976**, *69*, 3–5. [CrossRef] [PubMed]
16. Yang, Q.; Ren, T.; Yang, S.; Li, X.; Chi, Y.; Yang, Y.; Gu, J.; Hu, C. Synthesis and Pharmacokinetic Study of Three Gemfibrozil Salts: An Exploration of the Structure–Property Relationship. *Cryst. Growth Des.* **2016**, *16*, 6060–6068. [CrossRef]
17. Rubins, H.B.; Robins, S.J.; Collins, D.; Fye, C.L.; Anderson, J.W.; Elam, M.B.; Faas, F.H.; Linares, E.; Schaefer, E.J.; Schectman, G. Gemfibrozil for the secondary prevention of coronary heart disease in men with low levels of high-density lipoprotein cholesterol. *New Engl. J. Med.* **1999**, *341*, 410–418. [CrossRef]
18. ICH Guidelines, Q 3A(R); Impurities is New Drug Product. The Quality Guide Lines for Active Pharmaceutical Ingredients Related to Impurities According to the International Conference of Harmonization. 2022. Available online: http://www.ich.org (accessed on 1 January 2023).
19. Madasu, S.B.; Vekariya, N.A.; Velladurai, H.; Islam, A.; Sanasi, P.D.; Korupolu, R.B. Improved Process for Preparation of Gemfibrozil, an Antihypolipidemic. *Org. Process Res. Dev.* **2013**, *17*, 963–966. [CrossRef]
20. Hitesh, T.K.; Mahantesh, J.; Vikrant, M.P.; Somanath, N.S.; Pravin, M.N.; Ameya, M.T.; Sudhir, N. An Improved Process for the Preparation of a Key Intermediate of Gemfibrozil. WIPO (PCT) Patent Application No. WO2019069321A1, 11 April 2019.
21. Nunna, R.; Jayanna, N.; Ramachandran, D. An Improved New Path to Synthesize Gemfibrozil. *Asian J. Chem.* **2015**, *27*, 925. [CrossRef]
22. Friedland, S.N.; Leong, A.; Filion, K.B.; Genest, J.; Lega, I.C.; Mottillo, S.; Poirier, P.; Reoch, J.; Eisenberg, M.J. The cardiovascular effects of peroxisome proliferator-activated receptor agonists. *Am. J. Med.* **2012**, *125*, 126–133. [CrossRef] [PubMed]
23. Zhao, M.; Shi, J.; Li, W.; Guan, C.; Sun, C.; Peng, Y.; Zheng, J. Metabolic activation of gemfibrozil mediated by cytochrome P450 enzymes and sulfotransferases. *Chem. Res. Toxicol.* **2022**, *35*, 1257–1266. [CrossRef]

24. Boyd, M.R.; Grygiel, J.J.; Minchin, R.F. Metabolic activation as a basis for organ-selective toxicity. *Clin. Exp. Pharmacol. Physiol.* **1983**, *10*, 87–99. [CrossRef] [PubMed]
25. Ogilvie, B.W.; Zhang, D.; Li, W.; Rodrigues, A.D.; Gipson, A.E.; Holsapple, J.; Toren, P.; Parkinson, A. Glucuronidation converts gemfibrozil to a potent, metabolism-dependent inhibitor of CYP2C8: Implications for drug-drug interactions. *Drug Metab. Dispos.* **2006**, *34*, 191–197. [CrossRef]
26. Sallustio, B.C.; Foster, D. Reactivity of gemfibrozil 1-o-beta-acyl glucuronide. Pharmacokinetics of covalently bound gemfibrozil-protein adducts in rats. *Drug Metab. Dispos.* **1995**, *23*, 892–899.
27. U.S. Department of Health and Human Services. *Registry of Toxic Effects of Chemical Substances (RTECS, Online Database)*; National Toxicology Information Program, National Library of Medicine: Bethesda, MD, USA, 1993.
28. Omura, M.; Itonaga, Y.; Komatsu, H.; Mangen, Z.; Hirata, M.; Tanaka, A.; Inoue, N. The acute toxicity of allyl chloride by subcutaneous injection in mice. *Fukuoka Igaku Zasshi = Hukuoka Acta Med.* **1993**, *84*, 427–432. [PubMed]
29. U.S. Environmental Protection Agency. *Health and Environmental Effects Profile for Allyl Chloride*; Environmental Criteria and Assessment Office, Office of Health and Environmental Assessment, Office of Research and Development: Cincinnati, OH, USA, 1986.
30. Ingle, S.U.; Patil, P.P.; Barhate, S.D.; Umarkar, A.R. Stability indicating rp–hplc method for determination of gemfibrozil in pharmaceutical formulation. *World J. Pharm. Res.* **2015**, *4*, 1625–1635.
31. Vishnupriya, S.; Narenderan, S.; Vishnu, K.; Babu, B.; Meyyanathan, S. A novel analytical liquid chromatography-tandem mass spectrometry method for the estimation of Gemfibrozil in bulk and pharmaceutical formulations. *J. Appl. Pharm. Sci.* **2019**, *9*, 97–101.
32. Ulu, S.T. LC determination of gemfibrozil in tablets. *Chromatographia* **2006**, *64*, 447–451. [CrossRef]
33. Branch, S.K. Guidelines from the international conference on harmonisation (ICH). *J. Pharm. Biomed. Anal.* **2005**, *38*, 798–805. [CrossRef] [PubMed]

separations

MDPI

Article

Improved Quantitative Approach for Monitorization of Gangliosides Structural Diversity in Fungal Cell Factories by LC-MS/MS

Javier-Fernando Montero-Bullón *, Javier Martín-González, Gloria Muñoz-Fernández, Alberto Jiménez and José Luis Revuelta *

Metabolic Engineering Group, Department of Microbiology and Genetics, University of Salamanca, 37007 Salamanca, Spain
* Correspondence: jfmonbul@usal.es (J.-F.M.-B.); revuelta@usal.es (J.L.R.)

Abstract: Gangliosides are glycolipids occurring in higher animals, with a sphingoid core in the form of ceramide, bound to a glycan moiety including several units of sialic acid. Gangliosides are involved in important (patho)-physiological processes as components of cell membranes in humans, which has led to intensive study and interest in production strategies. Their structural variability depends on the combination of a sphingoid base, a fatty acyl chain, and an attached oligosaccharide. The combinatorial diversity differs and grows exponentially in synthetic biology approaches, e.g., use of microbial cell factories. A specific analytical platform accounting for this complexity is not available to date. However, quantification of the intermediates of the whole biosynthetic route is needed to boost projects on biotechnological ganglioside production. In this study, a fast high-throughput quantitative LC-MS/MS methodology was developed to cover analysis of gangliosides, with a wider structural perspective adapted to fungal organisms. This work was achieved using metabolically engineered strains that further allowed to test detection in biological complex matrixes. Ganglioside backbones—hitherto uncharacterized—with the five most common fungal sphingoid bases and both simple and hydroxylated fatty acids were subjected to characterization. The addition of glycans to the polar head was also successfully monitored with up to 4 units—corresponding to GD3 which bears two sialic acid units and furthermore represents the common precursor for the whole ganglio-series. This platform represents an improved methodology to study the biochemical diversity associated to gangliosides for natural and metabolically engineered biosynthetic pathways.

Keywords: LC; MS/MS; mass spectrometry; liquid chromatography; glycosphingolipids; ceramides; gangliosides; GM3; GD3

Citation: Montero-Bullón, J.-F.; Martín-González, J.; Muñoz-Fernández, G.; Jiménez, A.; Revuelta, J.L. Improved Quantitative Approach for Monitorization of Gangliosides Structural Diversity in Fungal Cell Factories by LC-MS/MS. *Separations* **2022**, *9*, 432. https://doi.org/10.3390/separations9120432

Received: 24 November 2022
Accepted: 8 December 2022
Published: 12 December 2022

1. Introduction

Ganglio-series sphingolipids, or gangliosides, are glycosphingolipids occurring in higher animals, consisting of a ceramide core and an anchored glycan moiety containing neuraminic acid. Gangliosides accumulate predominantly in central and peripheral nervous system—up to 10% of total lipid content, but they occur in all tissues in vertebrate animal cells [1]. Gangliosides have not been observed in lower animals and other kingdoms such as plants, fungi, or bacteria. They accumulate in the outer leaflet of the cell membrane, whereby they form part of the glycocalyx, constituting lipid rafts and microdomains of functional relevance. They operate as antigens/receptors of specific molecules, mediators in cell-to-cell interactions, and modulators of the charge density at the membrane surface. Accordingly, they show physiological and patho-physiological roles in mammalian cells [2] and ganglioside metabolism is implicated in human health and disease [3]. Recent studies pinpoint diagnostic and therapeutic properties in neurodegenerative diseases [4,5], cancer [6], or autoimmune diseases [7]. Monosialogangliosides and disialogangliosides with short saccharide forms—such as GM3 and GD3 herein studied—have shown beneficial effects in infant neurologic development [8,9].

Obtention in laboratory of isolated gangliosides has been essential for applied and fundamental studies. Gangliosides have also attracted interest as high-added value products due to their biological properties [10]. Synthetic methodologies are mainly based on chemical procedures [11], semisynthetic assembly [12], or more recently, complete enzymatic in vitro approaches [13,14]. However, scarce work has been performed to produce gangliosides via bioprocessing in the classical microbial biotechnological workhorses. It seems plausible, through synthetic biology, to engineer metabolism of microorganisms towards production of gangliosides, e.g., in eukaryotic fungal species. This work should account for the particular structural diversity associated to the variability of the sphingoid base, fatty acyl chain and attached oligosaccharide [15]. Remarkably, the dissimilar and more extensive sphingolipid structural basis associated to organisms such as fungi is well known [16]. An adapted analytical quantitative platform becomes of the utmost importance when dealing with these potential microbial producers. In fact, this may be one of the main factors that hampers, so far, biotechnological production of gangliosides. An efficient analysis for the putative biosynthetic pathway must consider several structural details.

The common glycan moiety pattern for mammalian gangliosides is based on a binding lactose upon the hydroxy group in carbon number one of the sphingoid base, with one or more sialic acid units linked to this core. A variable number of sialic acids and other neutral monosaccharides including glucose, galactose, and N-acetyl-galactose are sequentially added in the most complex forms [15]. In this context, monosialogangliosides GM3 and disialogangliosides GD3 are especially relevant. These species can be found ubiquitously but they predominate in extra-neural tissues. Enriched quantities of GM3 and GD3 are present in tissues as digestive system, kidney, heart, or liver—in the range of nanomols per milligram of dry weight [17], but most importantly, they represent the precursors of most of gangliosides structural diversity [15]. Regarding the sphingoid backbone, fatty acyl chains bound to the sphingoid base vary in length depending on the organism—from 16 to up to 26 carbons, and unsaturations—one or none most commonly [18]. The sphingoid base is most often C18-sphingosine (d18:1), as corresponds to higher animals sphingolipid biosynthesis [15]. Nonetheless, recent studies reveal the importance of gangliosides based on non-canonical sphingoid bases such as 3-ketosphinganine [19] or eicosphingosine [20]. In the context of synthetic biology, an emphasis on revisiting structural diversity is needed. In the case of fungal microorganisms, other sphingoid bases such as sphinganine (d18:0), 4-hydroxysphinganine (t18:0), sphingadienine (d18:2), or 4-methylsphingadienine (dm18:2) must be considered to evaluate production [16]. Moreover, the existence of alpha-hydroxylated fatty acids must not be overlooked [21].

Available analytical methodologies to monitor ganglioside profiles mainly rely on mass spectrometry hyphenated to liquid chromatography. They focus on the most common structures, namely sphingosine-containing gangliosides [17,22,23]. They also encompass analysis of the most complex forms including isomeric resolution [24,25]. With respect to glycosphingolipids in fungal species, some works have already elucidated the glycosphingolipid structural base [26] and help to propose plausible biosynthetic options. An approach that considers the analysis of putative gangliosides species in non-mammalian engineered organisms remains unexplored to date.

In this work, we expanded the analytical coverage of the structural diversity of gangliosides with a focus on the sphingoid base and fatty acyl variability in fungal organisms. We designed the analytical platform to offer a comprehensive and technically transferable analysis. Data from LC, MS and MS/MS have been gathered to rationalize the detection of ganglioside species based on non-canonical bases and chemical modifications such as extension of the aliphatic chains, hydroxylation, and degree of saturation. This was accomplished making use of metabolically engineered fungal strains (unpublished results) to generate an adequate expanded structural diversity and demonstrate detection in biological complex matrixes. The fast high-throughput quantitative LC-MS/MS methodology developed provides quantitative values of fungal glycosphingolipids—hitherto uncharacterized—with the five most common sphingoid bases and both simple and hydroxylated fatty acids. Sequential addition of monosaccharides to the bare ceramide was successfully analyzed — up to 4 units corresponding to GD3, which bears two sialic acid monomers. This monitorization platform represents an improved quantitative methodology to study the chemical diversity associated to the ganglio-series glycosphingolipids for natural and metabolically engineered biosynthetic pathways.

2. Materials and Methods

2.1. Chemicals

Internal standards D-erythro-sphingosine (C17 base), N-heptadecanoyl-D-erythro-sphingosine, N-(2′-(R)-hydroxyheptadecanoyl)-D-erythro-sphingosine, D-glucosyl-b-1,1-N-heptadecanoyl-D-erythro-sphingosine, D-lactosyl-b-1,1′-N-Heptadecanoyl-D-erythro-Sphingosine, and deuterated C18:0 GM3 Ganglioside-d5 from Avanti® Polar Lipids, Inc. (Alabaster, AL, USA) were used. HPLC grade chloroform, methanol, acetonitrile, formic acid, and ammonium formate were purchased from Fisher Scientific Ltd. (Leicestershire, UK). Milli-Q water was used for all experiments, filtered through a 0.22 µm filter and obtained using a Milli-Q Millipore system (Synergy®, Millipore Corporation, Billerica, MA, USA).

2.2. Standard Preparations

Six commercial standards of high purity (<99%) were dissolved at 200 ppb concentration: D-erythro-sphingosine (C17 base), N-heptadecanoyl-D-erythro-sphingosine, N-(2′-(R)-hydroxyheptadecanoyl)-D-erythro-sphingosine, D-glucosyl-b-1,1-N-heptadecanoyl-D-erythro-sphingosine, and D-lactosyl-b-1,1′-N-Heptadecanoyl-D-erythro-Sphingosine and deuterated C18:0 GM3 Ganglioside-d5, hereafter referred as Sph(17:1), Cer(d18:1/17:0). Cer(d18:1/h17:0), HexCer(d18:1/17:0), Hex2Cer(d18:1/17:0) and GM3(d18:0/18:0-d5), respectively. They were used without further purification in methanol:choloroform:water (4:1:1). The solution was kept at −20 °C until use. For method development, this mixture was injected for analysis. The standards mixture was also used as extraction solvent for biological samples, in order to allow the use of internal standards in known concentrations.

2.3. Extraction of Ceramides and Gangliosides in Biological Samples

Approximately 10 milligram of lyophilized mycelia was suspended in 1 mL of extraction solvant in 2 mL plastic tubes containing 1.4 mm ceramic beads (Precellys—Bertin Techologies, Montigny-le-bretonneux, France). Lyophilizate weight was annotated to correct absolute quantitative values. A benchtop Minilys homogenizer (Bertin Technologies) was used for mycelia lysis with three rounds of 30 s at 4000 rpm agitation. Samples were kept in ice during the process and between disruption rounds. After lysis, samples were centrifuged at 17,000 rpm for two minutes in a benchtop centrifuge. Supernatant was collected in LC-MS vials for analysis.

2.4. LC-MS/MS Analysis of Ceramides and Glycoceramides

Sample separation for ceramides and glycoceramides was performed using an Ascentis Express C8 Solid Core column (30 mm × 2.1 mm, 2.7 µm, 90 Å) from Sigma-Aldrich (St. Louis, MO, USA) in a Vanquish Flex UHPLC with flow rate set at 300 µL/min and column compartment temperature 30 °C. Binary gradient used was 70–100% B in 10 min, 10% B for 5 min, 1000–70% B in 1 min, and re-equilibration at 70% B for 4 min. The solvents used formic acid 0.1% (A) and acetonitrile (B).

The Orbitrap Q-Exactive Focus (Thermo Fisher Scientific, Waltham, MA, USA) mass spectrometer was operated with electrospray (ESI) voltage −2.75 kV/+3.5 kV, capillary temperature, 250 °C; sheath gas flow 40 units, auxiliary gas flow 10 units, spare gas 1.5 units, probe heater 300 °C). MS analysis was acquired in both polarities with resolving power 70,000 (full width half maximum), m/z = 250–1250 using an automatic gain control (AGC) target of 10^6. Targeted analysis was performed through parallel reaction monitoring

(PRM) with an inclusion list for the species monitored and a m/z window of 1 unit. Two different runs to split the number of transitions for an efficient detection—methods otherwise identical—were used (a) for ceramides and glucosyl-ceramides and (b) their lactosyl ceramides counterparts. MS/MS spectra were obtained using higher-energy collisional dissociation (HCD) fragmentation (stepwise 20, 25, and 30% normalized collision energy), AGC target of 10^6 and maximum injection time of 75 ms at a resolution of 17,500. Data acquisition was carried out using Xcalibur data system (V3.3, Thermo Fisher Scientific, Waltham, MA, USA). 10 µL of standards solution or biological samples were injected per run.

2.5. LC-MS/MS Analysis of Monosialo and Disialo Gangliosides

Sample separation for monosialo- and disioalogangliosides was performed using an Ascentis Express C8 Solid Core column (30 mm × 2.1 mm, 2.7 µm, 90 Å) from Sigma-Aldrich (St. Louis, MO, USA) in a Vanquish Flex UHPLC with flow rate set at 300µL/min and column compartment temperature 30 °C. Binary gradient used was 50–100% B in 10 min, 100% B for 5 min, 100–50% B in 1 min and re-equilibration for 5 min at 50%B. The solvents used were 0.1% formic acid (A) and acetonitrile:methanol:water (79:19:2; $v/v/v$), 20 mM ammonium formate, 20 mM formic acid (B).

The Orbitrap Q-Exactive Focus mass spectrometer (Thermo Fisher Scientific, Waltham, MA, USA) was operated with electrospray voltage −2.75 kV/+3.5 kV, capillary temperature, 250 °C; sheath gas flow 40 units, auxiliary gas flow 10 units, spare gas 1.5 units, probe heater 300 °C). MS analysis was acquired in both polarities with resolving power 70,000 (full width half maximum), m/z = 250–1250 using an automatic gain control (AGC) target of 10^6. Targeted analysis was performed through parallel reaction monitoring (PRM) with an inclusion list for the species monitored and a m/z window of 1 unit. MS/MS spectra were obtained using higher-energy collisional dissociation (HCD) fragmentation (stepwise 20, 25, and 30% normalized collision energy), AGC target of 10^5, and maximum injection time of 50 ms at a resolution of 17,500. Data acquisition was carried out using the Xcalibur data system (V3.3, Thermo Fisher Scientific, Waltham, MA, USA). A measure of 10 µL of standards solution or biological samples was injected per run.

2.6. Quantitative LC-MS/MS Analysis

Quantification was performed by integrating the area under the curve for the corresponding MS/MS fragment ions chromatographic peak chosen for each species, using Quan Browser software (Thermo Fisher Scientific, Waltham, MA, USA). Signals were normalized versus the associated internal standard: Sph(17:1) for free bases, Cer(d18:1/17:0) for ceramides. Cer(d18:1/h17:0) for hydroxylated fatty acid ceramides, HexCer(d18:1/17:0) for glucosyl ceramides, Hex2Cer(d18:1/17:0) for lactosyl ceramides and GM3(d18:0/18:0-d5) for monosialo- and disialogangliosides. The nanograms of each species was extrapolated from the nanograms of internal standards injected (2 nanograms), and further normalized versus the milligrams of mycelia used for extraction.

3. Results

3.1. Optimizatin of MS Ionizatin Parameters

A preliminary analysis based on ESI-MS spectra was performed to evaluate ionization efficiency in both polarities, including the presence of adducts. The corresponding standards mimicking ceramides, glycoceramides—with one or two hexoses, and the free sphingoid bases, showed $[M + H]^+$ ions that clearly outperform other ionization forms (Supplementary Figure S1a). Other ionization forms comprising salt adducts (e.g., sodium or potassium adducts) were not abundant or induced signal suppression. Analysis in negative voltage was rendered useless due to very low sensitivity for these species. In the case of gangliosides, both $[M + H]^+$ and $[M - H]^-$ ions of the corresponding standard yielded an intense comparable signal that dominates the spectrum in each polarity mode

(Supplementary Figure S1b). Eventually, positive mode was selected for the whole analysis, also due to optimization on MS/MS transitions as detailed in the next sections.

3.2. LC-MS/MS Method Development for Ceramides and Glycoceramides

According to the efficient formation of [M + H]+ ions for ceramides and glycoceramides, MS/MS fragmentation produces an intense ion based on the protonated sphingoid base. The doubly dehydrated sphingoid base ion is the most intense ion for all structures, except in the case of the free sphingoid base where it competes with simple dehydration (Supplementary Figure S2). Therefore, $[M + H]^+$ ion based on the doubly dehydrated sphingoid base was chosen for MS/MS transitions. It was optimal to completely elucidate the structures, since it allowed assignation of the sphingoid base and the fatty acid bound, with increased sensitivity for all species in the method (Table 1). To note, the presence of a hydroxylated fatty acid leads to an intense fragment cation associated to water loss from the $[M + H]^+$ precursor ion (Supplementary Figure S2). This may slightly hamper detection by decreasing sensitivity through competence with the production of sphingoid base fragments. In return, this trait can be considered to confirm this kind of species in a MS/MS scan. In the case of glycosylated ceramides, the sequential neutral loss of the hexose units could be observed providing structural information (Supplementary Figure S2).

Table 1. Most abundant observed MS/MS fragments for the five internal standards used in the method development of ceramides and glycoceramides analysis.

Internal Standard	MS/MS Fragment Ions
Sph (d18:1/17:0)	- NL of 2xH_2O, $[M + H]^+$ = 250.252 (equivalent to 264.268 in C18 sphingoid base) - NL of H_2O, $[M + H]^+$ = 268.263
Cer(d18:1/17:0)	- NL of H_2O, $[M + H]^+$ = 534.523 - NL of fatty acyl chain and H_2O $[M + H]^+$ = 264.268
Cer(d18:1/h17:0)	- NL of water, $[M + H]^+$ = 550.518 - NL of 2xH_2O, $[M + H]^+$ = 532.508 - NL of fatty acyl chain and water: $[M + H]^+$ = 264.268
HexCer(d18:1/17:0)	- NL of water, $[M + H]^+$ = 696.577 - NL of 1 hexose, $[M + H]^+$ = 534.524 - NL of fatty acyl chain, hexose, and water: $[M + H]^+$ = 264.268
Hex2Cer(d18:1/17:0)	- NL of water, $[M + H]^+$ = 858.629 - NL of 1 hexose, $[M + H]^+$ = 696.576 - NL of 2 hexoses, $[M + H]^+$ = 534.523 - NL of fatty acyl chain, 2 hexoses, and H_2O: $[M + H]^+$ = 264.268

The most abundant fragment ion for each species is underlined. Annotation is based on the corresponding neutral loss (NL) of different parts of the structure. Experimental m/z values in the MS/MS spectrum are included (Supplementary Figure S2), and match theoretical values with precision within 10 ppm.

Some in-source fragmentation events were found as the main issue regarding method development. These undesired interferences are observed for certain species and can mislead assignation. They are neutral losses from part of the structure in the process of ionization, forming structures that behave as a new precursor that represents a confounding equivalent to other species. Specifically, the water molecule in hydroxylated fatty acids or monosaccharides in glucosyl and lactosyl ceramides are labile and readily removed in the ionization process. Therefore, hydroxylated fatty acids can be confounded with their counterpart without hydroxylation but one more unsaturation. It is also the case of glycosylated species and their non-glycosylated counterparts with one more unsaturation. However, the method developed allowed discrimination of interferences by avoiding coelution in the chromatogram (Supplementary Figure S3).

Free sphingoid bases, ceramides, and glycoceramides with up to 2 neutral monosaccharides were successfully distinguished by LC separation in a unique gradient, with elution order based on their hydrophobicity (Figure 1). Hence, lactosyl-ceramides eluted in the shortest retention times, followed by glucosyl ceramides and non-glycosylated ceramides. Sphingoid bases eluted in the beginning of the run due to their high hydrophilicity but showed a well-defined peak shape that allowed quantification. The inclusion of a hydroxy group in the fatty acyl chain shortens retention times versus the non-hydroxylated fatty acids.

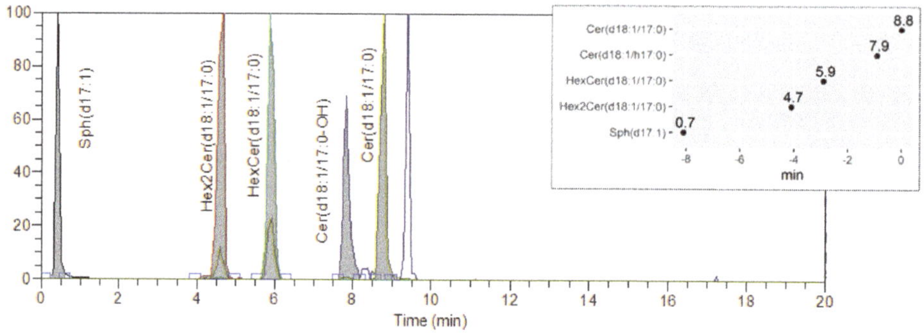

Figure 1. Annotated LC-MS/MS chromatogram for internal standards used for the method development of ceramides and glycoceramides analysis. The inset image presents retention time variations due to the structural differences, i.e., sequential addition of hexoses and hydroxylation of the fatty acyl chain, as well as the free sphingoid base retention time.

3.3. LC-MS/MS Method Development for Gangliosides

Seeking specific MS/MS transitions to unequivocally detect ganglioside structures, we opted for a positive polarity for method development. Even if negative ions provide a good intensity, fragment cations became of great advantage for structural discrimination. Product ions in the negative mode MS/MS spectrum mainly corresponded to the uninformative dehydrated sialic acid anion (Supplementary Figure S4). In the positive MS/MS, the cation corresponding to the sialic acid was observed but disregarded for the same reason. The doubly dehydrated sphingoid base remained the most prominent and structurally relevant fragment cation (Figure 2) and accordingly chosen for MS/MS transitions. Nonetheless, the MS/MS spectrum further provided structurally useful transitions such as the sequential neutral loss of the monosaccharides. (Figure 2).

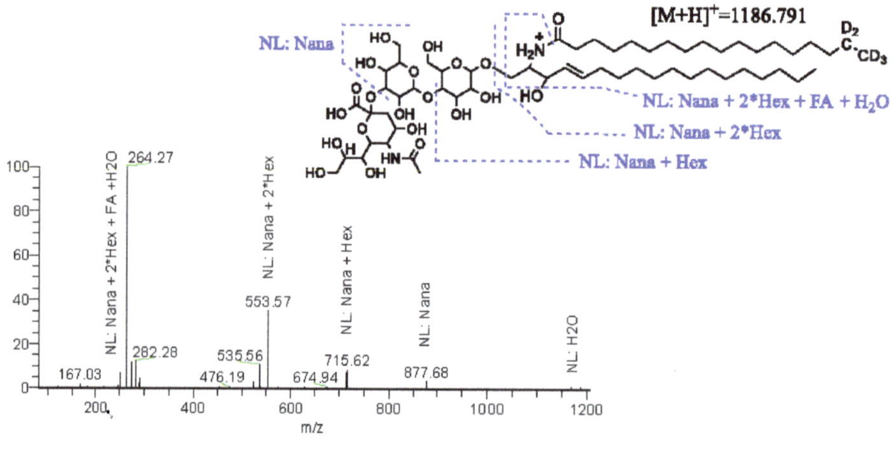

Figure 2. Structure and fragmentation of the deuterated GM3(d18:1/18:0-d5) standard. Schematic representation of the molecular structure and cleavage sites. Annotation is based on the corresponding neutral losses (NL) of different parts of the structure generating fragment ions.

From the chromatographic point of view, the method used for ceramides and glycoceramides was incompatible with detection of gangliosides. A new solvent system with elevated ionic strength—ammonium formate and formic acid in concentration 20 mM—was determined as best performing. After adaptation of the gradient, good peak shapes were obtained for the ganglioside standard. In this new method, the GM3 standard showed the shortest retention regarding ceramides and glycosylated ceramide standards (Figure 3), as expected due to the increased hydrophilicity upon addition of the sialic acid residue.

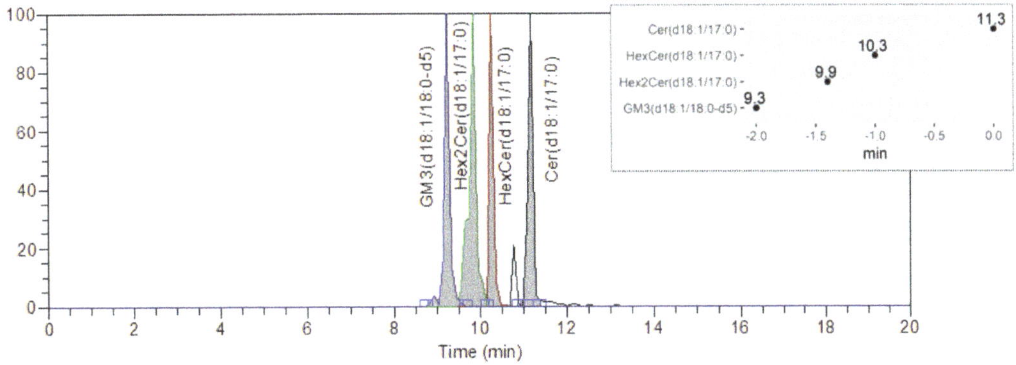

Figure 3. Annotated LC-MS/MS chromatogram for deuterated internal standard GM3(d18:17/18:0-d5) in the method developed for ganglioside analysis, compared with the other standards. The inset image presents variation due to the structural differences regarding the sequential addition of monosaccharides to the glycan moiety, one or two hexoses, and the subsequent sialic acid.

3.4. Expanded Characterization of Species upon Biological Samples

Following method development with standards, detection of species with a wider variety of sphingoid bases was undertaken by utilizing extracts from metabolically engineered organism, that allowed analysis of structures with the bases d18:1, d18:0, d18:2, dm18:2, and t18:0. The monitorization upon detection of the sphingoid base product ion, taking into account the chemical modification, allowed unequivocal assignation of the species. Chromatographic methods developed were efficient for separation of the five corresponding sphingoid backbones (Figure 4b), in both simple ceramides and their glycosylated equivalents. Considering ceramide structures with the same fatty acyl chain, addition of hydroxy groups or unsaturations shortened retention times, while saturation or addition of a methyl group delays elution (Figure 4a).

The increased number in structures generated by the variety in sphingoid bases implied that, even when chromatographic optimization was maximized to the best of our ability/setup capabilities, some of the species could still not be separated. This conveyed detection to the positive mode and more specifically to the use of sphingoid base fragments, whereas the use of sialic acid fragment anions or other potential transitions did not allow discrimination. Such is the case of GM3(d18:1/18:0) and GM3(dm18:2/h16:0)—both isobaric species in terms of m/z value and eluting at similar retention times, that would be indistinguishable if information on the sphingoid base is not available (Supplementary Figure S5). This handicap affects this and other pairs of species, but it was overcome in all cases throughout the developed method either by chromatographic separation or MS/MS information.

Disialogangliosides, namely GD3, were also characterized thanks to their presence in biological samples of the metabolically engineered organism. Same ionization forms and MS/MS transitions relying on the sphingoid base were proven efficient. Other fragments based on the sequential loss of monosaccharides and the protonated sialic acid helped to ascertain structures. The chromatographic behavior observed was a shift, of the whole group of GD3 species, to shorter retention times versus the monosialo- counterparts, due to the addition of a second hydrophilic sialic acid residue (Figure 5). Within the GD3 group, retention times depend on traits previously explained.

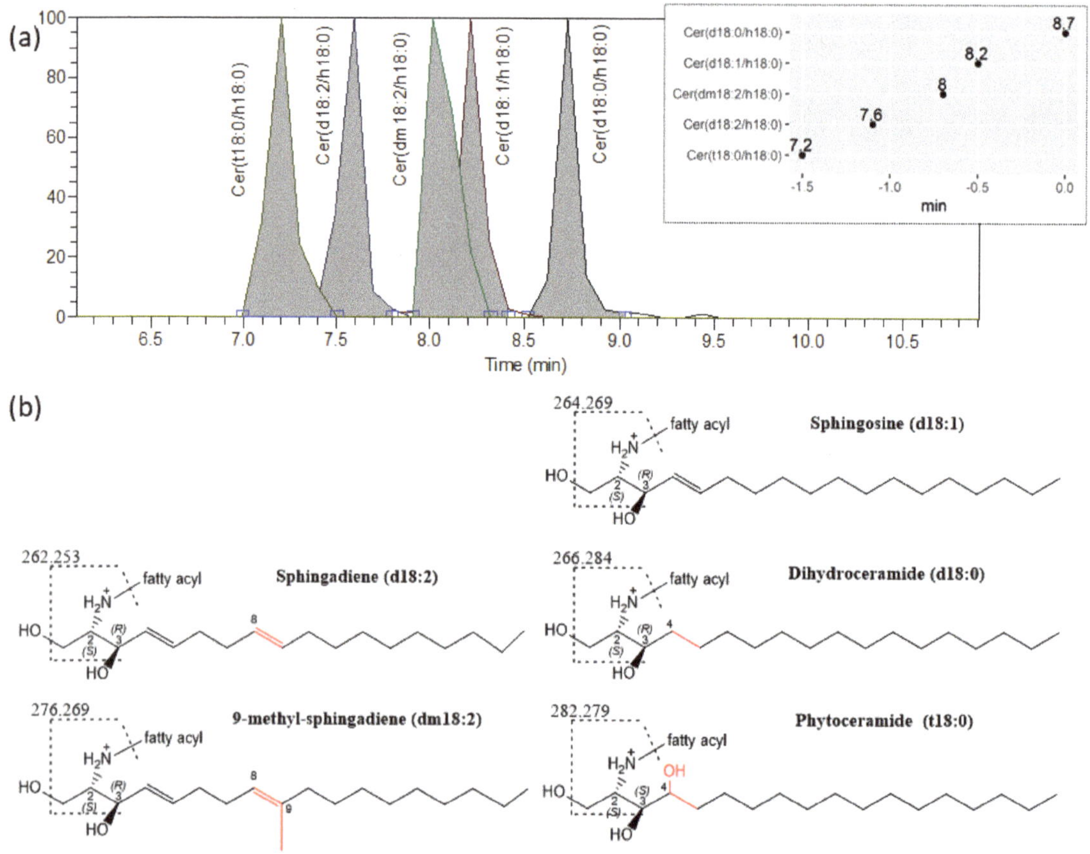

Figure 4. (**a**) Zoomed LC-MS/MS chromatogram for ceramide species containing five different sphingoid bases found in a fungal biological matrix (d18:1, d18:2, d18:0, dm18:2, and t18:0). All ceramide species plotted contain the same fatty acid, allowing variation in retention time due to the sphingoid base to be observed (inset image). (**b**) Structure and MS/MS fragments used for each of the five structures with different sphingoid bases. Mass of the fragment indicated beside the discontinuous line, which represents the most abundant MS/MS fragmentation cleavage.

3.5. Method Validation in Biological Samples

Our method was successfully tested in biological matrixes by analyzing extracts from fungal mycelia. Both wild-type and a mutant organism were analyzed, the latter potentially containing all of the repertory of species targeted in this work. Based on the known naturally occurring array of fatty acids in the wild-type organism—regarding length and degree of unsaturation, all combinations for the corresponding ceramide, glycosylated ceramide, and ganglioside structures were considered. A white list containing more than one hundred MS/MS transitions was generated. We observed that methods in our instrumental setup with more than 50 MS/MS transitions to be monitored led to poor chromatographic power and efficient detection was restored by adjusting number of transitions in unfolded methods.

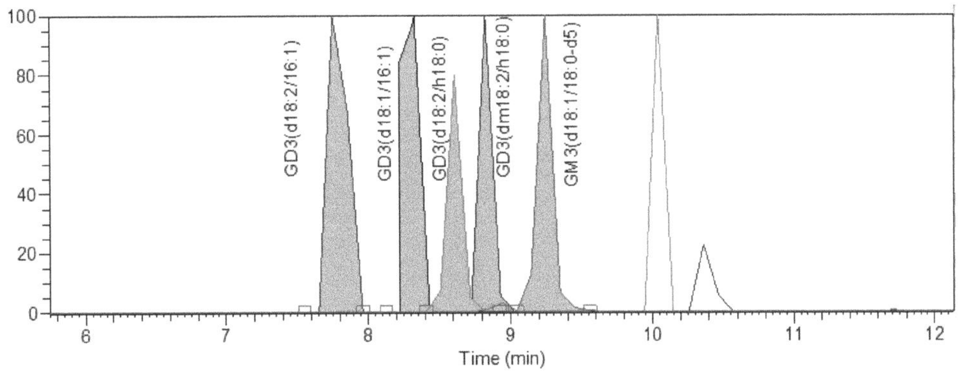

Figure 5. LC-MS/MS chromatogram for disialoganglioside GD3 species containing three different sphingoid bases (d18:1, d18:2, dm18:2) and both hydroxylated and non-hydroxylated fatty acids. GM3 internal standard is included as elution time reference. Comparison of retention times is not straightforward due to the different fatty acids attached, but GD3 species elute as a group before the GM3 internal standard.

Biological replicates showed reproducibility in quantification—extrapolated from the use of internal standards—and retention times (Supplementary Figure S6). Methods developed for ceramides and glycoceramides monitorization provided quantitative values for 42 species in the wild-type organism, including the five sphingoid types and occasionally alpha-hydroxylation in the fatty acyl chain. Concentrations observed ranged between values close to the observed limit of detection—around 1 nanogram of the species per milligram of mycelia, and a few hundreds of nanograms per milligram of mycelia. Species with two hexoses and the corresponding gangliosides were not observed in the wild type, serving therefore as negative control. Analysis of the mutant strain provided quantitative values in a number of species, with a greater structural variety as explained in the previous sections. Hence, methods developed for lactosyl ceramides and GM3 and GD3 gangliosides also yielded detection of species in this case. However, only structures with d18:1, d18:2, and dm18:2 sphingoid backbone were detected and quantified within these groups (unpublished results). These structures predominantly included C18 saturated alpha-hydroxylated acids, in comparison to the alternative unmodified fatty acids, chain lengths, and number of double bonds, present to a lesser extent up to absolute quantification.

4. Discussion

Gangliosides are lipids that form part of membranes in higher animals and own important functionalities rather than just structural roles. They are recognized as mediators of physiological and pathophysiological processes in humans. We believe that production of well-defined ganglioside structures can be of interest for fundamental and applied research. To date, most of the synthetic processes developed are based merely on extraction or in vitro preparation. However, current state-of-the-art in synthetic biology suggests it must be possible to produce these metabolites in microbial cell factories. The most plausible option seems to be, given current knowledge on gangliosides synthesis and natural occurrence, the use of well-established eukaryotic biotechnological chassis. Hence, fungi such as *Pichia pastoris, Saccharomyces cerevisiae, Eremothecium gossypii* or *Yarrowia lipolytica* ought to become important key players regarding ganglioside production. From a biosynthetic point of view, in such potential microbial factories, it is essential to consider both the heterologous design of the characteristic glycan moiety but also the available fungal structural basis as starting point to build up gangliosides.

Nevertheless, a structurally precise and quantitative method to facilitate this work must be at hand as a prerequisite. High-throughput sensitive and fine structural information provided by LC-MS/MS technique can meet this challenge. In this work we considered for analysis the five most important sphingoid bases that can be found in this context, namely sphingosine, sphinganine, 4-hydroxy-sphinganine, 4,8-sphingadienine and 9-methyl-4,8-sphingadienine [16]. We also considered the plausibility of fatty acid hydroxylation in alpha position. Accordingly, combination of fatty acids and sphingoid bases leads to exponential increment of the number of structures. Ambiguities for identification, as highlighted in the results section, must be overcome. Even when high resolution MS helped in this work—a few parts per million of mass deviation and resolution of up to 70,000—methods developed are aimed to allow the platform to be used in middle and low-end MS instruments. For this purpose, chromatographic separation and the selection of structurally relevant fragment ions for MS/MS were carefully chosen. All MS/MS transitions were established relying on sphingoid-base related fragments that allow a better structural elucidation. Hence, positive polarity was required for monitorization, including gangliosides—traditionally analyzed as anions [17,22–25]. Efficient ionization in positive polarity is due to proton addition to the secondary amine in the sphingoid bases. However, gangliosides also ionize intensely in negative mode due to the presence of sialic acid units, readily deprotonated. We found positive polarity is the best option, since sensitivity remains high and MS/MS analysis is crucially improved. To note, an Orbitrap HCD collision cell was employed and energy levels needed to be tuned to maximize fragment ion signals. For chromatography, reversed stationary phases, which have demonstrated good chromatographic performance for sphingolipids [17,22–25], were used for separation with some adaptations. For all free sphingoid bases, ceramides, glycoceramides, and gangliosides, a short C8 column was able to retain species and produce appropriate peak shapes within a 20 min run, that includes washing and equilibration. However, an unique chromatographic method to cover all species was not possible. Retention times heavily increased for the ganglioside internal standard when using the same method than for the rest of species. This is probably the reason why even less retentive phases have been previously used [23]. It was solved by using an elevated ionic strength buffering solvent, which most probably avoids secondary interactions via hydrogen bonds with the residual silanols in the stationary phase. This method for gangliosides could not be used, vice versa, for the rest of species, since it provided insufficient chromatographic separation.

We undertook method development on standards commercially available that can successfully represent all structural groups of the study in terms of mass spectrometry and chromatography behavior. Odd chain fatty acid structures containing heptadecanoic acid for ceramides—Cer(d18:1/17:0), and glycoceramides—HexCer(d18:1/17:0), Hex2Cer(d18:1/17:0); hydroxylated heptadecanoic acid for alpha-hydroxylated ceramides—Cer(d18:1/h17:0), an odd chain sphingoid base (d17:1) for free bases, and deuterated species in the case of gangliosides—pentadeuterated GM3 with sphingosine and octadecanoic acid GM3 (d18:1/18:0-d5), were selected. Their use allowed to face the main challenges in the development of a LC-MS/MS method and succeed in the obtention of a method that avoids interferences when combined with the choice of MS/MS transitions. Remarkably, they are all non-endogenous species for the biological cases of study, allowing eventual internal standard correction. Hence, they helped towards reliable quantification by compensating well-known technical caveats such as in-source fragmentation events, ionization efficiency and biological matrix ion suppression, providing good reproducibility. Given that ionization efficiency mainly depends on the polar head—that accommodates the charge—the selection of one internal standard for each structural polar head trait provides full correction. Furthermore, variation in fatty acyl chain length and unsaturation should still keep polar head-related species grouped in terms of elution times, therefore with similar matrix influence and ion suppression that can be corrected by the internal standards chosen. Accordingly, we consider that methods developed should be suitable for species with fatty acyl chains ranging within structural variety in fungal samples (from 14 to up to 26 carbons). In order to offer wider information and due to the lack of commercial standards, we employed genetically engineered fungal strains to expand the coverage of the analytical platform. The use of this platform allowed to separate and identify structures in a subsequent step, with the most common fungal sphingoid bases and the possibility of fatty acid hydroxylation, regarding the biological matrix of study. Furthermore, we were able to characterize up to GM3 monosialogangliosides and GD3 disialogangliosides with these backbones for the first time. These are fundamental species since they represent the main precursors of all ganglio-series structural diversity [15]. The design of these strains is out of the scope of this article (unpublished results), while being crucial to provide an expanded set of structures for method development and allow demonstration of the capabilities of the method in biological complex matrixes.

5. Conclusions

In this study, an LC-MS/MS method was developed for absolute quantification of ceramides and gangliosides with a wide structural diversity and applied for detection in biological matrixes. Five different sphingoid bases were characterized as part of the backbone: (d18:1), (d18:0), (t18:0), (d18:2), and (dm18:2). Fatty acyl chains with 16, 18, and 20 carbons and none or one unsaturation were also characterized (namely 16:0, 16:1, 18:0, 18:1, 20:1), including alpha-hydroxylated saturated fatty acids (namely h16:0, h18:0). Glycosylated counterparts with glucose, lactose, and subsequent sialic acid units were finally characterized, specifically GM3 monosialogangliosides and GD3 disialogangliosides.

The setup was designed to perform with high sensitivity, high chromatographic selectivity, and high structural discrimination allowing quantification of a broad variety of relevant gangliosides and their precursors. This analytical platform will allow to assess biotechnological production of gangliosides in fungal microbial factories. Nonetheless, it may further help to improve LC and MS analysis of gangliosides and disclose new biological implications or utilities for this important type of glycosphingolipids, with a wider structural perspective.

Supplementary Materials: The following supporting information can be downloaded at: https://www.mdpi.com/article/10.3390/separations9120432/s1, Supplementary Figure S1: (a) Zoomed region of the MS spectrum of the commercial standards used for method development in this work. The [M + H]+ is observed as the predominant ion without significant contribution of other adducts, that should be observed along the m/z range represented. (b). Zoomed region of the MS spectrum of GM3 standard in both positive—upper panel, and negative polarity-lower panel. Similar intensities are observed, according to the absolute signal scale used, to this end, along the y axis. Supplementary Figure S2: MS/MS spectrum in positive polarity of the five commercial standards used in method development for ceramides and glycoceramides. Schematic molecular structures, with fragmentation patterns, are depicted on the right and correlated with the masses annotated in the spectrum., matching theoretical values within a 10 ppm tolerance threshold. Supplementary Figure S3: LC-MS/MS chromatogram for the commercial standard Cer(d18:1/17:0). Correct signal appears at 8.8 min (shaded). The other ghost peaks belong to in-source fragmentation of other structures. Hex2Cer(d18:1/17:0) interferes in the chromatogram through in-source loss of the 2 hexoses of the glycan moiety (4.7 min). HexCer(d18:1/17:0) generates a ghost peak by release of its hexose. (5.9 min). Cer(d18:1/h17:0) easily decomposes by loss of a water molecule from the hydroxylated fatty acid, in such a way that the 2nd isotopic peak is present in the chromatogram of the non-hydroxylated counterpart. Supplementary Figure S4: MS/MS spectrum in negative polarity of the GM3(d18:1/18:0-d5) standard used in method development for ganglioside analysis. Mass annotated in the spectrum matches theoretical values, within 10 ppm tolerance, of the fragment ion consisting of the deprotonated sialic acid unit. Supplementary Figure S5: (a) LC-MS/MS chromatogram in positive polarity, based on the sphingoid base fragment ion, for the isobaric species GM3(d18:1/18:0) and GM3(d18:2/h16:0), that elute at similar retention times. (b) Chimeric MS/MS spectrum with information for both species, given that precursor ions have similar masses and retention times. Quantification is only possible when using MS/MS fragments that allow unequivocal identification, such as the annotated ions corresponding to the sphingoid base. Supplementary Figure S6: Quantitative LC-MS/MS analysis of ceramides (Cer) and glucosyl ceramides (HexCer) in a biological matrix extracted from the wild-type fungus *Eremothecium gossypii*. Quantitative values were extrapolated using internal standards for each group of molecules and calculated versus the initial amount of the mycelia of fungus obtained. A total of 42 species in the wild-type organism were detected, including the five sphingoid types and occasionally alpha-hydroxylation in the fatty acyl chain. Concentrations observed ranged between values close to the observed limit of detection, around one nanogram of the species per milligram of mycelium, and a few hundreds of ng per mg of mycelium.

Author Contributions: Conceptualization, J.L.R. and A.J.; methodology, J.L.R. and J.-F.M.-B.; formal analysis, J.-F.M.-B.; investigation, J.-F.M.-B., G.M.-F. and J.M.-G.; writing—J.-F.M.-B.; review and editing, J.L.R., A.J., G.M.-F. and J.M.-G.; project administration, J.L.R., A.J. and J.-F.M.-B.; funding acquisition, J.L.R., A.J. and J.-F.M.-B. All authors have read and agreed to the published version of the manuscript.

Funding: J.-F.M.-B. was supported by a post-doctoral contract (Juan de la Cierva FJC2019-041035-I) funded by the Spanish Ministerio de Ciencia e Innovación (MCIN/AEI/10.13039/501100011033). G.M.-F. was supported by a predoctoral contract (PRE2018-084931) from the Spanish Ministerio de Economía y Competitividad. J.M.-G. was recipient of USAL predoctoral fellowship from the University of Salamanca. This work was supported by the funding: grant PID2020-118200RB-I00 from the Spanish Ministerio de Ciencia e Innovación and "Programa propio C1" from University of Salamanca (18.K202/463AC01).

Acknowledgments: We thank Silvia Domínguez Barrios and Marta Santos López for excellent technical help and the development of metabolically engineered strains. We also thank NUCLEUS platform at University of Salamanca, specifically the Elemental analysis, Chromatography and Mass spectrometry Service, for the support.

Conflicts of Interest: The authors declare no conflict of interest.

References

1. Posse de Chaves, E.; Sipione, S. Sphingolipids and gangliosides of the nervous system in membrane function and dysfunction. *FEBS Lett.* **2010**, *584*, 1748–1759. [CrossRef] [PubMed]
2. Lopez, P.H.; Schnaar, R.L. Gangliosides in cell recognition and membrane protein regulation. *Curr. Opin. Struct. Biol.* **2009**, *19*, 549–557. [CrossRef] [PubMed]
3. Sandhoff, R.; Schulze, H.; Sandhoff, K. Ganglioside metabolism in health and disease. *Prog. Mol. Biol. Transl. Sci.* **2018**, *156*, 1–62. [PubMed]
4. Fazzari, M.; Lunghi, G.; Chiricozzi, E.; Mauri, L.; Sonnino, S. Gangliosides and the treatment of neurodegenerative diseases: A long italian tradition. *Biomedicines* **2022**, *10*, 363. [CrossRef] [PubMed]
5. Magistretti, P.J.; Geisler, F.H.; Schneider, J.S.; Andy Li, P.; Fiumelli, H.; Sipione, S. Gangliosides: Treatment avenues in neurodegenerative disease. *Front. Neurol.* **2019**, *10*, 859. [CrossRef]
6. Qamsari, E.S.; Nourazarian, A.; Bagheri, S.; Motallebnezhad, M. Ganglioside as a therapy target in various types of cancer. *Asian Pacific J. Cancer Prev.* **2016**, *17*, 1643–1647. [CrossRef]
7. Cutillo, G.; Saariaho, A.H.; Meri, S. Physiology of gangliosides and the role of antiganglioside antibodies in human diseases. *Cell. Mol. Immunol.* **2020**, *17*, 313–322. [CrossRef]
8. Skolnick, J.; Chou, C.; Miklavcic, J. Insights into novel infant milk formula bioactives. *Nutr. Diet. Suppl.* **2020**, *12*, 11–19. [CrossRef]
9. Gurnida, D.A.; Rowan, A.M.; Idjradinata, P.; Muchtadi, D.; Sekarwana, N. Association of complex lipids containing gangliosides with cognitive development of 6-month-old infants. *Early Hum. Dev.* **2012**, *88*, 595–601. [CrossRef]
10. Jiang, C.; Ge, J.; He, B.; Zeng, B. Glycosphingolipids in filamentous fungi: Biological roles and potential applications in cosmetics and health foods. *Front. Microbiol.* **2021**, *12*, 690211. [CrossRef]
11. Imamura, A.; Kiso, M. Chemical synthesis of gangliosides. *Methods Mol. Biol.* **2018**, *1804*, 293–310. [PubMed]
12. Schwarzmann, G. Labeled gangliosides: Their synthesis and use in biological studies. *FEBS Lett.* **2018**, *592*, 3992–4006. [CrossRef] [PubMed]
13. Yu, H.; Gadi, M.R.; Bai, Y.; Zhang, L.; Li, L.; Yin, J.; Wang, P.G.; Chen, X. Chemoenzymatic total synthesis of GM3 gangliosides containing different sialic acid forms and various fatty acyl chains. *J. Org. Chem.* **2021**, *86*, 8672–8682. [CrossRef] [PubMed]
14. Vilcaes, A.A.; Garbarino-Pico, E.; Daniotti, J.L.; Demichelis, V.T. Ganglioside synthesis by plasma membrane-associated sialyltransferase in macrophages. *Int. J. Mol. Sci.* **2020**, *21*, 1063. [CrossRef] [PubMed]
15. Merrill, A.H.; Wang, M.D.; Park, M.; Sullards, M.C. (Glyco)sphingolipidology: An amazing challenge and opportunity for systems biology. *Trends Biochem. Sci.* **2007**, *32*, 457–468. [CrossRef]
16. Barreto-Bergter, E.; Sassaki, G.L.; de Souza, L.M. Structural analysis of fungal cerebrosides. *Front. Microbiol.* **2011**, *2*, 239. [CrossRef]
17. Iwamori, M.; Shimomura, J.; Tsuyuhara, S.; Nagai, Y. Gangliosides of various rat tissues: Distribution of ganglio-N-tetraose-containing gangliosides and tissue-characteristic composition of gangliosides. *J. Biochem.* **1984**, *95*, 761–770. [CrossRef]
18. Guillou, H.; Zadravec, D.; Martin, P.G.P.; Jacobsson, A. The key roles of elongases and desaturases in mammalian fatty acid metabolism: Insights from transgenic mice. *Prog. Lipid Res.* **2010**, *49*, 186–199. [CrossRef]

19. Ordóñez, Y.F.; González, J.; Bedia, C.; Casas, J.; Abad, J.L.; Delgado, A.; Fabrias, G. 3-Ketosphinganine provokes the accumulation of dihydroshingolipids and induces autophagy in cancer cells. *Mol. Biosyst.* **2016**, *12*, 1166–1173. [CrossRef]
20. Ardail, D.; Popa, I.; Alcantara, K.; Pons, A.; Zanetta, J.P.; Louisot, P.; Thomas, L.; Portoukalian, J. Occurrence of ceramides and neutral glycolipids with unusual long-chain base composition in purified rat liver mitochondria. *FEBS Lett.* **2001**, *488*, 160–164. [CrossRef]
21. Proštenik, M.; Burcar, I.; Častek, A.; Ćosović, Č.; Golem, J.; Jandrić, Z.; Kljaić, K.; Ondrušek, V. Lipids of higher fungi. III. The fatty acids and 2-hydroxy fatty acids in some species of Basidiomycetes. *Chem. Phys. Lipids* **1978**, *22*, 97–103. [CrossRef]
22. Lagutin, K.; Mackenzie, A.; Bloor, S.; Scott, D.; Vyssotski, M. HPLC-MS, GC and NMR profiling of bioactive lipids of human milk and milk of dairy animals (cow, sheep, goat, buffalo, camel, red deer). *Separations* **2022**, *9*, 145. [CrossRef]
23. Sørensen, L.K. A liquid chromatography/tandem mass spectrometric approach for the determination of gangliosides GD3 and GM3 in bovine milk and infant formulae. *Rapid Commun. Mass Spectrom.* **2006**, *20*, 3625–3633. [CrossRef] [PubMed]
24. Lee, J.; Hwang, H.; Kim, S.; Hwang, J.; Yoon, J.; Yin, D.; Choi, S.I.; Kim, Y.H.; Kim, Y.S.; An, H.J. Comprehensive profiling of surface gangliosides extracted from various cell lines by LC-MS/MS. *Cells* **2019**, *8*, 1323. [CrossRef]
25. Park, J.Y.; Shrestha, S.A.; Cha, S. Isomer separation and analysis of amphiphilic polysialogangliosides using reversed-phase liquid chromatography-mass spectrometry. *J. Sep. Sci.* **2021**, *44*, 1824–1832. [CrossRef] [PubMed]
26. Barreto-Bergter, E.; Pinto, M.R.; Rodrigues, M.L. Structure and biological functions of fungal cerebrosides. *An. Acad. Bras. Cienc.* **2004**, *76*, 67–84. [CrossRef]

separations

Article

Phytochemical Content and Anticancer Activity of Jamaican *Dioscorea alata* cv. White Yam Extracts

Kenroy Wallace [1,2,*], Racquel Wright [3,4,*], Melisa Williams-Longmore [1], Sasha-Gay Wright [1,5] and Helen Asemota [1,3]

1 Department of Basic Medical Sciences, Biochemistry Section, Faculty of Medical Sciences, The University of the West Indies, Mona, Kingston, Jamaica; melisa_williams@live.com (M.W.-L.)
2 College of Natural and Applied Science Allied Health and Nursing, Northern Caribbean University, Mandeville, Jamaica
3 Biotechnology Centre, Faculty of Science & Technology, The University of the West Indies, Mona, Kingston, Jamaica
4 Faculty of Engineering and Applied Technology, Caribbean Maritime University, Kingston, Jamaica
5 Faculty of Engineering, The University of the West Indies, Mona, Kingston, Jamaica
* Correspondence: kenroywallace@hotmail.com (K.W.); racq.wright@gmail.com (R.W.)

Abstract: *Dioscorea* spp. is known for its myriad medicinal properties. *D. alata*, specifically crude extracts, have displayed potent anticancer properties. However, the chemical constituents of these extracts have not been examined. The aim of this study is to determine the chemical composition and antioxidant characteristics of the active extracts from *D. alata* tuber. Chemoinformatic profiling of the Jamaican *Dioscorea alata* cultivar white yam tuber was generated by a sequential Soxhlet extraction of dried milled tuber, producing five crude extracts: hexane (E-1), diethyl ether (E-2), acetone (E-3), ethanol (E-4) and water (E-5). The analytes within the five extracts were dissolved in 0.1% DMSO and their anticancer activity was determined using DU145 prostate cancer cells. Both the acetone and the ethanolic extract were able to induce greater than 50% cell death at 50 μg/mL. The order of growth inhibition of the extracts in DU-145 cell is E3 (IC_{50}, 10.81 μg/mL) > E-4 (IC_{50} 24.17 μg/mL) > E-1 (IC_{50} > 100 μg/mL) ≥ E-2 (IC_{50} > 100 μg/mL) ≥ E-5 (IC_{50} > 100 μg/mL). Phytochemical screening of both E-3 and E-4 revealed the presence of all major classes of secondary metabolites except tannins. Resins were also absent in the E-3 extract. Phenolic quantification indicated that E-3 and E-4 possessed GAEs of 31 ± 1.1 and 72 ± 1.8 mg per g of sample, respectively. Inversely, E-3 displayed greater antioxidant capability with IC_{50} of 82.9 μg/mL compared to E-4 (166.9 μg/mL); however, neither was comparable to citric acid (33.6 μg/mL). The extract E-3 was further isolated by HPLC into 11 fractions. Fractions 4 and 5 possessed potent cell growth inhibitory effects. GCMs of fractions 4 and 5 showed they possessed numerous saturated fatty acids with pharmacological relevance. The presence of these compounds shows potential for exploitation of *D. alata* extracts for pharmacological purposes.

Keywords: *Dioscorea alata*; anticancer; phytochemical; antioxidants

Citation: Wallace, K.; Wright, R.; Williams-Longmore, M.; Wright, S.-G.; Asemota, H. Phytochemical Content and Anticancer Activity of Jamaican *Dioscorea alata* cv. White Yam Extracts. *Separations* **2024**, *11*, 44. https://doi.org/10.3390/separations11020044

Academic Editor: Liangliang Liu

Received: 29 June 2023
Revised: 23 October 2023
Accepted: 24 October 2023
Published: 30 January 2024

1. Introduction

Yams are monocotyledonous plants belonging to the family Dioscoreaceae with the *Dioscorea* being the largest genus. They are known for their production of tubers, a storage organ composed mostly of starch. Yams are mainly found in tropical regions and are considered important staple crops in West Africa, Southeast Asia and the Caribbean [1,2]. Despite the large number of *Dioscorea* species, only about 11 species are cultivated or economically important, including the very popular *Dioscorea alata* or white yam [2].

Most commonly, yam is consumed boiled, baked, or fried [2]. Occasionally, yam is ingested in the form of raw yam, cooked soup, powder, or flour [2]. When compared to cassava or sweet potato, yam boasts a greater caloric value and contains relatively high levels of protein, vitamin C, and vitamin E [3]. Mucin, dioscin, dioscorin, allantoin, choline, polyphenols, diosgenin, and vitamins including carotenoids and tocopherols are only a few of the bioactive substances found in yam tubers [3] Mucilage of the yam tuber contains soluble glycoprotein and dietary fiber [3].

Dioscin, diosgenin, dioscorin, and anthocyanins are a few of the bioactive substances that have been linked to the anticancer effects of the *Dioscorea* species. Even though these compounds are prolific and anti-cancer—anti-proliferative, based on our results they are not the cytotoxic agents responsible for the antiproliferative activity in *D. alata* acetonic and ethanolic extracts.

These substances have been demonstrated to have strong cytotoxic effects against numerous cancer cell lines, including those from the breast, lungs, colon, and prostate, among others [4–6]. There are various cultivars of *D. alata* with the white yam cultivar (*D. alata cv* white yam), being very popular and widely eaten in Jamaica. Other cultivars of *D. alata* grown in Jamaica, such as the purple yam, commonly called St. Vincent, Dark Night, and Moonshine yam, have been shown to possess anticancer activities in vitro and in vivo [2]. The properties observed are thought to be due mainly to the presence of potent antioxidants, namely brightly coloured anthocyanins, which are responsible for the purple color in these yams [1]. It is possible that other cultivars may possess similar properties. We postulated that *D. alata* cv. white yam grown in Jamaica possesses anti-proliferative properties due to its phytochemical content. Studies have shown that many illnesses, including certain cancers and atherosclerosis, are linked to redox imbalance/oxidative stress [7,8]. Oxidative stress results from an imbalance in the reactive oxidative species (ROS) in cells. ROS are formed as by-products of metabolism but can also be generated from external sources. When there is a high concentration of ROS in cells, this may lead to uncontrolled cellular oxidation which can result in cellular damage and apoptosis. Conversely, studies have also indicated the stimulative effect of antioxidants in cancer cell growth [9].

Antioxidants are chemicals that are designed to reduce or neutralize ROS. Antioxidants are generated internally in the body; however, antioxidant sources are not limited only to compounds produced in the body as they can be obtained exogenously [10,11]. There has therefore been an effort to extract antioxidants from plants to determine their benefits. Yams possess a variety of compounds including saponins, polyphenols and vitamins; these components are thought to contribute to their bioactivity in some efficacy studies. The literature shows that these components have displayed antioxidant properties in other plants.

The effects of antioxidants on cancer cells are complex and can depend on several factors, such as the type of antioxidant, the type of cancer, and the stage of the cancer. Some studies have suggested that antioxidants may be beneficial in preventing cancer development and reducing cancer-related mortality. For example, a meta-analysis of randomized controlled trials found that supplementation with beta-carotene, vitamin C, and vitamin E was associated with a decreased risk of cancer incidence and mortality [12]. Additionally, several preclinical studies have suggested that certain antioxidants, such as curcumin and resveratrol, have anticancer properties and can inhibit the growth and metastasis of cancer cells [13,14].

Some studies have demonstrated that antioxidants can stimulate the growth of cancer cells in vitro and in vivo, as well as promote metastasis [15]. One potential mechanism by which antioxidants may promote cancer growth is through their ability to reduce oxidative stress in cancer cells, which may make cancer cells less vulnerable to chemotherapy and radiation therapy. These treatment options rely on the generation of ROS to induce cell death [16].

Studies have shown that yam plants possess antioxidant properties particularly in brightly colored cultivars such as the St. Vincent yam. There has been limited research on the antioxidant content of yams with white coloration such as the white yam. This is the first research known to the authors which focuses on the anticancer properties of yams with white pigmentation in Jamaica. This study therefore seeks to screen for phytochemical components present in the *D. alata*, cv white yam and identify the possible compounds contributing to the anticancer properties observed. The result of this study will provide the foundations for further investigation into the chemical compositions of the medicinal extracts of *D. alata*.

2. Materials and Methods

2.1. Collection of Dioscorea Alata Tuber Samples

Dioscora alata cv white yam tubers were gathered in the Jamaican parish of St. Catherine and shipped to the University of the West Indies' Biotechnology Center in Mona, Jamaica. The tubers underwent a 6-h sun-drying process after a tap-water wash. After being dried at 50 °C to a constant weight, the tubers were ground into a fine powder and kept in sample bottles at room temperature.

2.2. Preparation of Extracts

The milled sample (1.2 kg) was placed in a large thimble constructed with steel mesh wire and filter paper. A continuous extraction was performed via Soxhlet extraction using 4 L of 100% hexane, diethyl ether, acetone, ethanol and water sequentially. Each solvent was used under reflux for 1 week before removal and subsequent addition of the next solvent. Each extract was concentrated using a rotary evaporator at 120 rpm at the solvents' melting point. The water extract was vacuum filtered and dried using a lyophilizer. The five extracts were dried and stored at room temperature.

2.3. Phytochemical Analysis

The presence of the following phytochemicals was assessed using the acetone and ethanolic extracts: Phenols, alkaloids, tannins, sterols, glycosides, flavonoids, saponins, terpenoids, carboxylic acid and resins. Presence or absence was indicated by a plus (+) or minus (−) sign, respectively. Strong reactivity (that is, a large phytochemical content) was indicated by two pluses (++).

2.3.1. Test for Phenols

The E-3 and E-4 solutions were treated with five drops of 5% (*w*/*v*) glacial acetic acid and 5% (*w*/*v*) sodium nitrite solution. Muddy brown precipitate developed in the test tube; hence, the presence of phenols was confirmed [17,18].

2.3.2. Test for Alkaloids

A small quantity of extract was warmed in 2% hydrochloric acid for 2 min. The mixture was filtered, and a few drops of Mayer's reagent and Wagner's reagent were added separately. The Mayer's reagent produced creamy-white colored precipitation giving a positive result. The Wagner's reagent produced a reddish-brown precipitate, which indicated the presence of alkaloids [17,18].

2.3.3. Test for Tannins

A small quantity of extract was boiled in 5 mL of 45% ethanol solution. The mixture was then cooled and filtered. Two drops of ferric chloride ($FeCl_3$) were added to 1 mL of filtrate. A greenish black or dark blue color indicated the presence of tannins [17–19].

2.3.4. Test for Glycosides

Extract (2 mL) was mixed with glacial acetic acid (1 mL), one drop of ferric chloride and one drop of concentrated sulphuric acid. The presence of a reddish-brown color at the junction of two layers with the upper layer having a bluish green color would confirm the presence of glycosides [18].

2.3.5. Test for Flavonoids

Extracts (10 mg) were boiled in 5 mL of ethyl acetate for 3 min. The mixture was filtered and the filtrate was mixed in 1 mL of diluted ammonium chloride solution (1%) by shaking. The layers were allowed to separate. A yellow coloration indicated the presence of flavonoids [17,20].

2.3.6. Test for Sterols

Extract (2 mL) was mixed with 2 mL of chloroform. To the mixture, 2 mL of acetic anhydride was added followed by two drops of concentrated sulphuric acid to the mixture via the side of the test tube. If sterol was present, there would be a red, then blue and finally green colour change [21].

2.3.7. Test for Saponins

Extract (0.2 mg) was dissolved in boiling distilled water. The extract was allowed to cool. The extract was then shaken vigorously and the formation of froth indicated the presence of saponin. The procedure was repeated using 5 g of dried yam material. Following the boiling in distilled water, the extract was allowed to cool and the filtrate was collected. The filtrate was then shaken vigorously for the formation of froth. Formation of steady froth indicated the presence of saponins [18,22,23].

2.3.8. Test for Terpenoids

Extract (2 mL) was mixed with 2 mL of chloroform. To this mixture 3 mL of concentrated H_2SO_4 was added and a layer of reddish-brown coloration was formed at the interface, thus indicating the presence of terpenoids [17,19]

2.3.9. Test for Resins

Acetonic extract (1 mL) was treated with few drops of acetic anhydride. Sulfuric acid (2 mL) was added to the mixture. A color change to yellow orange was observed; this indicated the presence of resins [24].

2.3.10. Test for Carboxylic Acid

Acetonic extract (0.5 mL) of extract was treated separately with 1 mL of saturated sodium bicarbonate solution (0.3 g solid to 4.5 mL distilled water). The liberation of carbon dioxide (visualized by effervescence or bubbling in the sample) confirmed the presence of carboxylic acid [24,25].

2.4. *Cell Culture of DU145 Cells*

All cell lines used in this study were obtained from the American Type Culture Collection (ATCC; Rockville, MD, USA). DU145 prostate cancer cells were cultured at the Environmental Toxicology Department at Southern University, Baton Rouge in an RPMI-1640 medium. Media were supplemented with 10% (v/v) heat inactivated fetal bovine serum (50 mL) and 1% (v/v) penicillin-streptomycin (5 mL). Cells were harvested to perform various assays once 80–85% confluency was attained. Cells were incubated at 37 °C with 5% CO_2.

2.5. *Cell Viability of DU145 Cells*

Cells were seeded in 96-well plates at 5×10^3 cell/well and allowed to grow for 24 h. Then, extracts were applied to the cells in five separate twofold dilutions, ranging from 100 g/mL to 6.5 g/mL. Each extract's final dimethyl sulfoxide (DMSO) content in the growing medium did not go above 0.5% (v/v). Cell viability and cytotoxicity were measured using 10% (v/v) CellTiter 96® aqueous one solution reagent after cells had been exposed to the extract for 24 h. The absorbance was measured at 490 nm with an ELX800UV universal microplate reader following a 60-min incubation with (4,5-dimethyl-2-yl)-5-(3-

carboxymethoxyphenyl)-2-(4-sulfophenyl)-2H-tetrazolium. A GraphPad Prism 8 was used to analyze the findings [26].

2.6. Antioxidant Capacity of Extracts Using 2,2-Diphenyl-1-picrylhydrazyl (DPPH)

Standards and extracts were diluted in methanol to the proper quantities in order to determine each compound's antioxidant capability within a linear range. Totals of 100, 200, 400, 800, and 1000 μg/mL concentrations of the extracts were used. The ascorbic acid standard was made daily and diluted to concentrations of 2.5, 5, 10, 20, 40, 80, 160, and 320 μg/mL. A quantity of 0.1 mM DPPH in methanol was also utilized. The DPPH (1000 μL each) was reacted with the controls and extracts (200 μL each) in test tubes for 30 min in the dark. Additionally, a blank was created and incubated alongside the samples using methanol in place of the sample. Following incubation, the samples' absorbance was measured at 517 nm using a Unico scanning spectrophotometer (Unico, Dayton, NJ, USA) [27].

The mean optical density of the sample was utilized to determine the DPPH reduction (inhibition), representing the percentage of DPPH that was neutralized by the antioxidants present in the added samples.

The equation below was used to calculate this percentage:

$$\frac{A_o - A_s}{A_o} \times 100 \quad (1)$$

A_O: absorbance blank
A_S: sample absorbance

The findings were organized into a table and illustrated through graphs. The IC_{50} values were determined based on the DPPH reduction percentages and represent the concentration of the sample (in μg/mL) needed to neutralize 50% of the DPPH in the assay.

2.7. Polyphenol Quantification

Standards containing 1 μg, 2 μg, 4 μg, 6 μg, 8 μg and 10 μg of gallic acid dissolved in a volume of 200 μL were prepared. To each well, 20 μL of standard was added in quadruplicates. Some 80 μL of 7% calcium carbonate was added to each well containing 20 μL of standards and unknown samples. The Folin–Ciocalteu reagent was diluted 1:10 in deionized water and 100 μL was added to each well. The absorbance was measured at 700 nm after the addition of the Folin–Ciocalteu reagent. The total phenolic content of the extract was expressed as mg gallic acid equivalents (GAE) per gram of sample in dry weight (mg/g) [28].

The total phenolic content was calculated according to the equation:

$$C = \frac{cV}{m} \quad (2)$$

C: Total phenolic content mg GAE/g dry extract
c: Concentration of gallic acid derived from calibration curve in mg/mL
V: Volume of extract in mL
m: Mass of extract in grams

2.8. High Performance Liquid Chromatography (HPLC)

The E-3 extract was purified prior to HPLC separation using solid phase extraction possessing a carbon 18 column (C18-E, 55 μm, 70A, 8B-S001-LEG). Methanol (10%) was used to wash the column; then, 25%, 75%, and 100% methanol was used to elute it. The samples were subsequently dried using a combination of rotary evaporation and straightforward distillation. Before being injected into the HPLC apparatus, the 100% methanol elute was dissolved in 20% DMSO and filtered through a 0.25 μm filter disc.

The E-3 extract was prepared by performing a 1:100 dilution of E-3 stock solution (1.46×10^5 µg/mL); therefore, the sample injected had a concentration of 1.46×10^3 µg/mL The experiment was carried out using an Agilent HPLC 1200 series and Chemstation for LC systems Rev. B.04.03 SP2 with a diode array detector. The method employed a Phenonmenex-250 × 21.20 nm, 10 u, C18(2), 100 Å column. The chromatography was monitored at 210 nm, 230 nm, and 280 nm. The mobile phase contained two solvents (A: distilled deionized water; B: acetonitrile); the gradient profile is described in Table 1 below.

Table 1. Gradient profile method of HPLC separation of E-3.

Time (min)	Flow Rate (mL/min)	Deionized Water (%) A	Acetonitrile (%) B
0	1	75	25
5	1	50	50
10	1	25	75
20	1.5	0	100

Sample separation for 30 min. Column washed and regenerated with 95% acetonitrile (5 min, 1 mL/min) followed by 100% acetonitrile (5 min, 1.5 mL/min).

The fractions collection was conducted according to the table below (Table 2).

Table 2. Fraction collection of HPLC separation for E3.

Fractions	Time (min)	Fractions	Time (min)
0	3	6	2
1	3.5	7	2
2	3.5	8	2
3	2	9	2
4	2	10	2
5	2	11	5

Fraction collection was initiated at 3 min into the HPLC run. The Fr. 1 and Fr. 2 fractions were collected within a time interval of 3.5 min. The following eight fractions had a time interval of 2 min. The final fraction had a time interval of 5 min. The fractionation protocol was repeated 13 times. Fractions were dried using a vacuum concentrator and dissolved in 20% DMSO. DU145 cells were treated with 10 µL of each fraction and their cell growth inhibition activity was assessed using an MTT assay.

2.9. Gas Chromatography Mass Spectrometry

The analysis was conducted using an electron impact mode at 70 eV. The splitless mode was employed with an injection purge duration of 0.75 min. The injector temperature was set to 250 °C, while the transfer line temperature was maintained at 280 °C. The column used was a DB-1701 with dimensions of 30 m length, 0.25 mm inner diameter, and 0.25 mm film thickness. A ramped temperature program was employed beginning at 80 °C and held for 2 min, followed by an increase of 20 °C per minute for a duration of 10 min. Subsequently, the temperature was held for an additional 10 min. A solvent delay of 3 min was implemented, and helium was used as the carrier gas at a flow rate of 1.2 mL min^{-1}. The operations were conducted in a Scan mode.

3. Results

Natural products commonly found in our diet present promising possibilities for identifying potent anticancer compounds. *Dioscorea* spp. have been extensively studied by various researchers and have been recognized for their diverse nutritional and medicinal properties [29,30].

3.1. Dioscorea alata Solvent Extraction

D. alata contains a wide array of active and non-active compounds distributed throughout its different parts. To extract these compounds, solvents with varying polarities were utilized. The data obtained from these extractions allowed us to create a fingerprint of the bioactive analytes present in *D. alata*. For the extraction process, *D. alata* (18.802 g) from Jamaica, W. I., was sequentially extracted using hexane (E-1), diethyl ether (E-2), acetone (E-3), ethanol (E-4), and water (E-5) through Soxhlet extraction for 24 h at their respective boiling points. Subsequently, a combination of simple distillation and rotary evaporation was employed to remove the solvents, and the gravimetric method was used to determine the dry weights of the extracted materials. Solid–liquid extraction of *D. alata* with water for 24 h resulted in the largest number of extractable analytes (5.12% 0.96/18 g). We obtained (1.68% 0.32 g/18 g), (1% 0.19 g/18 g) and (0.29% 0.05/18 g) for hexane (E-1), diethyl ether (E-2), acetone (E-3), ethanol (E-4) or water (E-5), respectively. The initial analytical extraction was used to conduct a cytotoxicity screen using DU145 cells (Figure 1). The screen showed that extracts E-3 and E-4 displayed the lowest cell viability in DU145 cells when comparing all the extracts.

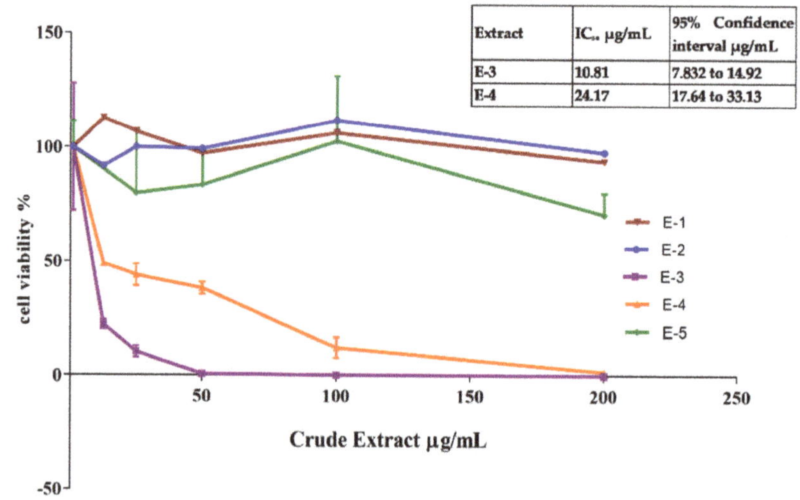

Figure 1. Cytotoxic effect of Jamaican Dioscorea alata crude extracts on DU145 cells. E−1—Hexane, E−2—diethyl ether, E−3 acetone, E−4 ethanol and E−5 water extract. Wells were seeded with 10,000 cells and were treated with crude extracts. Cells were harvested after 24 h and their viability was measured using a CellTiter 96® non-radioactive cell proliferation assay (MTS). Assays were performed in quadruplicate.

3.2. Preliminary Phytochemical Assessment of Acetonic and Ethanoic Extracts of Dioscorea alata

White yams have been attributed various medicinal properties owing to the existence of specific secondary metabolites in the tuber [31]. The bioactive compounds found in plants are secondary metabolites that can elicit pharmacological or toxicological effects in humans and animals. The Jamaican *Dioscorea alata* extracts were subjected to phytochemical evaluation. Various plant constituents present were identified in the acetone and ethanol extract using both qualitative and quantitative methods. The qualitative phytochemical assessment revealed the presence of various classes of secondary metabolites present within the acetonic and ethanolic extract (Table 3). Both extracts tested positive for the presence of phenols, alkaloids, sterols, alkaloids, glycosides, terpenoids, and phytosteroids. Both extracts tested negative for tannins. The ethanolic extract tested positive for resins; however, the acetonic extract tested negative. Both extracts possessed compounds that are known to exhibit antioxidant properties.

Table 3. Phytochemical Screening of E-3 and E-4 Extract from *Dioscorea alata*.

Phytochemicals	Acetone Extract (E-3)	Ethanol Extract (E-4)
Phenols	++	++
Alkaloids	+	+
Tannins	−	−
Sterols	+	++
Glycosides	++	++
Flavonoids	+	+
Saponins	++	+
Terpenoids	+	++
Carboxylic acid	++	+
Resins	−	+
Steroids	++	++

(+) indicates present, (++) indicates strongly present, (−) indicates absence.

To corroborate the phytochemical screen, the total phenol content of E-3 and E-4 extracts was determined. The Folin–Ciocalteu assay method, with a Gallic acid standard curve, was used to determine the total phenolic content of the extract. The content was expressed as milligrams of Gallic acid equivalents (GAE) per gram of extract. Table 4 reveals that the total phenolic content in E-4 was approximately twice as high as E-3, with values of 72 ± 1.8 mg GAE per g and 31 ± 1.1 mg GAE per g, respectively.

Table 4. Total Phenol Content in Acetonic and Ethanolic extracts of Jamaican *Dioscorea alata*.

Extracts	Gallic Acid Equivalents (GAE) mg/g
E-3	31 ± 1.1
E-4	72 ± 1.8

3.3. *Dioscorea alata Acetonic and Ethanolic Extracts Possess Antioxidant Properties*

Free radical scavenging ability when assessed showed that E-3 was more potent than E-4 (Figure 2). The order of increasing antioxidant potency of the yam extract was E-3, IC_{50} of 83 μg/mL, then E-4 with an IC_{50} value of 169.4 μg/mL (Table 5). It was observed that the DPPH reagent was reduced with increased concentration of both extracts. Ascorbic acid displayed more than double the potency of acetonic extract with an IC_{50} value of 33.6 ± 0.1 μg/mL and was five times more potent than that of E-4, an IC_{50} value of 167 μg/mL. The data highlighted the likelihood that the protective properties of *Dioscorea alata* tuber may not be due to its free radical scavenging abilities.

Table 5. IC_{50} Values of Free radical Scavenging Properties of *Dioscorea alata* Acetonic and Ethanolic extracts.

Sample	IC_{50} (μg/mL)	95% Confidence Interval
Ascorbic Acid	33.6	31.98 to 35.34
E-3	82.9	70.56 to 93.95
E-4	166.9	140.4 to 204.3

The E-3 extract displayed greater cytotoxic activity towards DU145 cells as well as more favorable antioxidant capabilities. Further investigations focused on identifying the compounds present within the E-3 extract. The E-3 extract was subjected to solid phase extraction. It was washed with 10% acetone and eluted with 25, 75 and 100% methanol (MeOH). The 100% MeOH elute, in a single dose assessment, inhibited 97.9% cell growth of the DU145 cells as compared to the crude extract's 98.5% cell growth inhibition. The resulting sample was further separated into 10 fractions via high performance liquid chromatography (HPLC) (Figure 3). The HPLC protocol was developed using the time fractionation method.

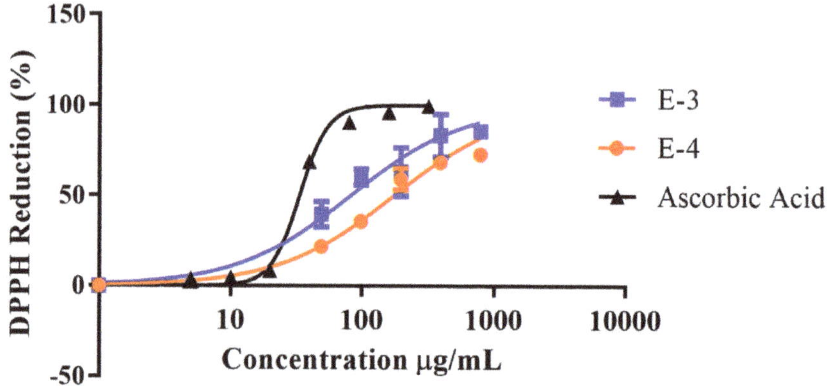

Figure 2. Antioxidant screening of bioactive acetonic and ethanolic extract. In vitro antioxidant screening of E-3 and E-4, using vitamin C (ascorbic acid) as a positive control. Assay was done in triplicate.

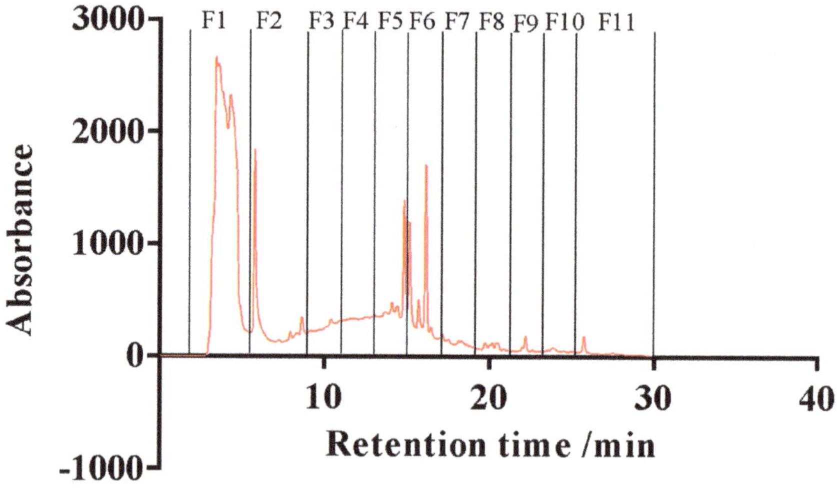

Figure 3. High performance liquid chromatography (HPLC) and fractionation design for E-3 extract gradient run of sample with detection using 240 nm HPLC fractionation design. Major peaks detected using time split.

Major peaks were collected in different fractions; this was enhanced by using two detection wavelengths, 210 nm and 230 nm. The protocol was designed using three different time intervals. Three time intervals were used; fractionation was initiated at a 2 min retention time. The first two fractions were collected within a time interval of 3.5 min. This was used to remove the major peak observed within the retention time of 3.4 min and 8.7 min detected using 230 nm and 210 nm. The following eight subsequent fractions had a time interval of 2 min. The final fraction had a time interval of 5 min, which collected all the smaller peaks at the end of the chromatogram.

Each HPLC fraction was concentrated and evaluated for proliferative activity using an MTT viability assay in DU145 cancer cells (Figure 4). Six fractions displayed positive inhibitory activity against DU145 cells; however, fraction 4 (Fr4) and fraction 5 (Fr5) possessed the greatest inhibitory effect at 92% and 93% growth inhibition. This was comparable to the load and docetaxel (positive control), which had a growth inhibitory activity of 94% and 97%, respectively. Both Fr4 and Fr5 were further assessed for their chemical composition using gas chromatography mass spectrometry (GC-MS).

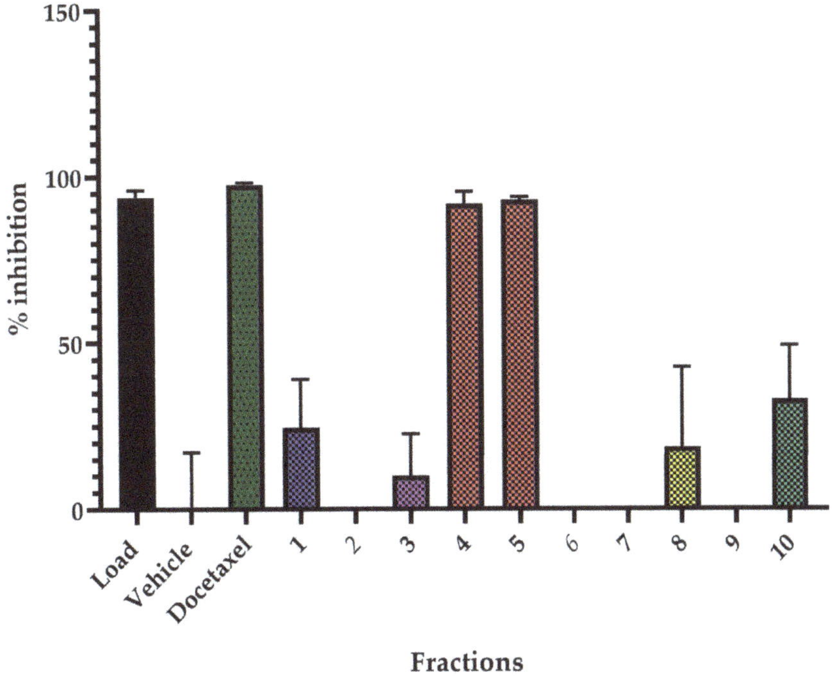

Figure 4. Single dose assessment of anticancer properties of HPLC fractions on DU145 cells. Fractions were assayed using DU145 cells with 5000 cell/well. Cells were harvested after 24 h and their viability was measured using a CellTiter 96® non-radioactive cell proliferation assay (MTT). Numbers 1–10 refer to the fractions from the HPLC.

The compounds identified in Fraction 4 (Fr4), and Fraction 5 (Fr5) are listed in Tables 5 and 6. Figure 5 highlights the gas chromatogram of the extract. The compounds were identified through the NIST08 L database as listed in Tables 5 and 6. The major compound identified by GC-MS analysis was azelaic acid RT 9.85 min in Fr4 and Fr5. However, the best matches from the database were hexadecanoic acid at RT 10.81 minutes in Fraction 4 with an abundance of 6.73% and RT 10.81 min with an abundance of 8.58% in Fraction 5. Other components also identified in the fractions are highlighted in Tables 6 and 7. Octanoic acid was present in both fractions but was more abundant in Fraction 4. Fractions 4 and 5 possessed the highest azelaic acid content. They also contained nonanoic acid RT 6.96 min, dodecanoic acid RT 8.79 min, hexadecanoic acid RT 10.81 min, and octadecanoic acid RT 9.33 min in relatively great abundance. Additionally, 1,2-Bis(trimethylsiloxy)ethane RT 5.24 min was identified in large quantities in Fraction 4.

Table 6. GC-MS analysis of the HPLC Fraction 4 of the E3 Extract of *Dioscorea alata*.

Retention Time (min)	Abundance %	Name of Compounds	Molecular Mass g/mol
5.24	9.59	1,2-Bis(trimethylsiloxy)ethane	206.43
6.26	2.8	Octanoic acid, trimethylsilyl ester	216.39
6.91	3.62	Butanedioic acid, bis(trimethylsilyl) ester	262.45
6.96	6.37	Nonanoic acid, trimethylsilyl ester	230.42
7.55	8.57	4H-1-Benzopyran-4-one, 5,7-dihydroxy-2-methyl-	192.17
8.79	5.82	Dodecanoic acid, trimethylsilyl ester	272.5
8.96	2.71	Benzoic acid, 4-[(trimethylsilyl)oxy]-, trimethylsilyl ester	282.48
9.85	16.18	Azelaic acid, bis(trimethylsilyl)ester	332.58
10.81	6.73	Hexadecanoic acid, trimethylsilylester	328.6
11.7	1.9	Octadecanoic acid, trimethylsilylester	356.7
12.31	2.74	Dodecanamide	199.33
11.7	1.9	Heptadecanoic acid, trimethylsilylester	342.6

Table 7. GC-MS analysis of the HPLC Fraction 5 of E3 Extract of *Dioscorea alata*.

Retention Time (min)	Abundance %	Name of Compounds	Molecular g/mol
4.76	3.34	Propanoic acid, 2-(methoxyimino)-,trimethylsilyl ester	189.28
5.25	3.24	Ethanedioic acid, bis(trimethylsilyl) ester	234.4
5.34	1.22	Dithioerythritol, tetrakis(trimethylsilyl)-	443
5.42	1.45	Butanoic acid, 2-[(trimethylsilyl)amino]-, trimethylsilyl ester	247.48
6.14	2.61	Trimethylsilyl ether of glycerol	308.64
6.26	0.92	Octanoic acid, trimethylsilyl este	216.39
6.91	4.96	Butanedioic acid, bis(trimethylsilyl) ester	262.45
6.96	3.04	Nonanoic acid, trimethylsilyl este	230.42
7.55	3.54	Anthracene, 1-methyl-	192.25
8.79	3.3	Dodecanoic acid, trimethylsilyl ester	272.5
8.96	1.73	Benzoic acid, 4-[(trimethylsilyl)oxy]-, trimethylsilyl ester	282.48
9.33	1.75	Octanedioic acid, bis(trimethylsilyl) ester	318.56
9.85	9.74	Azelaic acid, bis(trimethylsily) ester	332.58
9.96	1.55	Arabinofuranose, 1,2,3,5-tetrakis-O-(trimethylsilyl)-	150.13
10.21	0.98	Trimethylsilyl ether of glycerol	308.64
10.34	1.38	.alpha.-D-Glucopyranoside, methyl2,3,4,6-tetrakis-O-(trimethylsilyl)-	194.18
10.81	8.58	Hexadecanoic acid, trimethylsilylester	328.6
11.38	1.05	Dodecanamide	199.33
11.64	0.89	Oleic acid	282.5
11.7	4.03	Octadecanoic acid, trimethylsilylester	356.7
12.23	3.39	.beta.-D-Galactopyranose, 1,2,3,4,6-pentakis-O-(trimethylsilyl)-	541.1
12.3	3.54	Benzeneethanamine, 2-fluoro-.beta.,3,4-trihydroxy-N-isopropyl-	229.25
14.29	1.42	1H+R6:R27ido [1,2-a]quinoline-2-carboxylic acid, 5,6-dihydro-1-oxo-, methyl ester	256.26

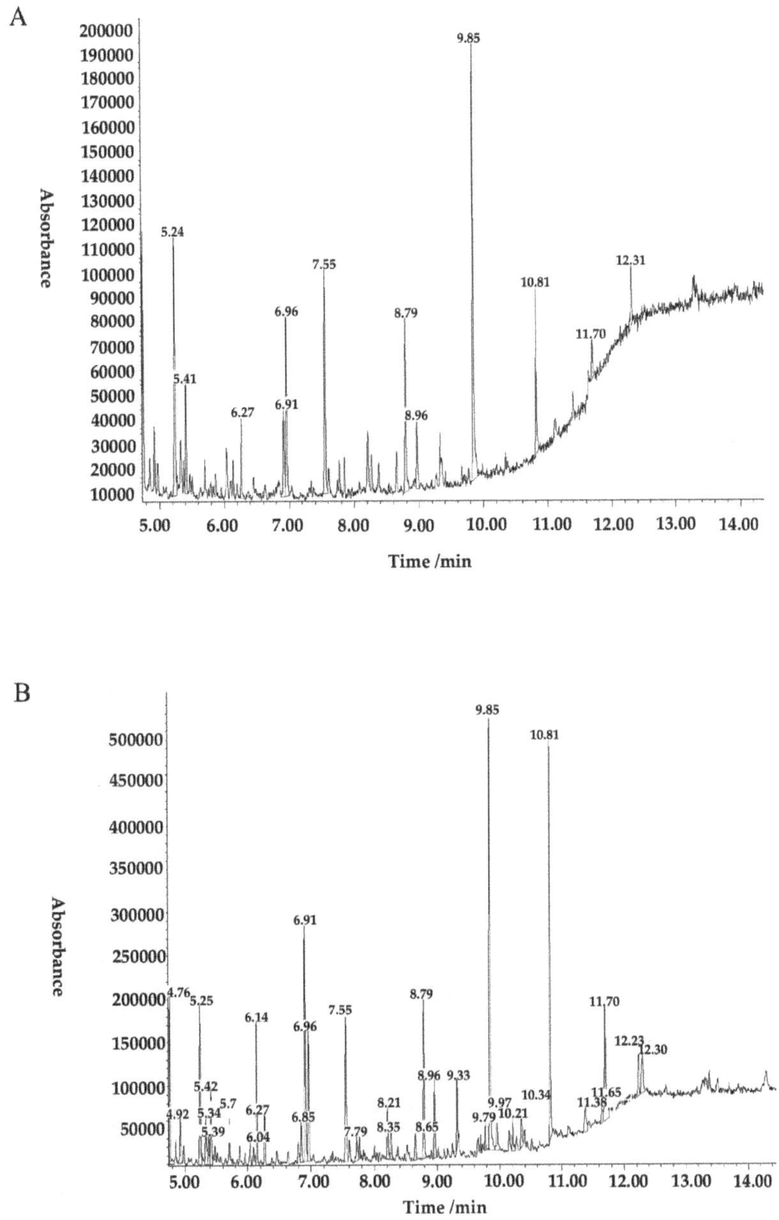

Figure 5. Chromatogram of gas chromatography for Fraction 4 (**A**) and Fraction 5 (**B**).

4. Discussion

D. alata is a widely consumed staple crop in Jamaica and the Caribbean; however, there are limited studies locally of its medicinal properties, particularly its antiproliferative capacity. Previous studies have shown the anti-proliferative capacity of the purple yam (*D alata* cultivar) but not of the generally consumed water yam. This study sought to evaluate the antiproliferative potential of components form fibers of locally grown *D. alata* cv water yams [26]. Sequential continuous extraction using solvents of increasing polarities allowed for separation of compounds in the yam tubers. This preliminary level of extraction has proven highly beneficial over the years (in natural compound isolation) because it allows for an initial separation of components for further analysis. The resulting extracts were tested for their cell inhibition activity against DU-145, an androgen-insensitive prostate cancer cell line. Androgen-insensitive prostate cells are the most difficult to test and cytotoxic compounds towards this cellular model would prove extremely beneficial. Extracts E3 and E4 were the most promising extracts as seen in Figure 1. These extracts displayed potent toxicity against androgen-insensitive prostate cancer cells (DU145) producing IC_{50} values of 10.81 μg/mL by E-3 and 24.17 μg/mL by E-4.

Despite the higher phenol content in E-4, E-3 (IC_{50} of 82.9 μg/mL) displayed superior antioxidant activity to E-4 (IC_{50} of 166.9 μg/mL). The use of solvents with varying polarity in the solid–liquid extraction allowed for the solubilizing of cytotoxic compounds using acetone (E-3) and ethanol (E-4). The extracts when screened for their phytochemical content had several phytochemicals of interest. Given the polarity of the extraction solvents used, many phytochemicals were polar in nature including phenolic compounds such as flavonoids, phenols, and tannins. The general classes of phytochemicals identified in our extracts, as seen in Table 2, have also been identified in various *Dioscorea* tuber species and cultivars from various locations [32,33]. Terpenoids, alkaloids, sterols, steroids, and glycosides were also present in the acetonic (E3) and ethanolic (E4) extracts [26,34]. Qualitative analysis additionally indicated a strong presence of carboxylic acid, which can be associated with high fatty acid content present within the tuber [35]. Phenolic compounds such as flavonoids have been implicated as antiproliferative agents. They are also known as antioxidants as they are able to scavenge free radicals [36]. The antioxidant capacity of the extracts E3 and E4 were evaluated as well as their phenol content. The total phenol content of E-3 and E-4 extracts was expressed in Gallic acid equivalents (GAE) and showed that E-4 possessed greater phenol content (72 ± 1.8 mg per g of sample) than E-3 (31 ± 1.1 mg per gram of sample). Even though the phenol and Gallic acid content of E4 was greater than E3, the antioxidant capacity was greater than that of E3. This may indicate that the phenolics present in the extracts may not be responsible for the antiproliferative activity displayed by E-3 and E-4 extracts.

Extract E3 (acetonic extract) was the most potent cytotoxic extract, as seen in Figure 1; and was subjected to SPE and HPLC fractionation and subsequently to cytotoxic testing. A total of 11 fractions was fractionated and evaluated for cellular inhibitory activity with Fraction 4 (Fr4) and Fraction 5 (Fr5) showing the greatest potency (over 90% inhibition).

The lack of discernible peaks in the HPLC chromatograms of Fr4 and Fr5 (Figure 3) suggests the potential presence of organic fatty acids. Due to the absence of significant chromophores, these fatty acids cannot be detected at UV wavelengths [37]. Therefore, to determine the chemical content of the fractions, GCMS was used. Fr4 and Fr5 were treated with BSFTA to increase the volatility of the samples by substituting –NH, –OH, –COOH, and –SH bonds with a dimethysilyl group. This derivatization reduced hydrogen bonding, which further reduced the polarity, increased the thermal stability and improved separation and detection by the GCMS system [38]. The major compounds identified are summarized in Tables 5 and 6 and the chromatograms are presented in Figure 5. The derivatives of some of the most abundant organic acid compounds in Fr4 and Fr5 were identified as octanoic acid (caprylic acid), propanedioic acid (malonic acid), butanedioic acid (succinic acid), nonanoic acid (pelargonic acid), dodecanoic acid (lauric acid), benzoic acid, hexadecenoic acid (palmitoleic acid), octadecanoic acid (stearic acid), azelaic acid and heptadecanoic acid (margaric acid).

Therapeutic relevant hexadecenoic acid has been shown to display a cell growth inhibitory effect against prostate cancer in vitro and in vivo [26,34]. Similar responses were observed with the treatment of E-3 extract against androgen-insensitive DU145 cells. Other relevant compounds include azelaic acid, ocatanoic, and octadecanoic acid, have been shown to display anticancer activity against leukemia, colon cancer, neuroblastoma human skin epidermoid (A-431), and human mammary gland adenocarcinoma cells [39–42]. Lauric acid prompts the demise of colon cancer cells by means of the downgrading of the epidermal growth factor receptor by 1.33- and 1.58-fold, leading to cell death [43].

The presence of these compounds suggests that the antiproliferative effects of the extract are possibly due to the individual or synergistic effects of these compounds.

Previous studies have shown the anticancer properties of dioscin, diosgenin, dioscorin, and anthocyanin. However, our research indicates that these compounds were not the primary agents responsible for the observed anticancer activity in Fractions 4 and 5. This finding supports the presence of other anticancer compounds in the *D. alata* tuber.

The diversity of bioactive compounds present in the *D. alata* tuber is an indicator of its potential use as a source of natural anticancer agents. Other studies indicate that these therapeutic relevant compounds are not localized to the tuber but are also found in the leaves and stems of *D. alata* plant [44]. Future studies will assess the effectiveness of the identified compounds on an in vivo prostate cancer model.

5. Conclusions

There is an ongoing demand for the development of alternative cancer treatment methods sourced from natural origins. Through the conducted experiments, it was discovered that compounds found in the *D. alata* tuber exhibit potent growth inhibitory effects against androgen-insensitive prostate cancer cells. Phytochemical screening successfully identified the chemical constituents of the *D. alata* tuber; however, no significant antioxidant properties were detected. The isolation of the bioactive compounds using techniques such as solid-phase extraction (SPE), high-performance liquid chromatography (HPLC), and gas chromatography-mass spectrometry (GC-MS) generated a chemical fingerprint that unveiled the presence of bioactive compounds within the *D. alata* tuber. These findings point towards a potential synergistic relationship between these compounds, contributing to the observed pharmacological properties.

Author Contributions: Conceptualization, K.W. and H.A.; methodology, K.W. and R.W.; software, K.W. and R.W.; validation, S.-G.W. and K.W.; formal analysis, K.W.; investigation, K.W., R.W. and S.-G.W.; resources K.W., H.A., M.W.-L. and S.-G.W.; data curation, K.W.; writing—original draft preparation, K.W.; writing—review and editing, S.-G.W., R.W. and M.W; visualization, KW., S.-G.W. and M.W.-L.; supervision, H.A.; project administration, K.W. and H.A.; funding acquisition, K.W. and H.A. All authors have read and agreed to the published version of the manuscript.

Funding: This research was funded by the University of the West Indies, Mona, Research and Publication grant.

Data Availability Statement: Data is contained within the article.

Acknowledgments: The authors express their gratitude to Wesley Gray from Southern University Baton Rouge for facilitating the cell culture aspect of this research and providing valuable guidance. Additionally, the authors would like to acknowledge the Office of Graduate Studies and Research and the Department of Basic Medical Sciences at the University of the West Indies for their funding support, which facilitated other components of this research.

Conflicts of Interest: The authors declare no conflicts of interest.

References

1. UWI Mona Discovers Anti Cancer Properties in Yams | Marketing and Communications Office, The University of West Indies at Mona. Available online: https://www.mona.uwi.edu/marcom/newsroom/entry/5775 (accessed on 6 January 2022).
2. Lebot, V.; Lawac, F.; Legendre, L. The Greater Yam (*Dioscorea Alata* L.): A Review of Its Phytochemical Content and Potential for Processed Products and Biofortification. *J. Food Compos. Anal.* **2023**, *115*, 104987. [CrossRef]
3. Bhandari, M.R.; Kasai, T.; Kawabata, J. Nutritional Evaluation of Wild Yam (*Dioscorea* Spp.) Tubers of Nepal. *Food Chem.* **2003**, *82*, 619–623. [CrossRef]
4. Srinivasan, S.; Koduru, S.; Kumar, R.; Venguswamy, G.; Kyprianou, N.; Damodaran, C. Diosgenin Targets Akt-Mediated Prosurvival Signaling in Human Breast Cancer Cells. *Int. J. Cancer* **2009**, *125*, 961–967. [CrossRef] [PubMed]

5. Hong, S.H.; Cha, H.-J.; Hwang-Bo, H.; Kim, M.Y.; Kim, S.Y.; Ji, S.Y.; Cheong, J.; Park, C.; Lee, H.; Kim, G.-Y.; et al. Anti-Proliferative and Pro-Apoptotic Effects of Licochalcone A through ROS-Mediated Cell Cycle Arrest and Apoptosis in Human Bladder Cancer Cells. *Int. J. Mol. Sci.* **2019**, *20*, 3820. [CrossRef] [PubMed]
6. Karar, J.; Maity, A. PI3K/AKT/mTOR Pathway in Angiogenesis. *Front. Mol. Neurosci.* **2011**, *4*, 51. [CrossRef] [PubMed]
7. Dufie, W.-M.F.; Oduro, I.; Ellis, W.O.; Asiedu, R.; Maziya-Dixon, B. Potential Health Benefits of Water Yam (*Dioscorea Alata*). *Food Funct.* **2013**, *4*, 1496–1501. [CrossRef]
8. Reuter, S.; Gupta, S.C.; Chaturvedi, M.M.; Aggarwal, B.B. Oxidative Stress, Inflammation, and Cancer: How Are They Linked? *Free Radic. Biol. Med.* **2010**, *49*, 1603–1616. [CrossRef]
9. Mendelsohn, A.R.; Larrick, J.W. Paradoxical Effects of Antioxidants on Cancer. *Rejuvenation Res.* **2014**, *17*, 306–311. [CrossRef]
10. Saha, S.K.; Lee, S.B.; Won, J.; Choi, H.Y.; Kim, K.; Yang, G.-M.; Dayem, A.A.; Cho, S.-G. Correlation between Oxidative Stress, Nutrition, and Cancer Initiation. *Int. J. Mol. Sci.* **2017**, *18*, E1544. [CrossRef]
11. Singh, U.; Jialal, I. Oxidative Stress and Atherosclerosis. *Pathophysiology* **2006**, *13*, 129–142. [CrossRef]
12. Bjelakovic, G.; Nikolova, D.; Simonetti, R.G.; Gluud, C. Antioxidant Supplements for Preventing Gastrointestinal Cancers. *Cochrane Database Syst. Rev.* **2008**, CD004183. [CrossRef] [PubMed]
13. Aggarwal, B.B.; Sundaram, C.; Malani, N.; Ichikawa, H. Curcumin: The Indian Solid Gold. *Adv. Exp. Med. Biol.* **2007**, *595*, 1–75. [CrossRef] [PubMed]
14. Singh, R.P.; Tyagi, A.K.; Dhanalakshmi, S.; Agarwal, R.; Agarwal, C. Grape Seed Extract Inhibits Advanced Human Prostate Tumor Growth and Angiogenesis and Upregulates Insulin-like Growth Factor Binding Protein-3. *Int. J. Cancer* **2004**, *108*, 733–740. [CrossRef] [PubMed]
15. Sayin, V.I.; Ibrahim, M.X.; Larsson, E.; Nilsson, J.A.; Lindahl, P.; Bergo, M.O. Antioxidants Accelerate Lung Cancer Progression in Mice. *Sci. Transl. Med.* **2014**, *6*, 221ra15. [CrossRef] [PubMed]
16. Ristow, M. Unraveling the Truth about Antioxidants: Mitohormesis Explains ROS-Induced Health Benefits. *Nat. Med.* **2014**, *20*, 709–711. [CrossRef] [PubMed]
17. Sheel, R.; Nisha, K. Qualitative Phytochemical Analysis for Isolation of Terpenes from Clerodendron Infortunatum Leaves. *ISOR J. Appl. Chem.* **2014**, *7*, 14–18. [CrossRef]
18. Ali, S.; Khan, M.R.; Irfanullah; Sajid, M.; Zahra, Z. Phytochemical Investigation and Antimicrobial Appraisal of Parrotiopsis Jacquemontiana (Decne) Rehder. *BMC Complement. Altern. Med.* **2018**, *18*, 43. [CrossRef] [PubMed]
19. Das, N.; Islam, M.E.; Jahan, N.; Islam, M.S.; Khan, A.; Islam, M.R.; Parvin, M.S. Antioxidant Activities of Ethanol Extracts and Fractions of Crescentia Cujete Leaves and Stem Bark and the Involvement of Phenolic Compounds. *BMC Complement. Altern. Med.* **2014**, *14*, 45. [CrossRef]
20. Gizaw, A.; Marami, L.M.; Teshome, I.; Sarba, E.J.; Admasu, P.; Babele, D.A.; Dilba, G.M.; Bune, W.M.; Bayu, M.D.; Tadesse, M.; et al. Phytochemical Screening and In Vitro Antifungal Activity of Selected Medicinal Plants against *Candida Albicans* and *Aspergillus Niger.* in West Shewa Zone, Ethiopia. *Adv. Pharmacol. Pharm. Sci.* **2022**, *2022*, e3299146. [CrossRef]
21. Shaikh, J.R.; Patil, M. Qualitative Tests for Preliminary Phytochemical Screening: An Overview. *Int. J. Chem. Stud.* **2020**, *8*, 603–608. [CrossRef]
22. Bankole, A.E.; Adekunle, A.A.; Sowemimo, A.A.; Umebese, C.E.; Abiodun, O.; Gbotosho, G.O. Phytochemical Screening and in Vivo Antimalarial Activity of Extracts from Three Medicinal Plants Used in Malaria Treatment in Nigeria. *Parasitol. Res.* **2016**, *115*, 299–305. [CrossRef]
23. Onwukaeme, D.N.; Ikuegbvweha, T.B.; Asonye, C.C. Evaluation of Phytochemical Constituents, Antibacterial Activities and Effect of Exudate of Pycanthus Angolensis Weld Warb (Myristicaceae) on Corneal Ulcers in Rabbits. *Trop. J. Pharm. Res.* **2007**, *6*, 725–730. [CrossRef]
24. Kumar, S.; Venkateshwar, C.; Samuel, G.; Rao, S.G. Phytochemical Screening of Some Compounds from Plant Leaf Extracts of Holoptelea Integrifolia (Planch.) and Celestrus Emarginata (Grah.) Used by Gondu Tribes at Adilabad District, Andhrapradesh, India. *Int. J. Eng. Sci. Invent.* **2013**, *2*, 65–70.
25. Siddique, A.B.; Rahman, S.M.M.; Hossain, M.A.; Hossain, M.A.; Rashid, M.A. Phytochemical Screening and Comparative Antimicrobial Potential of Different Extracts of Stevia Rebaudiana Bertoni Leaves. *Asian Pac. J. Trop. Dis.* **2014**, *4*, 275–280. [CrossRef]
26. Wallace, K.; Asemota, H.; Gray, W. Acetone Extract of *Dioscorea Alata* Inhibits Cell Proliferation in Cancer Cells. *Am. J. Plant Sci.* **2021**, *12*, 300. [CrossRef]
27. Wright, R.J.; Lee, K.S.; Hyacinth, H.I.; Hibbert, J.M.; Reid, M.E.; Wheatley, A.O.; Asemota, H.N. An Investigation of the Antioxidant Capacity in Extracts from Moringa Oleifera Plants Grown in Jamaica. *Plants* **2017**, *6*, 48. [CrossRef] [PubMed]
28. Durazzo, A.; Lucarini, M. Extractable and Non-Extractable Antioxidants. *Molecules* **2019**, *24*, 1933. [CrossRef]
29. Stennett, D.; Oladeinde, F.; Wheatley, A.; Bryant, J.; Dilworth, L.; Asemota, H. Effects of Dioscorea Polygonoides (Jamaican Bitter Yam) Supplementation in Normocholesterolemic and Genetically Modified Hypercholesterolemic Mice Species. *J. Food Biochem.* **2014**, *38*, 28–37. [CrossRef]
30. Grindley, P.B.; Omoruyi, F.O.; Asemota, H.N.; St. A. Morrison, E.Y. Asemota Effect of Yam (*Dioscorea cayenensis*) and Dasheen (*Colocassia esculenta*) Extracts on the Kidney of Streptozotocin-Induced Diabetic Rats. *Int. J. Food Sci. Nutr.* **2001**, *52*, 429–433. [CrossRef] [PubMed]

31. Chandrasekara, A.; Josheph Kumar, T. Roots and Tuber Crops as Functional Foods: A Review on Phytochemical Constituents and Their Potential Health Benefits. Available online: https://www.hindawi.com/journals/ijfs/2016/3631647/ (accessed on 14 September 2020).
32. Das, A.; Chaudhuri, D.; Ghate, N.; Chatterjee, A.; Mandal, N. Phytochemical Analysis, Antioxidant and Anticancer Potential of Leaf Extracts from Edible Greater Yam, Dioscorea Alata L., from North-East India. *Int. J. Phytopharm.* **2014**, *5*, 109–119.
33. Maithili, V.; Dhanabal, S.P.; Mahendran, S.; Vadivelan, R. Antidiabetic Activity of Ethanolic Extract of Tubers of Dioscorea Alata in Alloxan Induced Diabetic Rats. *Indian J. Pharmacol.* **2011**, *43*, 455–459. [CrossRef] [PubMed]
34. Zhu, S.; Jiao, W.; Xu, Y.; Hou, L.; Li, H.; Shao, J.; Zhang, X.; Wang, R.; Kong, D. Palmitic Acid Inhibits Prostate Cancer Cell Proliferation and Metastasis by Suppressing the PI3K/Akt Pathway. *Life Sci.* **2021**, *286*, 120046. [CrossRef] [PubMed]
35. Muzac-Tucker, I.; Asemota, H.N.; Ahmad, M.H. Biochemical Composition and Storage of Jamaican Yams (*Dioscorea* Sp). *J. Sci. Food Agric.* **1993**, *62*, 219–224. [CrossRef]
36. Sakthidevi, G.; Mohan, V.R. Total Phenolic, Flavonoid Contents and In Vitro Antioxidant Activity of Dioscorea Alata l. Tuber. *J. Pharm. Sci. Res.* **2013**, *5*, 115.
37. De Baere, S.; Eeckhaut, V.; Steppe, M.; De Maesschalck, C.; De Backer, P.; Van Immerseel, F.; Croubels, S. Development of a HPLC–UV Method for the Quantitative Determination of Four Short-Chain Fatty Acids and Lactic Acid Produced by Intestinal Bacteria during in Vitro Fermentation. *J. Pharm. Biomed. Anal.* **2013**, *80*, 107–115. [CrossRef] [PubMed]
38. Wettersten, H.I.; Ganti, S.; Weiss, R.H. Chapter Fourteen-Metabolomic Profiling of Tumor-Bearing Mice. In *Methods in Enzymology*; Galluzzi, L., Kroemer, G., Eds.; Cell-Wide Metabolic Alterations Associated with Malignancy; Academic Press: Cambridge, MA, USA, 2014; Volume 543, pp. 275–296.
39. Narayanan, A.; Baskaran, S.A.; Amalaradjou, M.A.R.; Venkitanarayanan, K. Anticarcinogenic Properties of Medium Chain Fatty Acids on Human Colorectal, Skin and Breast Cancer Cells in Vitro. *Int. J. Mol. Sci.* **2015**, *16*, 5014–5027. [CrossRef] [PubMed]
40. Aiyelaagbe, O.O.; Negi, A.S.; Hamid, A.A.; Luqman, S.; Kumar, S.B.; Kaneez, F. Chemical Constituents from Alafia Barteri Oliv. Leaves with Cytotoxic Activity. *J. Chin. Chem. Soc.* **2015**, *62*, 751–755. [CrossRef]
41. Mericli, F.; Becer, E.; Kabadayı, H.; Hanoglu, A.; Yigit Hanoglu, D.; Ozkum Yavuz, D.; Ozek, T.; Vatansever, S. Fatty Acid Composition and Anticancer Activity in Colon Carcinoma Cell Lines of *Prunus Dulcis* Seed Oil. *Pharm. Biol.* **2017**, *55*, 1239–1248. [CrossRef]
42. Atasever-Arslan, B.; Yilancioglu, K.; Bekaroglu, M.G.; Taskin, E.; Cetiner, E.A.S. Cytotoxic Effect of Extract from Dunaliella Salina against SH-SY5Y Neuroblastoma Cells. *Gen. Physiol. Biophys.* **2015**, *34*, 201–207. [CrossRef]
43. Sheela, D.L.; Narayanankutty, A.; Nazeem, P.A.; Raghavamenon, A.C.; Muthangaparambil, S.R. Lauric Acid Induce Cell Death in Colon Cancer Cells Mediated by the Epidermal Growth Factor Receptor Downregulation: An in Silico and in Vitro Study. *Hum. Human. Exp. Toxicol.* **2019**, *38*, 753–761. [CrossRef]
44. Dey, P.; Chaudhuri, T.K. Phytochemical Characterization of *Dioscorea alata* Leaf and Stem by Silylation Followed by GC-MS Analysis. *J. Food Biochem.* **2016**, *40*, 630–635. Available online: https://onlinelibrary.wiley.com/doi/abs/10.1111/jfbc.12235 (accessed on 24 January 2021).

separations

MDPI

Article

Efficient Extraction of Flavonoids from Lotus Leaves by Ultrasonic-Assisted Deep Eutectic Solvent Extraction and Its Evaluation on Antioxidant Activities

Liangliang Liu, Aiping Xiao, Yi Zhang and Shengwen Duan *

Institute of Bast Fiber Crops, Chinese Academy of Agricultural Sciences, Changsha 410205, China
* Correspondence: duanshengwen@caas.cn; Tel.: +86-731-88998516

Abstract: The discovery of a green extraction solvent for natural plants could promote related research. In this study, deep eutectic solvents (DES) were used as green solvents coupled with an ultrasound-assisted extraction method (UAE) to extract flavonoids from lotus leaves. Thirty-four different DES were performed and choline chloride/urea with 40% water was chosen as the most promising one, and the related parameters in the procedures were optimized, resulting in the highest extraction amount of flavonoids in lotus leaves. D-101 was selected from four macroporous resins to separate the flavonoids from DES. Moreover, DES could be recycled and efficiently reused four times with satisfactory performances. In addition, the lotus leaf flavonoids from the DES extract exhibited antioxidant activities in five kinds of assays including DPPH, ABTS, Fe^{3+} reducing, FRAP, and Fe^{2+} chelating. It also showed antibacterial activities on *Staphylococcus aureus* and *Escherichia coli* bacterial strains with minimal inhibitory concentrations at 1666 μg/mL and 208 μg/mL, respectively. In the HPLC analysis, the three main components in the DES extract were identified as astragalin, hyperoside, and isoquercitrin. In conclusion, the developed UAE-DES followed by macroporous resin treatment could become an efficient and environmentally friendly extraction and enrichment method for flavonoids from lotus leaves and other natural products.

Keywords: antioxidant; deep eutectic solvents; flavonoids; lotus leaves; ultrasound-assisted

Citation: Liu, L.; Xiao, A.; Zhang, Y.; Duan, S. Efficient Extraction of Flavonoids from Lotus Leaves by Ultrasonic-Assisted Deep Eutectic Solvent Extraction and Its Evaluation on Antioxidant Activities. *Separations* **2023**, *10*, 65. https://doi.org/10.3390/separations10020065

Academic Editor: Marcello Locatelli

Received: 3 January 2023
Revised: 13 January 2023
Accepted: 16 January 2023
Published: 17 January 2023

1. Introduction

Lotus leaves are the leaves of *Nelumbo nucifera* Gaertn., which are considered to be both foods and traditional medicines in China [1]. Lotus leaves have had a great vogue as drinks such as tea or in herbal formulations and have had a lot of recognition [2]. As reported by many papers, they contain a variety of alkaloids and flavonoids, which made them possess antioxidant, antiviral, anti-obesity, and bacteriostatic functions [3,4]. To develop better lotus-leaf-based products, extraction of lotus leaves could provide more specific materials (such as flavonoids) for research and industrial purposes [5]. Hence, it is necessary to apply a novel extraction method to promote research on lotus leaves.

As in Abbott's first report, transparent and uniform liquids called deep eutectic solvents (DES) could be formed by mixing ammonium salts and amides at low temperatures [6]. After a series of research, DES are typically composed of a hydrogen bond acceptor (HBA) and a hydrogen bond donor (HBD) by intermolecular hydrogen bonding [7]. The resulting DES showed a low melting point, which was lower than that of each pure component. Compared with traditional organic solvents, DES exhibited various characteristics such as low cost, simple preparation, less volatility, and non-toxicity, which made it widely used in the extraction of natural, biologically active components in recent years [8,9]. Choline chloride has often been applied as an HBA in DES for the extraction of flavonoids [10]. In Jalel Labidi's report, choline chloride/1,4-butanediol (75 wt%) showed the highest extraction ability for flavonoids (383 mg CE/g dried bark extract) and the highest antioxidant capacity [11]. In addition, a study optimized a ternary combination in DES with choline chloride/betaine hydrochloride/ethylene glycol with the mole ratio of 1:1:2 and 20% water content in a negative pressure cavitation-assisted extraction method for flavonoids from *Equisetum palustre* L. [12]. In Weng's research, the flavonoids in *Spirodela polyrrhiza* were rapidly extracted by a DES-based microwave-assisted extraction method using choline chloride/levulinic acid with the mole ratio of 1:2 as the DES [13].

The DES could be combined with some assisted extraction technology such as microwave, ultrasonic, and vortex technology [14,15]. These combination methods could improve extraction efficiency and capacity. Among them, ultrasound-assisted extraction (UAE) could promote the solvent to quickly penetrate the plant material and enhance the diffusion of compounds [16]. The instrument condition of UAE was relatively easy to control in the lab compared to the other assisted methods. In recent years, UAE-DES has been applied in the extraction of many natural plants [17–19]. However, these methods still needed selection and optimization for different plant samples due to their complexity and composition. Considering the many benefits of flavonoids and their abundance in lotus leaves and the fact that very little research has been devoted to the utilization of UAE-DES in lotus leaves, the present research aimed to extract flavonoids by an ultrasonic-assisted DES method.

For this purpose, choline chloride, benzyltriethylammonium chloride, betaine, and guanidine-hydrochloride-based DES were prepared to extract flavonoids from lotus leaves to the maximum extent with the assistance of ultrasonic technology (UAE-DES). Then, the flavonoid extract was separated by macroporous resin from the DES, and the antioxidant capacities and antibacterial activities of the extracted flavonoids were systematically studied with beneficial results. Through this study, an effective strategy was provided to extract flavonoids in lotus leaves, which had green and convenient advantages compared to traditional methods. The lotus leaf flavonoids also exhibited antioxidant activities, making them promising in the healthcare area.

2. Materials and Methods

2.1. Materials

Dried lotus leaves were bought from Mingxing Biotechnology Co., LTD. in Bozhou, Anhui. The dried leaves were pulverized and sieved in 40 mesh sieves before use. Choline chloride, benzyltriethylammonium chloride, betaine, guanidine hydrochloride, ethylene glycol, triethylene glycol, glycerol, 2-chloropropionic acid, malonic acid, lactic acid, formic acid, acetic acid, and urea were purchased from Macklin Chemical Reagent Co. (Shanghai, China). 2,2-Diphenyl-1-picrylhydrazyl (DPPH), Diammonium 2,2′-azino-bis(3-ethylbenzothiazoline-6-sulfonate) (ABTS), tri-2-pyridyl-s-triazine (TPTZ), and Folin–Ciocalteu were acquired from Sigma-Aldrich Chemicals (St. Louis, MO, USA). The *Staphylococcus aureus* and *Escherichia coli* bacterial strains were provided by Professor Qi Yang from the Institute of Bast Fiber Crops, Chinese Academy of Agricultural Sciences (Changsha, China). All of the other chemicals were of analytical grade.

2.2. Preparation of DES

Choline chloride, benzyltriethylammonium chloride, betaine, and guanidine hydrochloride were selected as the HBA to synthesize DES with different HBDs (ethylene glycol, triethylene glycol, glycerol, 2-chloropropionic acid, malonic acid, lactic acid, formic acid, acetic acid, urea, oxalic acid, and acetamide). The reagents were weighed with the molar ratio at 1:2 (0.2 mol:0.4 mol) and manually stirred at 80 °C for 2 h until a homogeneous liquid formed. When the DES cooled to room temperature, they were stored at ambient temperature.

2.3. Extraction of Flavonoids by the UAE-DES Method

A dried lotus leaf (1.0 g) was powdered and added to the DES (20.0 mL) in a centrifuge tube. The sample was ultrasonically treated by an ultrasonic cleaner (Kunshan Ultrasonic Instrument, Suzhou, China) for 5 min and then extracted in an oscillator at 60 °C under shaking for 60 min. After extraction, the mixture was centrifuged at 4000 r/min for 15 min. The supernatant was taken for further analysis.

To investigate the effect of factors on the extraction efficiency of the flavonoids, the detailed conditions including the ultrasonic time (0 to 30 min), extraction time (10 to 140 min), solid–liquid ratio (1:20 to 1:80), and extraction temperature (40 to 120 °C) were optimized in single-factor experiments. All of the experiments were carried out in triplicate and the values were expressed as means $\pm$ SD.

2.4. Determination of the Total Flavonoids Content

The total flavonoid content of the extract was determined by a colorimetric method with some modifications [20]. Briefly, 10 μL of DES extract was mixed with 25 μL of sodium acetate solution (1.0 mol/L), 25 μL of aluminum chloride solution (10%, wt%), and 1940 μL of ethanol solution (50%, *v*:*v*). The mixture was thoroughly mixed and incubated at 25 °C for 30 min. After that, the absorbance of the mixture was measured at 415 nm by a UV-2700 UV–Vis spectrophotometer (Shimadzu, Kyoto, Japan). Rutin was used as the standard in this analysis and the results were expressed as milligrams of rutin equivalent per gram of extract.

2.5. Recovery of Flavonoids from the DES Extract

The recovery of flavonoids from the DES extract was carried out by adsorption and desorption by macroporous resins. The pretreatment of macroporous resin was performed before usage as in Li's report [21]. Ten mL of DES extracts were put into a flask containing 2.0 g D101 macroporous resin. The adsorption was conducted in an oscillator under shaking (200 rpm) at 25 °C for 6 h. Then, the resin remained, was washed with water, and then desorbed with 10 mL of ethanol at 25 °C and 200 rpm for 2 h in an oscillator under shaking.

2.6. Evaluation of Antioxidant Activities

2.6.1. DPPH Assay

In brief, fresh DPPH solution (0.1 mmol/L) was prepared in ethanol. Two mL of sample solution (1–20 mg/mL) was mixed with DPPH solution and incubated in a 25 °C constant temperature water bath for 30 min. Then, the absorbance of the mixture at 517 nm was measured with a spectrophotometer. The scavenging rate of DPPH radicals was calculated as in the previously reported equation [22].

2.6.2. ABTS Assay

The ABTS assay was measured as previously described [22]. ABTS stock solution (7 mmol/L of ABTS solution mixed in 2.45 mmol/L of $K_2S_2O_8$ solution with equal volume) was incubated for 12 h in the dark and diluted for use. One mL of sample solution (1–20 mg/mL) was mixed with 2 mL of ABTS working solution and incubated in the dark for 6 min. After that, the absorbance of the mixture at 734 nm was measured. The scavenging rate of ABTS radicals was calculated as the equation reported by the reference.

2.6.3. Fe^{3+} Reducing Assay

The Fe^{3+} reducing ability of the extract was investigated by the reduction of $K_3[Fe(CN)_6]$ method [23]. One mL of the sample solution (1–20 mg/mL) was added into 2 mL of phosphate buffer (0.2 M, pH 6.6) and 2 mL of potassium ferricyanide solution (1%, *w*/*v*) and then incubated at 50 °C for 20 min. Then, 2 mL of water, 2 mL of trichloroacetic acid (10%, *w*/*v*), and 0.1 mL of $FeCl_3$ (0.1%, *w*/*v*) were respectively added into the mixture. The absorbance of the final solution was recorded at 700 nm in a spectrophotometer. Rutin was used as standard (0.1–0.5 mg/mL) and the results were expressed as milligrams of rutin equivalents concentration (mg RE/mL).

2.6.4. Ferric Ion Reducing Antioxidant Power (FRAP) Assay

According to Tungjai's report, TPTZ solution (10 mmol/L), sodium acetate solution (300 mmol/L), and $FeCl_3$ solution (20 mmol/L) were mixed in the proportion of 10:1:1 (*v:v:v*) as the working solution. Then, 0.1 mL of the sample solution (1–20 mg/mL) was mixed in 2 mL of water and 3 mL of the FRAP working solution and incubated at 37 °C for 50 min. After that, the absorbance of the mixture was measured at 596 nm. $FeSO_4$ solution was used for the calibration curve and the results were expressed as the FRAP value (mmol/L Fe^{2+}) [24].

2.6.5. Fe^{2+} Chelating Assay

The metal chelating ability of the sample was conducted according to Buravlev et al. with some modifications [25]. A measurement of 0.1 mL of $FeSO_4$ solution (2 mmol/L) and 1 mL of sample solution (1–20 mg/mL) were added to 4 mL of water. Two hundred μL of ferrozine solution (5 mmol/L) was added to start the reaction. The mixture was incubated at room temperature for 10 min and monitored the absorption at 562 nm. The results were given as the percentage of inhibition.

2.7. Antibacterial Activity

The diameter of the inhibition zone (DIZ) was determined by the paper disc diffusion method as previous reports have described [26]. Bacteria suspensions were cultured in a nutrient broth at 37 °C for 48 h and diluted to about 1×10^{-6} CFU. One hundred μL of bacterial suspensions was evenly spread on an agar plate. Then, sterilized filter paper discs (Φ = 6 mm) were soaked in samples (50 mg/mL and 100 mg/mL) and then carefully placed on the plates. The plates were incubated at 37 °C for 24 h, and the DIZ was measured by a vernier caliper. Rutin was used as a positive control, and a solvent without samples was used as a negative control. The DIZ values were expressed in millimeters and values < 6.00 mm were considered as no inhibition zone (NIZ).

The minimal inhibitory concentration (MIC) was determined by the micro-dilution broth method with some modification [26]. five hundred μL of nutrient broth and 50 μL of samples were mixed in a tube. Then, the bacterial solution (1×10^{-6} CFU) was added to the tube and incubated at 37 °C for 12 h. The absorbance at 600 nm of the bacterial suspensions was monitored by a spectrophotometer at predetermined time intervals. Rutin was used as the positive control and the nutrient broth was used as a negative control. The lowest concentration of the sample inhibiting the growth of bacteria was defined as the MIC value.

2.8. HPLC Analysis

The components analysis of the DES extract was carried out on an Agilent 1260 HPLC system (Agilent Technologies Inc., Santa Clara, CA, USA). The chromatographic separation was performed on an Agilent ZORBAX C18 column (150 mm × 4.6 mm i.d.; 5 μm, Santa Clara, CA, USA). The mobile phase consists of solvent A (0.1% *v*/*v* acetic acid solution) and solvent B (acetonitrile containing 0.1% *v*/*v* acetic acid) with gradient elution mode: 0–4 min, 10% B; 4–25 min, 10–20% B; and 25–30 min, 20–20% B. The flow rate, temperature, injection volume, and detection wavelength were controlled at 1.0 mL/min, 25 °C, 5 μL, and 254 nm, respectively.

2.9. Statistical Analysis

The statistical analysis was performed by SPSS V20 software (IBM SPSS Statistics 20, New York, NY, USA). The data were then compared using Duncan's multiple range tests at 5% significance levels, and one-way ANOVA with Tukey's post-hoc test [27,28]. Values followed by different letters in the same column are significantly different according to Duncan's test ($p < 0.05$).

3. Results and Discussion

3.1. Selection of DES

The extraction of active compounds using DES was easily influenced by many conditions related to the target compounds and the composition of DES. A lot of research used one or two HBAs for selection [29,30]. While to select appropriate DES on a larger scale for the full extraction and dissolution of flavonoids in lotus leaves, the extraction efficiencies of 34 DES belonging to four kinds of HBA were investigated. The extraction was completed in the same ultrasonic condition and molar ratio of HBA/HBD, and the results are recorded in Table 1. The highest flavonoid yield by DES was achieved when the DES was prepared by choline chloride and urea as in HBA and HBD. The DES system using choline chloride and urea often exhibited the best extraction capacity for flavonoids and polyphenols in natural products. Choline chloride–lactic acid had the best extraction effect among forty groups of DES in the extraction of flavonoids from *Dendrobium officinale* [31]. Choline chloride/levulinic acid (1:2, mol/mol) could be applied as the solvent for extracting the flavonoids in *Spirodela polyrrhiza* [13]. Choline chloride/1,4-butanediol exhibited the best extraction efficiency from seven different DES for extracting valuable bioactive compounds from Chinese wild rice [32]. The finding in this study was also in accordance with several reports [11,33,34]. Therefore, the optimal DES type for the extraction of flavonoids in lotus leaves was set as choline chloride and urea.

Table 1. The extraction efficiencies of flavonoids with different DES types.

No.	HBA	HBD	Molar Ratio	Extraction Efficiency (mg/g)
1	Choline chloride	Ethylene glycol	1:2	114.9 ± 2.5 d
2		Triethylene glycol	1:2	91.5 ± 1.2 h
3		Glycerol	1:2	49.6 ± 1.2 p
4		2-Chloropropionic acid	1:2	33.7 ± 0.1 rs
5		Malonic acid	1:2	36.3 ± 0.4 r
6		Lactic acid	1:2	54.1 ± 1.5 o
7		Formic acid	1:2	81.4 ± 0.7 l
8		Acetic acid	1:2	88.3 ± 0.8 ij
9		Urea	1:2	178.9 ± 2.8 a
10	Benzyltriethylammonium chloride	Ethylene glycol	1:2	82.1 ± 0.9 kl
11		Triethylene glycol	1:2	100.9 ± 1.7 g
12		Glycerol	1:2	92.2 ± 1.3 h
13		2-Chloropropionic acid	1:2	49.6 ± 0.4 p
14		Malonic acid	1:2	49.3 ± 0.5 p
15		Lactic acid	1:2	83.4 ± 1.4 kl
16		Formic acid	1:2	154.9 ± 2.7 b
17		Acetic acid	1:2	48.0 ± 0.9 p
18		Oxalic acid	1:2	107.4 ± 0.8 f
19		Acetamide	1:2	30.7 ± 0.2 st
20	Betaine	Glycerol	1:2	87.6 ± 0.6 ij
21		Malonic acid	1:2	73.3 ± 0.4 m
22		Lactic acid	1:2	97.7 ± 1.0 g
23		Formic acid	1:2	84.4 ± 0.5 g
24		Acetic acid	1:2	113.3 ± 1.1 d
25		Urea	1:2	59.3 ± 0.4 n
26		Oxalic acid	1:2	39.8 ± 0.1 q

Table 1. *Cont.*

No.	HBA	HBD	Molar Ratio	Extraction Efficiency (mg/g)
27	Guanidine hydrochloride	Ethylene glycol	1:2	99.3 ± 0.9 [g]
28		Triethylene glycol	1:2	105.2 ± 1.1 [f]
29		Glycerol	1:2	85.3 ± 0.7 [jk]
30		2-Chloropropionic acid	1:2	83.7 ± 0.4 [kl]
31		Lactic acid	1:2	111.4 ± 3.0 [e]
32		Formic acid	1:2	98.7 ± 1.0 [g]
33		Acetic acid	1:2	29.1 ± 0.8 [t]
34		Acetamide	1:2	138.9 ± 2.5 [c]

Values followed by different letters ([a–t]) in the same column are significantly different according to Duncan's test ($p < 0.05$).

3.2. Optimization of Extraction Parameters

3.2.1. Effect of the HBA/HDB Ratio

The variation of the HBA/HBD ratio would affect the viscosity and surface tension of the DES solution and the resulting different extraction capacities of flavonoids. To improve the extraction ability of DES, the extraction efficiencies of flavonoids with DES in different HBA/HBD ratios were investigated (Figure 1a). With the increasing HBA/HBD ratio, the extraction efficiencies of flavonoids first increased and then decreased. The highest extraction efficiency of 195.2 mg/g was obtained when the ratio was 1:2, which was then set as the optimal HBA/HBD ratio for preparing DES. This molar ratio (1:2) of HBA/HBD has often been selected in a lot of research and can be supported [29,35,36].

3.2.2. Effect of Water Content in DES

The water content in DES could change the viscosity and polarity of DES. The appropriate water content of DES would decrease the solvent viscosity, improve mass transfer efficiency, and increase the extraction yield of polyphenol and flavonoids [37]. In this study, the extraction efficiencies of various DES with water contents from 10% to 50% were conducted. With the results shown in Figure 1b, when the water content increased up to 40%, the extracted flavonoid amount improved, which was because of the reduction of the viscosity of solvents and the weakening of the hydrogen bonds [38]. When the water content surpassed this value, the extracted flavonoid amount started to drop off. Hence, 40% water content in DES was chosen since this was the highest extraction efficiency for the flavonoids. In Gan's research, 40% water content in DES was chosen for the extraction of polyphenols from green tea as well [39].

3.2.3. Effect of Ultrasonic Time

The influence of ultrasonic time ranged from 0 to 30 min on the extraction efficiency was monitored. It was observed in Figure 1c that the extraction amount of flavonoids increased with the increasing of the ultrasonic time before 10 min and then was not improved with longer ultrasonic time, which indicated that a dynamic balance was reached during the ultrasonic-assisted extraction. Thus, 10 min was chosen for the extraction in further experiments taking into account the extraction efficiency and energy saving.

3.2.4. Effect of the Solid–Liquid Ratio

The effect of the solid–liquid ratio (g/mL) on the extraction efficiency of total flavonoids was monitored. As indicated in Figure 1d, the extraction efficiency of flavonoids increased with an increasing solid–liquid ratio from 1:20 to 1:80 and then decreased slightly. As a result, a 1:80 solid–liquid ratio was selected for later study.

3.2.5. Effect of Temperature

As the literature has reported, high temperatures could reduce the viscosity and surface tension of DES as well, and then improve the mobility and solubility of compounds [40]. To obtain the optimal extraction ability, the effect of different temperatures in the range of 40–120 °C on the extraction efficiency of flavonoids was studied (Figure 1e). As the temperature increased from 40 °C to 60 °C, the extraction amount of flavonoids increased up to a maximum value, but a decrease was then observed at temperatures higher than 60 °C. The decrease in high temperature might be attributed to the elevated temperature increasing the oxidation of flavonoid compounds. This trend can be observed in Wang's and Weng's reports as well [13,41]. Accordingly, a moderate temperature (60 °C) was set for the extraction.

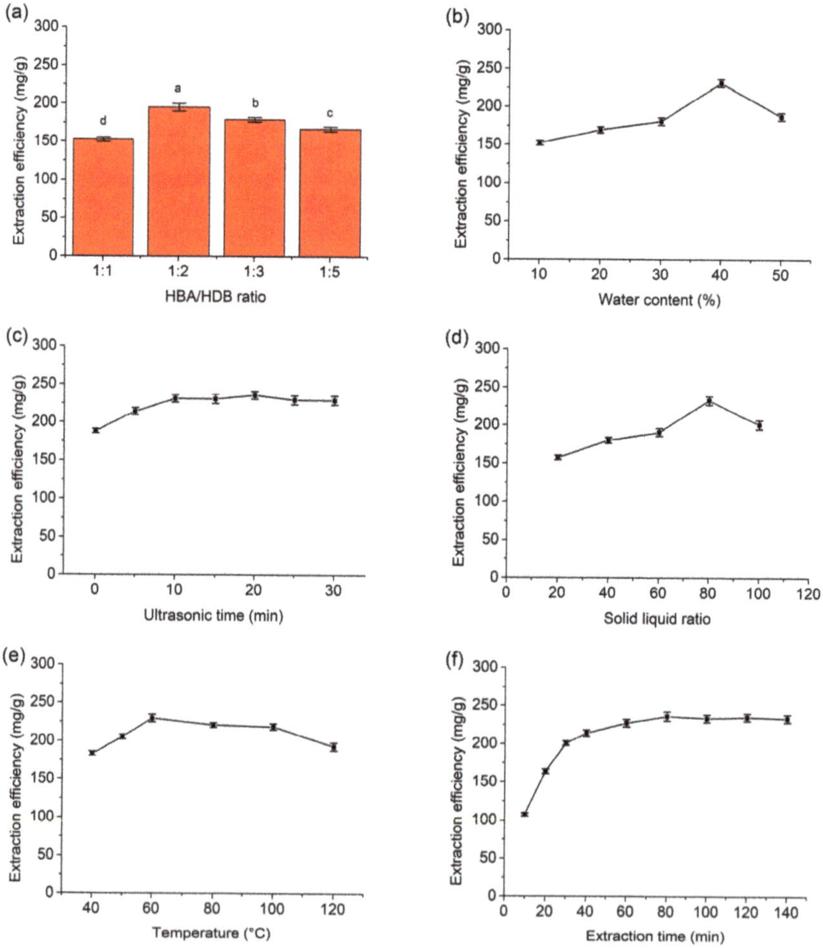

Figure 1. Optimization of extraction parameters in the UAE-DES method. (**a**) Effect of the HBA/HBD ratio on extraction efficiency; (**b**) effect of water content on extraction efficiency; (**c**) effect of ultrasonic time on extraction efficiency; (**d**) effect of the solid–liquid ratio on extraction efficiency; (**e**) effect of temperature on extraction efficiency; (**f**) effect of extraction time on extraction efficiency. Values followed by different letters (a–d) are significantly different according to Duncan's test ($p < 0.05$).

3.2.6. Effect of Extraction Time

To optimize the extraction equilibrium, the extraction of flavonoids in lotus leaves was performed at different times from 10 min to 140 min using the UAE-DES method. As indicated in Figure 1f, the extraction efficiency increased as the extraction time increased from 10 to 60 min. However, the value kept stable after 60 min, which could be considered as the extraction having achieved balance. Typically, long-term extraction was not suitable for natural active components, which would affect the chemical structures and activities [13]. Accordingly, the DES extraction time was confirmed at 60 min.

3.2.7. Comparison to Conventional Solvents

To evaluate the extraction efficiency of UAE-DES, conventional solvents were used for comparison in the same extraction condition and procedure [42]. The used conventional solvents included water, 50% methanol, 95% methanol, and 95% ethanol. Under the same treatment, 50% methanol solvent exhibited the highest extraction efficiency (246.9 mg/g), which was higher than that using DES (236.6 mg/g) by 4% (Figure 2), while water demonstrated the worst extraction efficiency for target compounds. Methanol was a good organic solvent in the extraction of natural products including flavonoids. However, its toxicity to the human body and harm to the environment could not be ignored. The extraction efficiency of DES is the second-best solvent in this comparison, which was slightly lower than that of 50% methanol. Considering the ease of recycling, low cost in preparation, and environmentally friendliness, DES were selected as the extraction solvent because of its satisfactory extraction ability and sustainable properties [43].

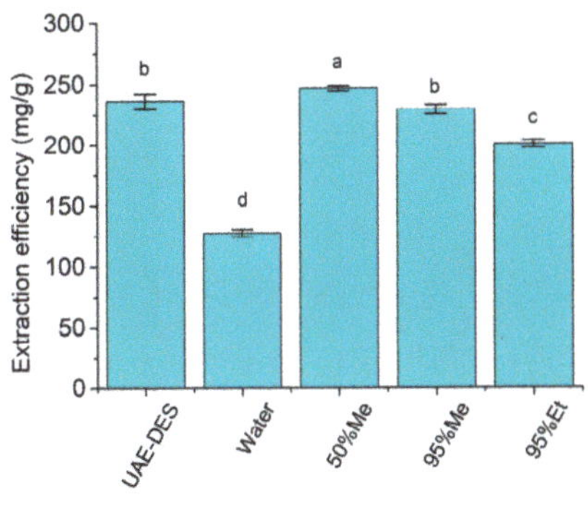

Figure 2. Extraction efficiencies of prepared DES and conventional solvents (50%me-50% methanol, 95%me-95% methanol, and 95%et-95% ethanol). Values followed by different letters (a–d) are significantly different according to Duncan's test ($p < 0.05$).

3.3. *Recovery of Flavonoids and Reusability of DES*

As studied in many pieces of research, macroporous resin has often been used to enrich natural active compounds, especially for flavonoids, and then target compounds could be eluted using organic solutions [44]. In this study, four different resins (HPD-500, AB-8, D-101, and NKA-9) were selected to investigate. Through the comparison, D-101 showed the best adsorption ability for flavonoids with 90% during the same conditions (Figure 3a). In many pieces of research, D-101 has also been selected for the collection of coumarins and flavonoids [45,46]. However, AB-8 and other kinds of resin have been used for the recovery of target compounds, which exhibited the importance of selection [29,47]. Then, different concentrations of ethanol solutions have been used for desorption. As shown in Figure 3b, the desorption ratio reached 95% when ethanol concentrations were higher than 60%, and the optimal desorption ratio (99.5%) was obtained by 100% ethanol. Moreover, 100% ethanol solution was easier for removing solvent. Therefore, D-101 was used for the recovery of flavonoids, and 100% ethanol was chosen as the optimal desorption solvent.

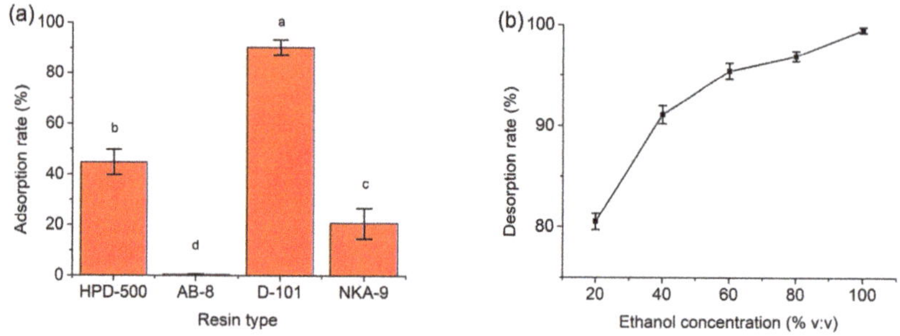

Figure 3. (**a**) The adsorption rate of four different macroporous resins (HPD-500, AB-8, D-101, and NKA-9); (**b**) The desorption rate obtained by different ethanol concentrations. Values followed by different letters (a–d) are significantly different according to Duncan's test ($p < 0.05$).

Since the flavonoids in the solvent were adsorbed by macroporous resin, the DES could be reused for the next round of extraction [48]. After the evaporation of water from DES through rotary evaporators, a new solvent was prepared based on the appropriate water content and applied for the next cycle of extraction. As shown in Figure 4, the extraction efficiency of DES for flavonoids was decreased with increasing reuse of recovered DES. The extraction efficiency of the fifth cycle was only 78% of the initial cycle. The results demonstrated that DES could be reused four times leading to cost savings.

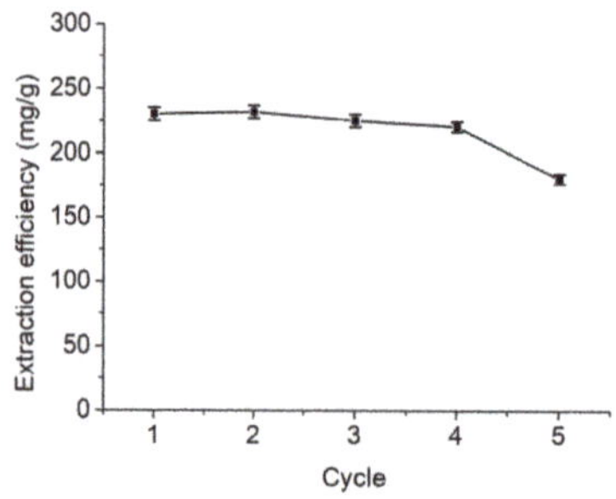

Figure 4. The recycling of recovered DES. One g of powdered dried lotus leaf was added to 20.0 mL of the DES. After extraction by the UAE-DES method, the DES extracts were adsorbed with 4.0 g of D101. After the adsorption procedure, the resin was desorbed and the DES were reused. The DES were evaporated to remove water and an appropriate amount of water was added for the next cycle of extraction.

3.4. Antioxidant Activities of the DES Extract

Antioxidant activity was an important bioactivity for natural components, and five assays were introduced to multi-aspect evaluate the antioxidant capacity [49]. The antioxidant activities including DPPH, ABTS, Fe^{3+} reducing, FRAP, and Fe^{2+} chelating of the DES extracts are presented in Figure 5. As a result, the DES extract ranged from 1 mg/mL to 20 mg/mL and expressed gradually increased antioxidant activities in different degrees. For the DPPH and ABTS radical scavenging abilities (Figure 5a), the DES extract showed the highest inhibition at 10 mg/mL (94.1% and 94.7%). These results showed that the DES extract had strong free radical scavenging activities, although it was lower than that of rutin. The iron chelating activity was significant of active compounds in the food and cosmetic industries, which meant the protection of the metabolism, skin, and environment against heavy metals [50]. The DES extract exhibited an increasing inhibition ability with increasing concentrations and reached 41.3% at 20 mg/mL (Figure 5a), while rutin showed no activity and only 0.11% at 0.5 mg/mL. In the Fe^{3+} reduction and FRAP assays, the values were expressed as the equivalent concentration of standards. The higher the values, the better the antioxidant activities the sample exhibited. The measured Fe^{3+} reduction and FRAP values reached 0.73 mg RE /mL and 0.71 mmol/L Fe^{2+} (Figure 5b), while the FRAP value of 0.1 mg/mL rutin was 0.74 mmol/L Fe^{2+}, which could be referred to as a reference. As some reports have shown, the ethanol extract of lotus leaves exhibited antioxidant activities in DPPH scavenging activity, ABTS scavenging activity, and the FRAP assay [51]. Compared to water extraction, methanol extraction had higher antioxidant activities [52]. In conclusion, the DES extract showed dose-dependent antioxidant activities in these five assays, and it could be further developed as a crude antioxidant in many areas.

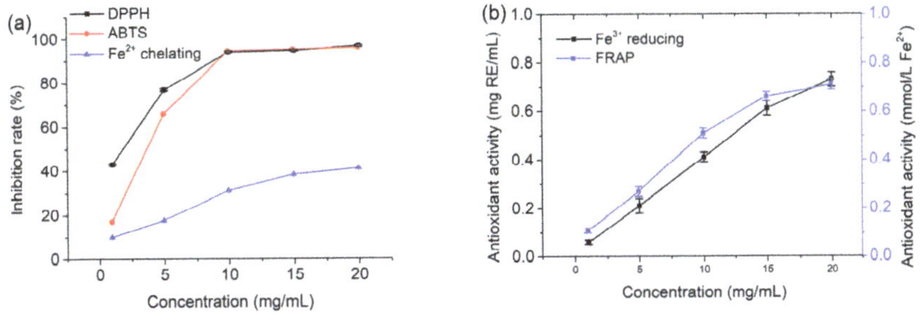

Figure 5. The antioxidant activities of DES extract in different concentrations. (**a**) DPPH, ABTS, and Fe^{2+} chelating assay. (**b**) Fe^{3+} reducing and FRAP assay.

3.5. Antibacterial Activity of the DES Extract

Antibacterial evaluation is a rapid, simple, and effective method to determine antimicrobial activity [53]. The antibacterial potentials of the DES extract against *Staphylococcus aureus* and *Escherichia coli* bacterial strains were investigated and are shown in Table 2. The DIZ of the DES extract at the concentration of 100 mg/mL was 8.49 ± 0.01 mm and 12.29 ± 0.01 mm for *S. aureus* and *E. coli*. It could be found that the antibacterial effect of the DES extract on two bacteria was confirmed, although the activities were lower than rutin. The antibacterial ability of the DES extract was further verified by the determination of the MIC value. As revealed in Table 2, the MIC values of the DES extract to *S. aureus* and *E. coli* were 1666 and 208 μg/mL, respectively, while the MIC values of rutin were 100 and 400 μg/mL, which was consistent with the results of DIZ. Antibacterial activity tests showed that the DES extract had a good inhibitory effect on *E. coli* at a low concentration, which made the potential usage of the DES extract an antibacterial agent [54]. The antibacterial testing of lotus leaf extracts is rare. Abderrahim Benslama reported that methanol extract has a moderate antibacterial activity against *Staphylococcus aureus*, *Micrococcus luteus*, and *Bacillus subtilis* [55]. The antibacterial ability of the DES extract in this study could expand the application range of lotus leaves.

Table 2. The antibacterial activities of DES extracts.

Antibacterial Properties	Samples	*S. aureus*	*E. coli*
DIZ (mm)	DES extract (50 mg/mL)	7.29 ± 0.01	9.02 ± 0.02
	DES extract (100 mg/mL)	8.49 ± 0.01	12.29 ± 0.01
	Rutin (1 mg/mL)	7.21 ± 0.01	7.11 ± 0.01
MIC (μg/mL)	DES extract	1666	208
	Rutin	100	400

3.6. HPLC Analysis of the DES Extract

For a better understanding of the extraction of DES, the composition of the DES extract was analyzed by HPLC. The typical chromatogram of the DES extract is shown in Figure 6. As shown in the Figure, the peaks were well separated, and three peaks at 17.3 min, 20.1 min, and 21.6 min could be found. In the previous study in our group, many flavonoid standards were acquired and analyzed [56]. Through the comparison of the UV spectra and retention time in the same HPLC analysis conditions between the DES extract and the standards, these three main peaks in the DES extract were identified as astragalin, hyperoside, and isoquercitrin. The structures and UV spectra of them are shown in the Supplementary File. The chromatograms of the three standards are also illustrated in Figure 6 (colorful lines). It could be found that the retention times of these standards were the same as those peaks in the DES extract. Astragalin was identified from the receptacles of N. nucifera as the main component in the inhibition of tyrosinase activity [57]. In other groups' reports, hyperoside and isoquercitrin were also identified from lotus leaves, which could confirm their existence in lotus leaves [2,58]. Based on these data and analyses, three peaks in the DES extract were identified as astragalin, hyperoside, and isoquercitrin, which made further utilization of the DES extract easier.

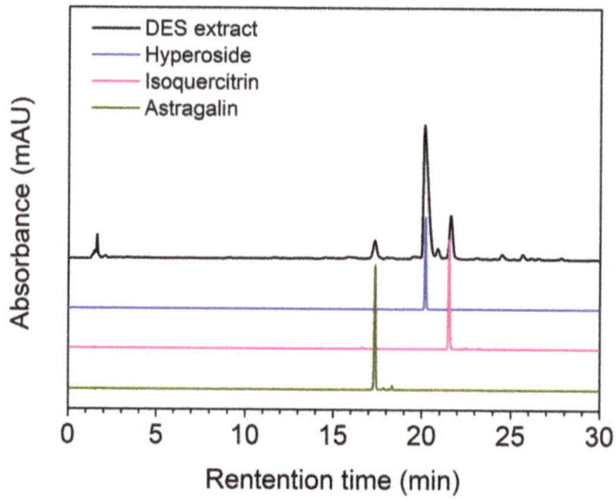

Figure 6. The chromatograms of the DES extract (black line), hyperoside (blue line), isoquercitrin (violet line), and astragalin (dark yellow line).

4. Conclusions

In this study, the flavonoids from lotus leaves were extracted using UAE-DES. The choline chloride/urea system was selected from 34 kinds of DES compositions, and the factors related to the preparation of the DES and the extraction process were optimized. Through the following purification by D-101 macroporous resin, the flavonoid extract was concentrated and the solvent could be recycled. The antioxidant and antibacterial tests confirmed that the extracted lotus leaf flavonoids had good activities in dose-dependent manners, which has promising applications in food additives and food preservations. Through the HPLC analysis of the DES extract, three peaks were positively identified as astragalin, hyperoside, and isoquercitrin by comparing the UV and retention time data. Based on the findings in this study, a proper deep eutectic solvent was optimized for the extraction of flavonoids from lotus leaves using UAE-DES, which gave a more eco-friendly method for the development of lotus-based products. To meet the challenge, further research in the future could be focused on research of specific flavonoids and the industrial application of DES in the separation of natural compounds.

Supplementary Materials: The following supporting information can be downloaded at: https://www.mdpi.com/article/10.3390/separations10020065/s1, Figure S1: The DIZ of DES extracts (w: Control, x: 50mg/mL DES extract, y: 100mg/mL DES extract, z: 1mg/mL Rutin); Figure S2: The structures of astragalin, hyperoside and isoquercitrin; Figure S3: The UV spectra of three peaks in DES extract, astragalin, hyperoside and isoquercitrin.

Author Contributions: Data curation, Y.Z.; formal analysis, Y.Z.; methodology, L.L. and S.D.; project administration, S.D.; writing—original draft, L.L.; writing—review and editing, A.X. All authors have read and agreed to the published version of the manuscript.

Funding: This research received no external funding.

Data Availability Statement: The data presented in this study are available on request from the corresponding author. The data are not publicly available due to security requirements in the institute of the corresponding author.

Acknowledgments: This work was supported by the Natural Science Foundation of Changsha, China (Grant No. kq2202323) and the Chinese Agricultural Science and Technology Innovation Project (No. ASTIP-IBFC05).

Conflicts of Interest: The authors declare no conflict of interest.

References

1. Liu, J.; Shi, K.; Shi, J.; Feng, Y.; Hao, C.; Peng, J.; Chen, S. A simple strategy to monitor the temporal and spatial distribution of alkaloids in sacred lotus leaves. *Biosci. Biotechnol. Biochem.* **2021**, *85*, 1332–1340. [CrossRef] [PubMed]
2. Chen, S.; Zheng, Y.; Fang, J.-B.; Liu, Y.-L.; Li, S.-H. Flavonoids in lotus (Nelumbo) leaves evaluated by HPLC–MSn at the germplasm level. *Food Res. Int.* **2013**, *54*, 796–803. [CrossRef]
3. Limwachiranon, J.; Huang, H.; Shi, Z.; Li, L.; Luo, Z. Lotus Flavonoids and Phenolic Acids: Health Promotion and Safe Consumption Dosages. *Compr. Rev. Food Sci. Food Saf.* **2018**, *17*, 458–471. [CrossRef]
4. Wang, Z.; Li, Y.; Ma, D.; Zeng, M.; Wang, Z.; Qin, F.; Chen, J.; Christian, M.; He, Z. Alkaloids from lotus (Nelumbo nucifera): Recent advances in biosynthesis, pharmacokinetics, bioactivity, safety, and industrial applications. *Crit. Rev. Food Sci. Nutr.* **2021**, *2021*, 1–34. [CrossRef] [PubMed]
5. Yaneva, Z.; Simeonov, E.; Rusenova, N.; Ivanova, D.; Nikolova, G.; Karamalakova, Y.; Chilev, C.; Beev, G. Flavonoids Extraction Kinetics, Antimicrobial Activity and Radical Scavenging Potential of Bulgarian Woundwort (*Solidago virgaurea* L.). *Separations* **2022**, *9*, 27. [CrossRef]
6. Abbott, A.P.; Capper, G.; Davies, D.L.; Rasheed, R.K.; Tambyrajah, V. Novel solvent properties of choline chloride/urea mixtures. *Chem. Commun.* **2003**, *39*, 70–71. [CrossRef]
7. Dheyab, A.S.; Kanaan, M.Q.; Hussein, N.A.; AlOmar, M.K.; Sabran, S.F.; Abu Bakar, M.F. Antimycobacterial Activity of Rosmarinus officinalis (Rosemary) Extracted by Deep Eutectic Solvents. *Separations* **2022**, *9*, 271. [CrossRef]
8. Dheyab, A.S.; Abu Bakar, M.F.; AlOmar, M.; Sabran, S.F.; Muhamad Hanafi, A.F.; Mohamad, A. Deep Eutectic Solvents (DESs) as Green Extraction Media of Beneficial Bioactive Phytochemicals. *Separations* **2021**, *8*, 176. [CrossRef]
9. Dwamena, A.K. Recent Advances in Hydrophobic Deep Eutectic Solvents for Extraction. *Separations* **2019**, *6*, 9. [CrossRef]
10. Amoroso, R.; Hollmann, F.; Maccallini, C. Choline Chloride-Based DES as Solvents/Catalysts/Chemical Donors in Pharmaceutical Synthesis. *Molecules* **2021**, *26*, 6286. [CrossRef]
11. Sillero, L.; Prado, R.; Welton, T.; Labidi, J. Extraction of flavonoid compounds from bark using sustainable deep eutectic solvents. *Sustain. Chem. Pharm.* **2021**, *24*, 100544. [CrossRef]
12. Qi, X.-L.; Peng, X.; Huang, Y.-Y.; Li, L.; Wei, Z.-F.; Zu, Y.-G.; Fu, Y.-J. Green and efficient extraction of bioactive flavonoids from Equisetum palustre L. by deep eutectic solvents-based negative pressure cavitation method combined with macroporous resin enrichment. *Ind. Crops Prod.* **2015**, *70*, 142–148. [CrossRef]
13. Hao, H.; Lin, L.; Liu, S.; Kang, Y.; Wang, Y.; Huang, J.; Weng, W. Deep Eutectic Solvent-Based Microwave-Assisted Extraction for the Chromatographic Analysis of Bioactive Flavonoids in Spirodela polyrrhiza. *J. Chromatogr. Sci.* **2022**, *60*, 501–510. [CrossRef] [PubMed]
14. Balanescu, F.; Busuioc, A.C.; Botezatu, A.V.; Gosav, S.; Avramescu, S.M.; Furdui, B.; Dinica, R.M. Comparative Study of Natural Antioxidants from Glycine max, Anethum graveolensand Pimpinella anisum Seed and Sprout Extracts Obtained by Ultrasound-Assisted Extraction. *Separations* **2022**, *9*, 152. [CrossRef]
15. Boli, E.; Prinos, N.; Louli, V.; Pappa, G.; Stamatis, H.; Magoulas, K.; Voutsas, E. Recovery of Bioactive Extracts from Olive Leaves Using Conventional and Microwave-Assisted Extraction with Classical and Deep Eutectic Solvents. *Separations* **2022**, *9*, 255. [CrossRef]
16. Tsvetov, N.; Sereda, L.; Korovkina, A.; Artemkina, N.; Kozerozhets, I.; Samarov, A. Ultrasound-assisted extraction of phytochemicals from Empetrum hermafroditum Hager. using acid-based deep eutectic solvent: Kinetics and optimization. *Biomass Convers. Biorefinery* **2022**, *12*, 145–156. [CrossRef]
17. Xue, H.; Li, J.; Wang, G.; Zuo, W.; Zeng, Y.; Liu, L. Ultrasound-Assisted Extraction of Flavonoids from *Potentilla fruticosa* L. Using Natural Deep Eutectic Solvents. *Molecules* **2022**, *27*, 5794. [CrossRef] [PubMed]
18. Gao, H.; Wang, Y.; Guo, Z.; Liu, Y.; Wu, Q.; Xiao, J. Optimization of ultrasound-assisted extraction of phenolics from Asparagopsis taxiformis with deep eutectic solvent and their characterization by ultra-high-performance liquid chromatography-mass spectrometry. *Front. Nutr.* **2022**, *9*, 1036436. [CrossRef]

19. Qin, G.; Lei, J.; Li, S.; Jiang, Y.; Qiao, L.; Ren, M.; Gao, Q.; Song, C.; Fu, S.; Zhou, J.; et al. Efficient, green extraction of two biflavonoids from Selaginella uncinata with deep eutectic solvents. *Microchem. J.* **2022**, *183*, 108085. [CrossRef]
20. Aouam, I.; El Atki, Y.; Taleb, M.; Taroq, A.; El Kamari, F.; Lyoussi, B.; Abdellaoui, A. Antioxidant Capacities and Total Phenolic Contents of Thymus riatarum. *Mater. Today Proc.* **2019**, *13*, 579–586. [CrossRef]
21. Sang, J.; Liu, K.; Ma, Q.; Li, B.; Li, C.-Q. Combination of a deep eutectic solvent and macroporous resin for green recovery of anthocyanins from Nitraria tangutorun Bobr. fruit. *Sep. Sci. Technol.* **2019**, *54*, 3082–3090. [CrossRef]
22. Zhang, L.; Liu, L.; Xiao, A.; Huang, S.; Li, D. Screening and analysis of xanthine oxidase inhibitors in jute leaves and their protective effects against hydrogen peroxide-induced oxidative stress in cells. *Open Chem.* **2020**, *18*, 1481–1494. [CrossRef]
23. Mansinhos, I.; Gonçalves, S.; Rodríguez-Solana, R.; Ordóñez-Díaz, J.L.; Moreno-Rojas, J.M.; Romano, A. Ultrasonic-Assisted Extraction and Natural Deep Eutectic Solvents Combination: A Green Strategy to Improve the Recovery of Phenolic Compounds from Lavandula pedunculata subsp. lusitanica (Chaytor) Franco. *Antioxidants* **2021**, *10*, 582. [CrossRef]
24. Huang, Q.; Zhang, H.; Xue, D. Enhancement of antioxidant activity of Radix Puerariae and red yeast rice by mixed fermentation with Monascus purpureus. *Food Chem.* **2017**, *226*, 89–94. [CrossRef] [PubMed]
25. Buravlev, E.V.; Shevehenko, O.G. 2-Hydroxy-3-isobornyl-5-methylbenzaldehyde derivatives: Synthesis and antioxidant activity in vitro. *Russ. Chem. Bull.* **2019**, *68*, 79–85. [CrossRef]
26. Xie, Y.; Chen, J.; Xiao, A.; Liu, L. Antibacterial Activity of Polyphenols: Structure-Activity Relationship and Influence of Hyperglycemic Condition. *Molecules* **2017**, *22*, 1913. [CrossRef] [PubMed]
27. Alanazi, A.K.; Alqasmi, M.H.; Alrouji, M.; Kuriri, F.A.; Almuhanna, Y.; Joseph, B.; Asad, M. Antibacterial Activity of Syzygium aromaticum (Clove) Bud Oil and Its Interaction with Imipenem in Controlling Wound Infections in Rats Caused by Methicillin-Resistant Staphylococcus aureus. *Molecules* **2022**, *27*, 8551. [CrossRef]
28. Naim, N.; Bouymajane, A.; Oulad El Majdoub, Y.; Ezrari, S.; Lahlali, R.; Tahiri, A.; Ennahli, S.; Laganà Vinci, R.; Cacciola, F.; Mondello, L.; et al. Flavonoid Composition and Antibacterial Properties of *Crocus sativus* L. Petal Extracts. *Molecules* **2023**, *28*, 186. [CrossRef] [PubMed]
29. Wang, G.; Cui, Q.; Yin, L.-J.; Zheng, X.; Gao, M.-Z.; Meng, Y.; Wang, W. Efficient extraction of flavonoids from Flos Sophorae Immaturus by tailored and sustainable deep eutectic solvent as green extraction media. *J. Pharm. Biomed. Anal.* **2019**, *170*, 285–294. [CrossRef] [PubMed]
30. Wang, X.; Wu, Y.; Li, J.; Wang, A.; Li, G.; Ren, X.; Yin, W. Ultrasound-assisted deep eutectic solvent extraction of echinacoside and oleuropein from Syringa pubescens Turcz. *Ind. Crops Prod.* **2020**, *151*, 112442. [CrossRef]
31. Sui, M.; Feng, S.; Liu, G.; Chen, B.; Li, Z.; Shao, P. Deep eutectic solvent on extraction of flavonoid glycosides from Dendrobium officinale and rapid identification with UPLC-triple-TOF/MS. *Food Chem.* **2023**, *401*, 134054. [CrossRef] [PubMed]
32. Zeng, J.; Dou, Y.; Yan, N.; Li, N.; Zhang, H.; Tan, J.-N. Optimizing Ultrasound-Assisted Deep Eutectic Solvent Extraction of Bioactive Compounds from Chinese Wild Rice. *Molecules* **2019**, *24*, 2718. [CrossRef] [PubMed]
33. Jakovljević, M.; Vladić, J.; Vidović, S.; Pastor, K.; Jokić, S.; Molnar, M.; Jerković, I. Application of Deep Eutectic Solvents for the Extraction of Rutin and Rosmarinic Acid from *Satureja montana* L. and Evaluation of the Extracts Antiradical Activity. *Plants* **2020**, *9*, 153. [CrossRef]
34. Pal, C.B.T.; Jadeja, G.C. Deep eutectic solvent-based extraction of polyphenolic antioxidants from onion (*Allium cepa* L.) peel. *J. Sci. Food Agric.* **2019**, *99*, 1969–1979. [CrossRef] [PubMed]
35. Zuo, J.; Ma, P.; Geng, S.; Kong, Y.; Li, X.; Fan, Z.; Zhang, Y.; Dong, A.; Zhou, Q. Optimization of the extraction process of flavonoids from Trollius ledebouri with natural deep eutectic solvents. *J. Sep. Sci.* **2022**, *45*, 717–727. [CrossRef]
36. Ali, M.C.; Chen, J.; Zhang, H.; Li, Z.; Zhao, L.; Qiu, H. Effective extraction of flavonoids from *Lycium barbarum* L. fruits by deep eutectic solvents-based ultrasound-assisted extraction. *Talanta* **2019**, *203*, 16–22. [CrossRef]
37. Huang, J.; Guo, X.; Xu, T.; Fan, L.; Zhou, X.; Wu, S. Ionic deep eutectic solvents for the extraction and separation of natural products. *J. Chromatogr. A* **2019**, *1598*, 1–19. [CrossRef]
38. Zheng, B.; Yuan, Y.; Xiang, J.; Jin, W.; Johnson, J.B.; Li, Z.; Wang, C.; Luo, D. Green extraction of phenolic compounds from foxtail millet bran by ultrasonic-assisted deep eutectic solvent extraction: Optimization, comparison and bioactivities. *LWT* **2022**, *154*, 112740. [CrossRef]
39. Luo, Q.; Zhang, J.-R.; Li, H.-B.; Wu, D.-T.; Geng, F.; Corke, H.; Wei, X.-L.; Gan, R.-Y. Green Extraction of Antioxidant Polyphenols from Green Tea (*Camellia sinensis*). *Antioxidants* **2020**, *9*, 785. [CrossRef]
40. Liu, Y.; Li, J.; Fu, R.; Zhang, L.; Wang, D.; Wang, S. Enhanced extraction of natural pigments from *Curcuma longa* L. using natural deep eutectic solvents. *Ind. Crops Prod.* **2019**, *140*, 111620. [CrossRef]
41. Zhang, H.; Hao, F.; Yao, Z.; Zhu, J.; Jing, X.; Wang, X. Efficient extraction of flavonoids from Polygonatum sibiricum using a deep eutectic solvent as a green extraction solvent. *Microchem. J.* **2022**, *175*, 107168. [CrossRef]
42. Zhu, S.; Liu, D.; Zhu, X.; Su, A.; Zhang, H. Extraction of Illegal Dyes from Red Chili Peppers with Cholinium-Based Deep Eutectic Solvents. *J. Anal. Methods Chem.* **2017**, *2017*, 2753752. [CrossRef] [PubMed]
43. Meng, Z.; Zhao, J.; Duan, H.; Guan, Y.; Zhao, L. Green and efficient extraction of four bioactive flavonoids from Pollen Typhae by ultrasound-assisted deep eutectic solvents extraction. *J. Pharm. Biomed. Anal.* **2018**, *161*, 246–253. [CrossRef] [PubMed]
44. Wang, X.-H.; Wang, J.-P. Effective extraction with deep eutectic solvents and enrichment by macroporous adsorption resin of flavonoids from *Carthamus tinctorius* L. *J. Pharm. Biomed. Anal.* **2019**, *176*, 112804. [CrossRef]

45. Wang, Y.; Hu, Y.; Wang, H.; Tong, M.; Gong, Y. Green and enhanced extraction of coumarins from Cortex Fraxini by ultrasound-assisted deep eutectic solvent extraction. *J. Sep. Sci.* **2020**, *43*, 3441–3448. [CrossRef]
46. Xia, G.-H.; Li, X.-H.; Jiang, Y.-H. Deep eutectic solvents as green media for flavonoids extraction from the rhizomes of Polygonatum odoratum. *Alex. Eng. J.* **2021**, *60*, 1991–2000. [CrossRef]
47. Zhang, X.; Su, J.; Chu, X.; Wang, X. A Green Method of Extracting and Recovering Flavonoids from Acanthopanax senticosus Using Deep Eutectic Solvents. *Molecules* **2022**, *27*, 923. [CrossRef]
48. Gao, M.; Wang, D.; Deng, L.; Liu, S.; Zhang, K.; Quan, T.; Yang, L.; Kang, X.; Xia, Z.; Gao, D. High-Crystallinity Covalent Organic Framework Synthesized in Deep Eutectic Solvent: Potentially Effective Adsorbents Alternative to Macroporous Resin for Flavonoids. *Chem. Mater.* **2021**, *33*, 8036–8051. [CrossRef]
49. Al-Azzawi, A.; Al Dibsawi, A.; Talath, S.; Wali, A.F.; Sarheed, O. Method Development: The Antioxidant and Antifungal Activity of the Isolated Component from the Ethanolic Extract of Tecoma stans Leaves Using Flash Chromatography. *Separations* **2022**, *9*, 317. [CrossRef]
50. Oliveira, G.; Marques, C.; de Oliveira, A.; de Almeida dos Santos, A.; do Amaral, W.; Ineu, R.P.; Leimann, F.V.; Peron, A.P.; Igarashi-Mafra, L.; Mafra, M.R. Extraction of bioactive compounds from *Curcuma longa* L. using deep eutectic solvents: In vitro and in vivo biological activities. *Innov. Food Sci. Emerg. Technol.* **2021**, *70*, 102697. [CrossRef]
51. Zhu, M.-Z.; Wu, W.; Jiao, L.-L.; Yang, P.-F.; Guo, M.-Q. Analysis of Flavonoids in Lotus (Nelumbo nucifera) Leaves and Their Antioxidant Activity Using Macroporous Resin Chromatography Coupled with LC-MS/MS and Antioxidant Biochemical Assays. *Molecules* **2015**, *20*, 10553–10565. [CrossRef] [PubMed]
52. Su, D.; Li, N.; Chen, M.; Yuan, Y.; He, S.; Wang, Y.; Wu, Q.; Li, L.; Yang, H.; Zeng, Q. Effects of in vitro digestion on the composition of flavonoids and antioxidant activities of the lotus leaf at different growth stages. *Int. J. Food Sci. Technol.* **2018**, *53*, 1631–1639. [CrossRef]
53. An, J.-Y.; Wang, L.-T.; Lv, M.-J.; Wang, J.-D.; Cai, Z.-H.; Wang, Y.-Q.; Zhang, S.; Yang, Q.; Fu, Y.-J. An efficiency strategy for extraction and recovery of ellagic acid from waste chestnut shell and its biological activity evaluation. *Microchem. J.* **2021**, *160*, 105616. [CrossRef]
54. Naseem, Z.; Zahid, M.; Hanif, M.A.; Shahid, M. Green extraction of ethnomedicinal compounds from Cymbopogon citratus Stapf using hydrogen-bonded supramolecular network. *Sep. Sci. Technol.* **2021**, *56*, 1520–1533. [CrossRef]
55. Benslama, A.; Harrar, A.; Gul, F.; Demirtas, I. Phenolic Compounds, Antioxidant and Antibacterial Activities of *Zizyphus lotus* L. Leaves Extracts. *Nat. Prod. J.* **2017**, *7*, 316–322. [CrossRef]
56. Tang, X.; Tang, P.; Liu, L. Molecular Structure–Affinity Relationship of Flavonoids in Lotus Leaf (Nelumbo nucifera Gaertn.) on Binding to Human Serum Albumin and Bovine Serum Albumin by Spectroscopic Method. *Molecules* **2017**, *22*, 1036. [CrossRef]
57. Jung, S.Y.; Jung, W.S.; Jung, H.K.; Lee, G.H.; Cho, J.H.; Cho, H.W.; Choi, I.Y. The mixture of different parts of Nelumbo nucifera and two bioactive components inhibited tyrosinase activity and melanogenesis. *J. Cosmet. Sci.* **2014**, *65*, 377–388.
58. Cho, W.-K.; Yang, H.J.; Ma, J.Y. Lotus (Nelumbo nucifera Gaertn.) leaf water extracts suppress influenza a viral infection via inhibition of neuraminidase and hemagglutinin. *J. Funct. Foods* **2022**, *91*, 105019. [CrossRef]

separations

MDPI

Article

Bioassay-Guided Isolation and Identification of Antibacterial Components against *Escherichia coli* from Industrial Hemp Leaves

Yafen Fu, Siyuan Zhu *, Shengwen Duan * and Liangliang Liu *

Institute of Bast Fiber Crops, Chinese Academy of Agricultural Sciences, Changsha 410205, China
* Correspondence: zhusiyuan@caas.cn (S.Z.); duanshengwen@caas.cn (S.D.); liuliangliang@caas.cn (L.L.); Tel.: +86-731-88998579 (S.Z.); +86-731-88998549 (S.D.); +86-731-88998525 (L.L.)

Abstract: Industrial hemp leaves have raised much interest in nutraceuticals and functional foods areas. To expand its application ranges, the antibacterial activities of industrial hemp leaf extract on *Escherichia coli*, *Staphylococcus aureus*, and *Bacillus cereus* were evaluated and the active components were screened. As a result, the industrial hemp leaf extract was found to have strong bacteriostatic effects on *E. coli* and *S. aureus*. Bioassay-guided fractionation and isolation from fractions active against *E. coli* were conducted. Two compounds, cannabidivarinic acid and cannabidiolic acid, were firstly recognized by analytical HPLC by comparing the retention times and UV spectra with standards and later isolated using preparative HPLC. Moreover, the antibacterial mechanisms of cannabidivarinic acid and cannabidiolic acid were investigated by testing the alkaline phosphatase activity, β-galactosidase activity, conductivity, proteins leakage, nucleic acid leakage, and scanning electron microscope observation. The results demonstrated that cannabidivarinic acid and cannabidiolic acid could destroy the cell wall and membrane of *E. coli*, resulting in the inhibition of enzyme activity and leakage of contents. They could damage the bacteria cell envelope as well. Presented results pointed out cannabidivarinic acid and cannabidiolic acid as promising natural bacteriostatic agents for the food, pharmaceutical, and cosmetic industry.

Keywords: antibacterial; cannabidiolic acid; cannabidivarinic acid; *Escherichia coli*; HPLC; industrial hemp

Citation: Fu, Y.; Zhu, S.; Duan, S.; Liu, L. Bioassay-Guided Isolation and Identification of Antibacterial Components against *Escherichia coli* from Industrial Hemp Leaves. *Separations* **2023**, *10*, 35. https://doi.org/10.3390/separations10010035

Academic Editor: Marcello Locatelli

Received: 13 December 2022
Revised: 4 January 2023
Accepted: 4 January 2023
Published: 6 January 2023

1. Introduction

Industrial hemp (IH) is an annual herbal plant of the cannabis species (Cannabaceae) which is cultivated for fiber, oil, and dietary supplements [1]. IH is widely cultivated worldwide and has a long planting history in China for its applications in textile and consumption. In recent years, various kinds of nutraceuticals and functional foods made from IH leaves (IHL) and seeds are presented [2]. The most active compounds in IHL are flavonoids, alkaloids, terpenes, and cannabinoids, which are responsible for diverse pharmacological functions [3]. As a group of unique and typical natural active compounds, cannabinoids have anti-pain activity closely related to various physiological activities such as obesity, dyslipidemia, hypertension, pain, and diabetes [4,5]. Specifically, cannabinol has inhibition on cancer cells [6]. Cannabidivarin could be applied in medicine, nutrition, and cosmetics, limiting the production of psychoactive and illegal cannabinoids [7]. In general, cannabinoids have great potential in scientific research and application development [8].

Escherichia coli (*E. coli*) is one of the bacteria that seriously endanger the healthy development of humans and animals [9]. Under certain conditions, it would cause diarrhea in humans and animals [10]. It also could cause the spoilage of food, which is the main topic in food preservation [11]. For a long time, its prevention and treatment mainly relied on antibiotics. With the concern of food safety and drug resistance, screening of natural sources, pathogenic bacteria sensitivity, and low toxicity active components to prevent and control human and animal diseases has become the current research hot spot and trend [12,13].

Research of the antibacterial activity of IHL extract (IHLE) and cannabinoids on *E. coli* is relatively rare. Ali et al. investigated seed oil, petroleum ether extract, and methanol extract of the whole IH for their antimicrobial activity against four bacterial strains including *E. coli*. The results show that the petroleum ether extract exhibited high activity against *E. coli* [14]. Novak investigated the activity of essential oils against potential pathogen microorganisms including *E. coli*. However, only one variety shows inhibition on *E. coli* [15]. Naveed reported the ethanol extract of the leaf of *Cannabis Sativa* exerted antibacterial activity against E. coli [16]. Another report shows that CBD could significantly enhance the antibacterial effect of erythromycin and rifampicin against *E. coli* [17].

Staphylococcus aureus is a gram-positive and aerobic bacterium and opportunistic pathogen [18]. The main sites of colonization for *S. aureus* are the skin and mucous membranes. It could cause a wide range of diseases including common problems such as skin infections, and some severe problems such as endocarditis, bloodstream infections, and osteomyelitis [19]. *Bacillus cereus* is a common foodborne gram-positive pathogenic bacteria existing in food, soil, and human skin [20]. It could produce cytotoxins and enzymes that cause diarrhea and vomiting, such as hemolysin BL, nonhemolytic enterotoxin, cytotoxin K, and vomitoxin. *B. cereus* has become a threat to the health of populations since about 1.4–12% of the world's food poisoning is caused by it [21]. Consequently, the finding of antibacterial ingredients for these bacteria is meaningful and helpful for human health.

2. Materials and Methods

2.1. Reagents

IHL was provided by the Institute of Bast Fiber Crops, Chinese Academy of Agricultural Sciences, which was collected in Yunnan province in 2019. Institute of Bast Fiber Crops, Chinese Academy of Agricultural Sciences provided *E. coli*, *S. aureus*, and *B. cereus*. The strains were cultured in a nutrient broth at 37 °C for 48 h with shaking. Ethanol, petroleum ether, ethyl acetate, *n*-butanol, and glutaraldehyde were bought from Sinopharm Chemical Reagent Co., Ltd. (Shanghai, China). Acetic acid and acetonitrile in HPLC grade were obtained from Macklin Inc. (Shanghai, China). Deionized water used in the experiments was supplied by an ELGA water purification system (ELGA Berkefeld, Veolia, Germany). All other chemicals were of analytical grade.

2.2. Preparation of IHLE

The workflow was shown in Figure 1. The collected IHL were dried in the shade and then ground into powder. A total of 25.0 g of IHL powders were weighed and poured into 250 mL of 70% ethanol solution. The solution was refluxed at 80 °C for 2 h two times and obtained extracts were combined and evaporated using a vacuum rotary evaporator to afford IHLE.

For preparing the antibacterial activity test, the IHLE was re-dissolved in 70% ethanol to adjust the concentration to certain values. The IHLE was re-dissolved in water to screen active antibacterial ingredients. In sequence, the suspension was extracted with petroleum ether, ethyl acetate, and *n*-butanol. Organic solvents and water were evaporated to obtain appropriate extracts. Four parts of IHLE were, respectively, redissolved in 70% ethanol to certain concentrations and marked as IHLE-p, IHLE-e, IHLE-b, and IHLE-w, which were diluted with water in further experiments.

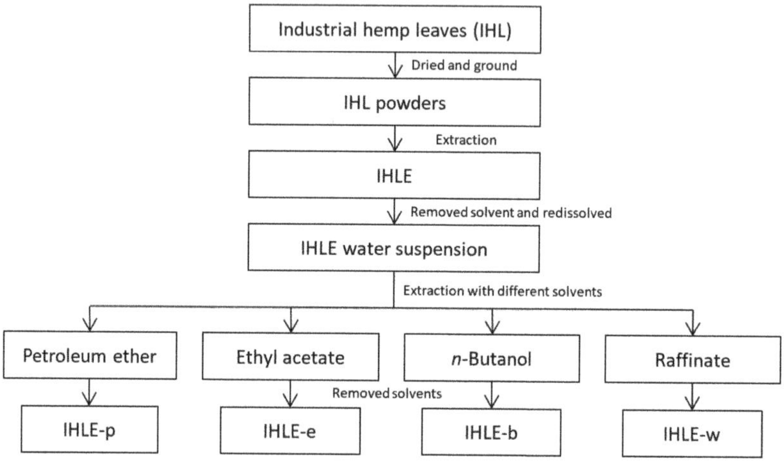

Figure 1. The workflow of extraction and fractionation of IHLE.

2.3. *Evaluation of Antibacterial Activity*

2.3.1. Determination of Bacterial Inhibition Zone

The diameter of the inhibition zone (DIZ) of extracts was determined by the paper disc diffusion method, as previous reports described [22]. Three bacteria suspensions (*E. coli*, *S. aureus*, and *B. cereus*) were cultured in a nutrient broth at 37 °C for 48 h and diluted to about 1×10^{-6} CFU. A total of 100 μL of bacterial suspensions was evenly spread on an agar plate. Then, sterilized filter paper discs (Φ = 6 mm) were soaked in samples with different concentrations and carefully placed on the plates. The plates were incubated at 37 °C for 24 h, and the DIZ was measured by a vernier caliper. A solvent without samples was used as a negative control.

2.3.2. Determination of Minimal Inhibitory Concentration (MIC)

The MIC was determined by the micro-dilution broth method with some modifications [23]. Extracts with concentrations of 100.0, 50.0, 25.0, 12.5, and 6.25 mg/mL were obtained by the double dilution method. A total of 500 μL of nutrient broth and 50 μL of samples were mixed in a tube. Then, the bacterial solution (1×10^{-6} CFU) was added to the tube and incubated at 37 °C for 12 h. The absorbance at 600 nm of the bacterial suspensions was monitored by a spectrophotometer at predetermined time intervals. The nutrient broth was used as a negative control. The lowest concentration of the sample inhibiting the growth of bacteria was defined as the MIC value.

2.3.3. Effect of Extracts on the Growth Curve

To study the effect of IHL extracts on the growth curve of bacteria [24]. The sample was added to the bacteria suspensions (1×10^{-6} CFU) and incubated at 37 °C and 100 rpm. The absorbance at 600 nm of the cultures was monitored by a spectrophotometer at predetermined time intervals. The sample using nutrient broth instead of the extract was set as a control.

2.4. *Analytical and Preparative HPLC Conditions*

The analysis of IHL extracts was carried out on the Agilent 1260 HPLC system including a quaternary pump, an autosampler, a thermostatic column compartment, and a diode array detector (Agilent Technologies Inc., Santa Clara, CA, USA). The chromatographic separation was performed on the Waters Sunfire C_{18} column was employed (250 mm × 4.6 mm i.d.; 5 μm, Milford, MA, USA). The mobile phase consisted of solvent A (0.1% *v*/*v* acetic acid solution) and solvent B (acetonitrile containing 0.1% *v*/*v* acetic acid) with gradient elution: 0–5 min, 5% B; 5–20 min, 5–15% B; and 20–50 min, 15–95% B; and 50–60 min, 95% B. The flow rate, temperature, injection volume, and detection wavelength were respectively controlled at 0.8 mL/min, 25 °C, 20 μL, and 220 nm. In the analysis of IHLE-pe, the elution mode was re-optimized as isocratic elution as 60% B for 40 min. The separation and isolation of target active compounds were performed by preparative HPLC (Sykam S500, Eresing, Germany) with an Innoval ODS-2 column (150 mm × 10 mm i.d.; 5 μm, Agela company, Tianjin, China). The elution mode was isocratic elution at 65% B for 25 min. The flow rate was 1.2 mL/min, and the injection volume was 200 μL. The temperature and detector were set as in previous methods.

2.5. Antibacterial Mechanism

2.5.1. Scanning Electron Microscope (SEM) Observation

An appropriate amount of bacterial suspension was added to the liquid culture containing IHL in MIC, and the bacteria were cultured under shaking at 37 °C and 160 r/min for 12 h. Then, 10 mL of suspension was centrifuged (4000 r/min, 15 min), and fixed for 12 h with 2.5% glutaraldehyde solution. The fixed cells were pelleted and washed with PBS for 15 min 3 times. The samples were then dehydrated by gradient ethanol of 30%, 50%, 70%, and 90% respectively, and dried in freeze drying. After the spraying with gold, the morphology and structure of the bacteria were observed by SEM [25].

2.5.2. Effect on the Cell Membrane

The effect of IHL extracts and active compounds on *E. coli* cell membrane was studied by measuring conductivity, protein concentration, nucleic acid leakage, and β-galactosidase (β-GAL) enzyme activity [26,27]. The bacteria were treated with MIC concentration of samples for 3 h. Then, the suspension was centrifuged at 4000 r/min for 15 min, and the conductivity of the supernatant was measured with a conductivity meter after 20-fold dilution (DDSJ-308A, Shanghai Precision Scientific Instrument Co., Ltd., Shanghai, China). The content of proteins in suspension was determined by bicinchoninic acid (BCA) protein quantification kit (Beijing Solarbio Science & Technology Co., Ltd., Beijing, China). The concentration of nucleic acid in the supernatant was expressed as the absorbance at 260 nm by UV–Vis spectrophotometer (UV-2700, Shimadzu, Kyoto, Japan). The activity of β-GAL was measured by the Micro β-GAL Assay Kit provided by Solarbio (Beijing Solarbio Science & Technology Co., Ltd., Beijing, China).

2.5.3. Effect on the Cell Wall

The cell wall integrity of bacteria was measured through an alkaline phosphatase (AKP) activity assay [28]. The bacteria were treated with MIC concentration of samples for 3 h. Then, the suspension was centrifuged at 4000 r/min for 15 min, and the supernatant was submitted for determination. The extracellular AKP level was determined by spectrophotometry method using the commercial AKP assay kit following the manufacturer's instructions (Nanjing Jiancheng Bioengineering Institute, Nanjing, China).

2.6. Statistical Analysis

All values were expressed as means ± SD of triplicate experimental studies. Statistical analysis was performed by SPSS V20 software (IBM SPSS Statistics 20, New York, NY, USA). The data were then compared using Duncan's multiple range tests at 5% significance levels, and one-way ANOVA with Tukey's post hoc test [29,30].

3. Results and Discussion

3.1. Evaluation of the Antibacterial Activity of IHLE

3.1.1. Screening of the Target Bacterial Strain

IHLE was acquired by traditional reflux extraction and then we screened its antibacterial activity in *E. coli*, *S. aureus*, and *B. cereus*. The DIZ of IHLE in various concentrations was determined by the disc diffusion method, which could reflect the sensitivity of bacterial strains [31]. The highest bacteriostatic effects were observed on *E. coli* and *S. aureus*

(Table 1), with DIZ values of 15.57 ± 1.65 mm and 14.86 ± 1.37 mm when the extract was 100 mg/mL, respectively. However, the IHLE was insensitive to *B. cereus* and had no inhibitory effect. With the decrease in the concentration of IHLE, the antibacterial ring significantly decreased. The minimum inhibitory concentrations of *E. coli* and *S. aureus* were 12.5 mg/mL. In previous investigation of antibacterial activity of *Cannabis sativa* extracts against *E. coli*, ethyl acetate leaf extract exhibited the highest activity, ranging from 7.50 ± 0.71 to 21.50 ± 0.71 mm, which is in agreement with our results for 70% ethanol extract [32].

Table 1. The DIZ of three strains of IHLE in different concentrations (mm).

Bacteria Strains	100 mg/mL	50 mg/mL	25 mg/mL	12.5 mg/mL	6.25 mg/mL
E. coli	15.57 ± 1.65 ^{a}	11.66 ± 2.87 ^{b}	10.74 ± 2.23 ^{b}	8.65 ± 2.09 ^{b}	- ^{c}
S. aureus	14.86 ± 1.37 ^{a}	10.46 ± 2.57 ^{b}	9.02 ± 3.09 ^{b}	8.60 ± 2.21 ^{b}	- ^{c}
B. cereus	-	-	-	-	-

Values followed by different letters ($^{a–c}$) in the same column are significantly different according to Duncan's test ($p < 0.05$).

3.1.2. Screening of the Antibacterial Fraction

To find the active ingredients in the extract, the antibacterial activities of four fractions of IHLE including IHLE-p, IHLE-e, IHLE-b, and IHLE-w on *E. coli* were detected. Table 2 showed the DIZ of IHLE-p, IHLE-e, IHLE-b, and IHLE-w on *E. coli*, and the concentration of samples was adjusted to be 20 mg/mL. As a result, IHLE-p and IHLE-e had apparent bacteriostatic effects on *E. coli*, whose DIZ values were 15.43 ± 1.01 mm and 16.18 ± 1.28 mm, respectively. Meanwhile, IHLE-b and IHLE-w had no inhibitory effect on *E. coli*. Therefore, IHLE-p and IHLE-e were used as the objects in subsequent screening research to explore the specific antibacterial active components in IHLE.

Table 2. The DIZ of *E. coli* under different fractions of IHLE (mm).

Samples	Concentration	DIZ
IHLE-p	20 mg/mL	15.43 ± 1.01 ^{a}
IHLE-e	20 mg/mL	16.18 ± 1.28 ^{a}
IHLE-b	20 mg/mL	- ^{c}
IHLE-w	20 mg/mL	- ^{c}
IHLE-pe1	4 mg/mL	- ^{c}
IHLE-pe2	4 mg/mL	11.40 ± 1.41 ^{b}
IHLE-pe3	4 mg/mL	12.76 ± 1.55 ^{b}
IHLE-pe4	4 mg/mL	- ^{c}

Values followed by different letters ($^{a–c}$) in the same column are significantly different according to Duncan's test ($p < 0.05$).

To screen the active components, HPLC analyses of fractions were conducted. According to the chromatograms of IHLE-p and IHLE-e in Figure 2, the main peaks of IHLE-p and IHLE-e were similar. Therefore, these two parts were combined as IHLE-pe, and the new fraction's HPLC parameters were re-optimized. Under the new optimized HPLC condition, the IHLE-pe was well separated in Figure 3. According to the retention time in the chromatogram, the IHLE-pe was divided into four parts marked as IHLE-pe1 (1–9 min), IHLE-pe2 (9–15 min), IHLE-pe3 (15–25 min) and IHLE-pe4 (25–40 min).

For screening active ingredients in IHLE-pe, the antibacterial activities of IHLE-pe1 to IHLE-pe4 on *E. coli* were subsequently detected. As shown in Table 2, the DIZ of IHLE-pe2 and IHLE-pe3 were respectively 11.40 ± 1.41 mm and 12.76 ± 1.55 mm, indicating these two parts contained active compounds on *E. coli*. However, there was no inhibition of IHLE-pe1 and IHLE-pe4, which showed there was no active compound in these samples. Hence, IHLE-pe2 and IHLE-pe3 contained antibacterial components. According to the chromatogram in Figure 3, two peaks could be observed in IHLE-pe2 and IHLE-pe3. These two peaks would be further isolated using preparative HPLC and identified.

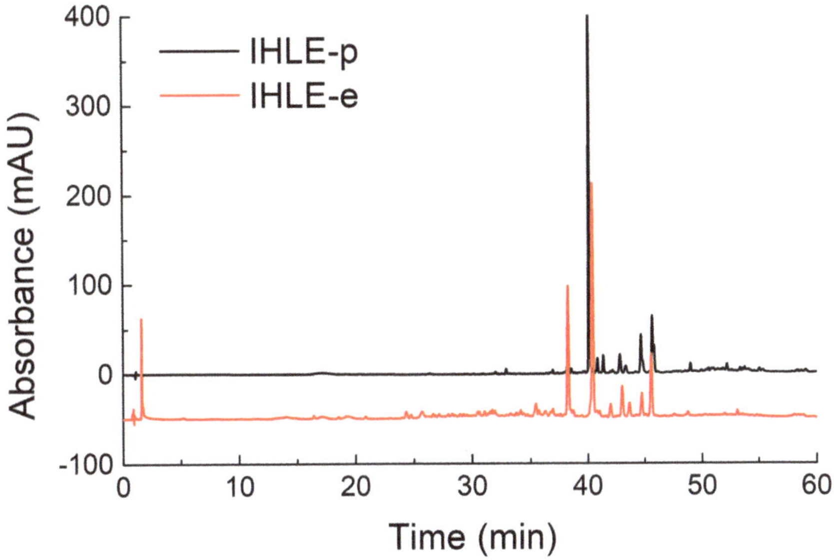

Figure 2. The chromatograms of IHLE-p and IHLE-e in HPLC analysis.

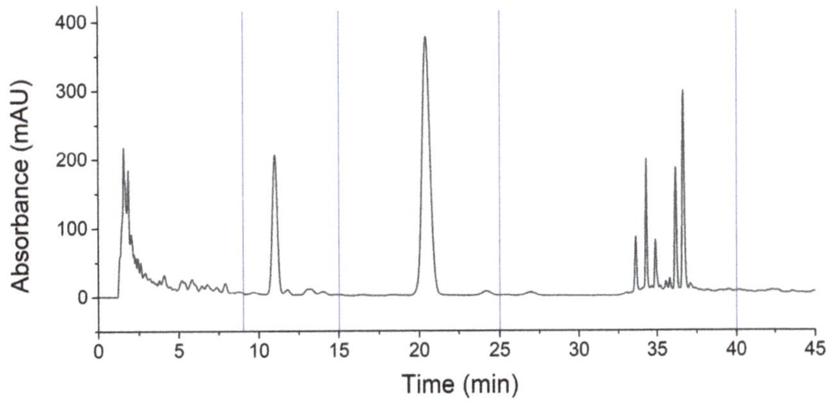

Figure 3. The chromatograms of IHLE-pe and the division of four fractions.

3.2. Identification of Antibacterial Active Components

Since fractions IHLE-pe2 and IHLE-pe3 exhibited significant antibacterial activity, and both possessed one peak each, according to HPLC analyses, antibacterial activity was ascribed to these two peaks (compounds). Peaks from IHLE-pe2 and IHLE-pe3 were identified as CBDVA and CBDA using standards and the same HPLC conditions, according to the same retention times and UV spectra [33–35]. Both compounds were later isolated using preparative HPLC. The structures of CBDA and CBDVA were shown in Figure 4. The chromatograms of CBDA and CBDVA were illustrated in Figure 5 together with the chromatograms of IHLE-pe, IHLE-pe2, and IHLE-pe3.

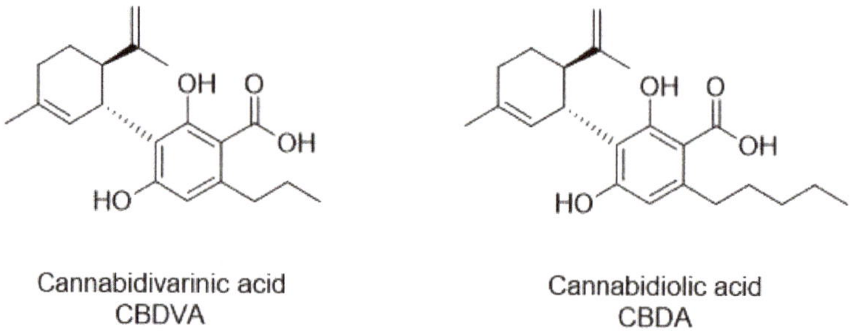

Figure 4. The chemical structure of CBDVA and CBDA.

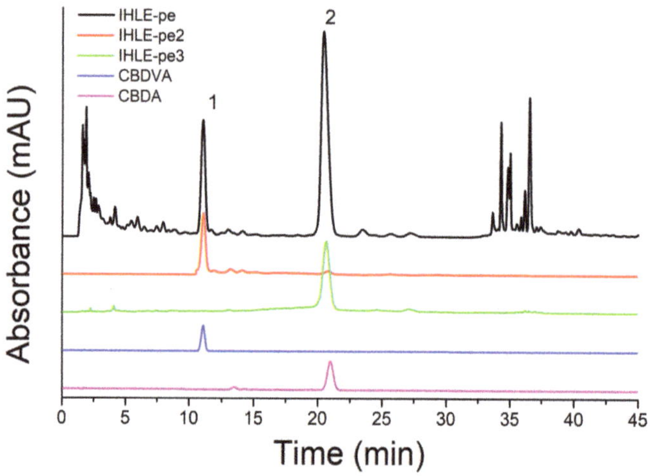

Figure 5. The chromatograms of IHLE-pe, IHLE-pe2, IHLE-pe3, CBDVA, and CBDA.

3.3. Antibacterial Activity of CBDA and CBDVA

3.3.1. MIC

The DIZ and MIC values of CBDA and CBDVA were also conducted. As indicated in Table 3, CBDVA has an inhibitory effect on *E. coli*. Up to 4.0 mg/mL CBDVA has an apparent inhibition with the DIZ of 11.19 ± 1.54 mm. Meanwhile, CBDA exhibited more obvious inhibition with the DIZ of 14.80 ± 1.35 mm at 1.0 mg/mL. As the concentration is diluted, the inhibitions of both CBDVA and CBDA decreased gradually, and the MIC of CBDVA and CBDA were at 1.0 mg/mL and 0.1 mg/mL. The results confirmed that CBDVA and CBDA had bacteriostatic effects on *E. coli*. Compared with that of IHLE, their activities were higher, and their bacteriostatic zones were more uniform and transparent.

Table 3. The DIZ of *E. coli* under different concentrations of CBDVA and CBDA (mm).

Samples	Concentration	DIZ
CBDVA	4 mg/mL	11.19 ± 0.84 [a]
CBDVA	3 mg/mL	9.17 ± 0.81 [b]
CBDVA	2 mg/mL	8.76 ± 1.35 [b]
CBDVA	1 mg/mL	8.51 ± 0.65 [b]
CBDVA	0.5 mg/mL	- [c]
CBDA	1.0 mg/mL	14.80 ± 1.35 [a]
CBDA	0.5 mg/mL	13.32 ± 1.46 [a]
CBDA	0.1 mg/mL	8.61 ± 0.80 [b]
CBDA	0.05 mg/mL	- [c]

Values followed by different letters ([a–c]) in the same column are significantly different according to Duncan's test ($p < 0.05$).

3.3.2. Growth Curves

To further investigate the antibacterial activity of CBDVA and CBDA, the growth curves of the *E. coli* in the presence of samples at different concentrations were monitored at 600 nm for 24 h. As shown in Figure 6, the growth of *E. coli* in the control group was in line with the typical S-type growth and reached stable 12 h of incubation [36]. When the concentrations of CBDVA and CBDA were at 1/2 MIC, the growth of *E. coli* was relatively slower than that of the control, and also reached the same stable level at 15 h. However, the growth of *E. coli* was completely inhibited in the treatments of CBDVA and CBDA at MIC and 2MIC within 24 h of culture, indicating the bactericidal effect of these two active compounds. These results indicated that IHLE, CBDVA, and CBDA treatment could reduce and inhibit the growth rate of *E. coli* in a dose-dependent manner, which is consistent with the growth inhibition effect of other bacteriostatic substances on pathogenic bacteria [37,38]. Based on these results, CBDVA and CBDA could be regarded as natural antibacterial agents.

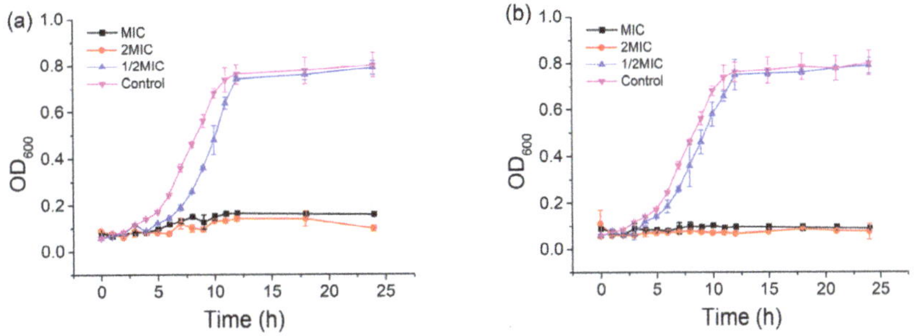

Figure 6. Growth curves of *E. coli* in the presence of (**a**) CBDVA and (**b**) CBDA at different concentrations.

3.4. Effect of CBDVA and CBDA on Cell Wall Integrity

The activity of AKP could not be detected in bacterial culture when the cell wall was unbroken [39]. However, AKP would leak into the extracellular environment when the cell wall is damaged, and AKP activity assay would reflect the damage to the cell wall [28]. When the *E. coli* was treated with CBDVA and CBDA at MIC for 12 h, the AKP activity increased with incubation time and finally reached six times that of control at 12 h (Figure 7). In comparison, the level of AKP activity remained in the control group. Therefore, it indicated that CBDVA and CBDA could damage the cell wall of *E. coli*.

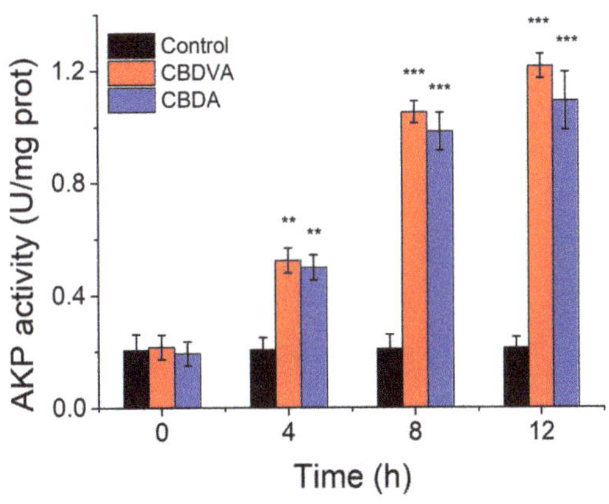

Figure 7. The AKP activities of *E. coli* in the presence of CBDVA and CBDA at different times. ** $p < 0.01$, *** $p < 0.001$ in comparison to control group.

3.5. *Effect of CBDVA and CBDA on the Cell Membrane*

3.5.1. Cell Membrane Permeability

The effect of CBDVA and CBDA on the cell membrane of *E. coli* was shown i Table 4. After the treatment of cannabinoids, the conductivity respectively increase from 0.567 mS/cm to 1.562 mS/cm and 1.414 mS/cm. This phenomenon shows the ce membrane was destroyed and small molecules such as sodium and potassium ions leake out of the cells [40].

Table 4. The analysis of cell membrane permeability of *E. coli* treated by CBDVA and CBDA at MIC.

Samples	Control	CBDVA Treatment	CBDA Treatment
Conductivity (mS/cm)	0.567 ± 0.004	1.562 ± 0.004 ***	1.414 ± 0.01 ***
Proteins leakage (mg/mL)	200.31 ± 0.090	270.12 ± 3.241 ***	268.31 ± 2.366 ***
β-GAL activity (OD_{405})	0.862 ± 0.034	0.133 ± 0.086 **	0.165 ± 0.086 **

** $p < 0.01$, *** $p < 0.001$ in comparison to control group.

3.5.2. Cell Membrane Integrity

The intracellular contents, such as proteins, small ions, and nucleic acid, would leak out when the cell membrane had a problem. The nucleic acid leakage was studied by monitoring the absorbance at 260 nm [41]. As shown in Figure 8, the absorbance slowly increased through the incubation. This trend might be due to the rupture of the cell membrane and then nucleic acid release into the medium in the presence of CBDVA and CBDA. Loss of intracellular nucleic acids would affect the replication, transcription, and translation of bacterial DNA resulting in cell death. The release of proteins was also an indicator of membrane damage [42]. The protein concentrations of the supernatant were shown in Table 4. After 12 h incubation, the leakage of protein was, respectively, 270 and 268 ug/mL, compared to 200 µg/mL in the control group. β-GAL could catalyze lactose into galactose and glucose, which then affects cell metabolism [43]. The activity of β-GAL could be determined by monitoring the absorption at 405 nm. The absorption value of the treatment groups was reduced by 84.6% and 80.9%, compared to the control group. It could be found that the presence of CBDVA and CBDA would destroy the membrane structure and inhibit intracellular enzyme activity [44]. The above results demonstrated that CBDVA and CBDA could destroy the membrane structure of *E. coli*, leading to the inhibition of enzyme activity and leakage of contents.

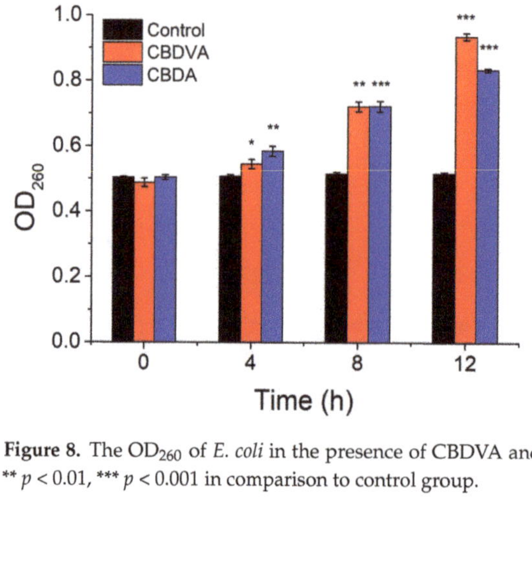

Figure 8. The OD_{260} of *E. coli* in the presence of CBDVA and CBDA at different times. * $p < 0.05$, ** $p < 0.01$, *** $p < 0.001$ in comparison to control group.

3.6. Morphological Transformation

The surface morphology of *E. coli* in the treatment of CBDVA and CBDA at MIC was visualized by SEM. The SEM images of the three groups were shown in Figure 9. The cells in the control group (6a and 6b) were rod-shaped, regular, and intact. Whereas the surface of cells showed evident morphological changes after treatment of CBDVA and CBDA. Many breakages and collapses of the cell could be observed, and the surface became coarse and multi-holed [45]. The observed results demonstrated CBDVA and CBDA exhibiting their bactericidal activities by affecting cell envelope integrity, consisting of the cell wall and membrane affections [46].

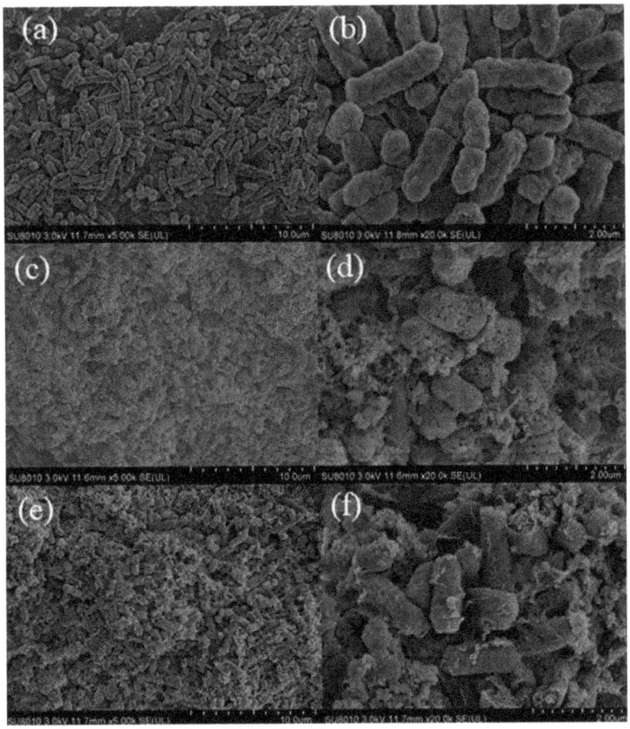

Figure 9. SEM images of *E. coli* treated with CBDA and CBDVA ((**a**,**b**): control group; (**c**,**d**): CBDA treatment; (**e**,**f**): CBDVA treatment).

4. Conclusions

In this work, antibacterial activity of IHLE against *E. coli* and *S. aureus* has been presented. Bioactivity-guided isolation revealed CBDVA and CBDA as active bacteriostatic compounds. Investigations of AKP activity, conductivity, proteins leakage, nucleic acid leakage, and β-GAL activity, CBDVA and CBDA exhibited the ability to destroy the cell wall and cell membrane of *E. coli*, leading to the inhibition of enzyme activity and leakage of contents. The morphological evaluation also showed these active components could damage the bacteria cell envelope. The discovery of bacteriostatic activity of CBDVA and CBDA against *E. coli* and *S. aureus* makes these compounds good candidates for use in the food, pharmaceutical and cosmetic industry as natural preservatives.

Author Contributions: Conceptualization, S.Z.; data curation, Y.F.; formal analysis, Y.F.; funding acquisition, S.D.; investigation, L.L.; project administration, S.D.; writing—original draft, Y.F. and L.L.; writing—review and editing, S.Z. All authors have read and agreed to the published version of the manuscript.

Funding: This research received no external funding.

Institutional Review Board Statement: Not applicable.

Informed Consent Statement: Not applicable.

Data Availability Statement: The data presented in this study are available on request from the corresponding author. The data are not publicly available due to security requirements in the institute of the corresponding author.

Acknowledgments: This work was supported by the Natural Science Foundation of Changsha, China (Grant No. kq2202323), and the Chinese Agricultural Science and Technology Innovation Project (No. ASTIP-IBFC05).

Conflicts of Interest: The authors declare no conflict of interest.

References

1. Brighenti, V.; Pellati, F.; Steinbach, M.; Maran, D.; Benvenuti, S. Development of a new extraction technique and HPLC method for the analysis of non-psychoactive cannabinoids in fibre-type *Cannabis sativa* L. (hemp). *J. Pharm. Biomed. Anal.* **2017**, *143*, 228–236. [CrossRef] [PubMed]
2. Salentijn, E.M.J.; Zhang, Q.; Amaducci, S.; Yang, M.; Trindade, L.M. New developments in fiber hemp (*Cannabis sativa* L.) breeding. *Ind. Crops Prod.* **2015**, *68*, 32–41. [CrossRef]
3. Liu, Y.; Xiao, A.-P.; Cheng, H.; Liu, L.-L.; Kong, K.W.; Liu, H.-Y.; Wu, D.-T.; Li, H.-B.; Gan, R.-Y. Phytochemical differences of hemp (*Cannabis sativa* L.) leaves from different germplasms and their regulatory effects on lipopolysaccharide-induced inflammation in Matin-Darby canine kidney cell lines. *Front. Nutr.* **2022**, *9*, 902625. [CrossRef] [PubMed]
4. Radwan, M.M.; Chandra, S.; Gul, S.; ElSohly, M.A. Cannabinoids, Phenolics, Terpenes and Alkaloids of Cannabis. *Molecules* **2021**, *26*, 2774. [CrossRef]
5. Papaseit, E.; Pérez-Mañá, C.; Pérez-Acevedo, A.P.; Hladun, O.; Torres-Moreno, M.C.; Muga, R.; Torrens, M.; Farré, M. Cannabinoids: From pot to lab. *Int. J. Med. Sci.* **2018**, *15*, 1286–1295. [CrossRef] [PubMed]
6. Almeida, C.F.; Teixeira, N.; Correia-da-Silva, G.; Amaral, C. Cannabinoids in Breast Cancer: Differential Susceptibility according to Subtype. *Molecules* **2022**, *27*, 156. [CrossRef]
7. Pattnaik, F.; Nanda, S.; Mohanty, S.; Dalai, A.K.; Kumar, V.; Ponnusamy, S.K.; Naik, S. Cannabis: Chemistry, extraction and therapeutic applications. *Chemosphere* **2022**, *289*, 133012. [CrossRef]
8. Chye, Y.; Kirkham, R.; Lorenzetti, V.; McTavish, E.; Solowij, N.; Yücel, M. Cannabis, Cannabinoids, and Brain Morphology: A Review of the Evidence. *Biol. Psychiatry Cogn. Neurosci. Neuroimaging* **2021**, *6*, 627–635. [CrossRef]
9. Lagha, R.; Ben Abdallah, F.; Al-Sarhan, B.O.; Al-Sodany, Y. Antibacterial and Biofilm Inhibitory Activity of Medicinal Plant Essential Oils against *Escherichia coli* Isolated from UTI Patients. *Molecules* **2019**, *24*, 1161. [CrossRef]
10. Yue, Y.; He, Z.; Zhou, Y.; Ross, R.P.; Stanton, C.; Zhao, J.; Zhang, H.; Yang, B.; Chen, W. *Lactobacillus plantarum* relieves diarrhea caused by enterotoxin-producing *Escherichia coli* through inflammation modulation and gut microbiota regulation. *Food Funct.* **2020**, *11*, 10362–10374. [CrossRef]
11. Gutiérrez-del-Río, I.; Fernández, J.; Lombó, F. Plant nutraceuticals as antimicrobial agents in food preservation: Terpenoids, polyphenols and thiols. *Int. J. Antimicrob. Agents* **2018**, *52*, 309–315. [CrossRef] [PubMed]
12. Roth, N.; Käsbohrer, A.; Mayrhofer, S.; Zitz, U.; Hofacre, C.; Domig, K.J. The application of antibiotics in broiler production and the resulting antibiotic resistance in *Escherichia coli*: A global overview. *Poult. Sci.* **2019**, *98*, 1791–1804. [CrossRef] [PubMed]
13. Corrêa, R.C.G.; Heleno, S.A.; Alves, M.J.; Ferreira, I.C.F.R. Bacterial Resistance: Antibiotics of Last Generation used in Clinical Practice and the Arise of Natural Products as New Therapeutic Alternatives. *Curr. Pharm. Des.* **2020**, *26*, 815–837. [CrossRef]
14. Ali, E.M.M.; Almagboul, A.Z.I.; Khogali, S.M.E.; Gergeir, U.M.A. Antimicrobial activity of *Cannabis sativa* L. *Chin. Med.* **2012**, *3*, 61–64. [CrossRef]
15. Novak, J.; Zitterl-Eglseer, K.; Deans, S.G.; Franz, C.M. Essential oils of different cultivars of *Cannabis sativa* L. and their antimicrobial activity. *Flavour Fragr. J.* **2001**, *16*, 259–262. [CrossRef]
16. Naveed, M.; Tahir, A.; Khan, I.; Ali, A.; Rehman, A.-U. In Vitro antibacterial activity of *Cannabis sativa* leaf extracts to some selective pathogenic bacterial strains. *Int. J. Biosci.* **2014**, *4*, 65–70.
17. Kosgodage, U.S.; Matewele, P.; Awamaria, B.; Kraev, I.; Warde, P.; Mastroianni, G.; Nunn, A.V.; Guy, G.W.; Bell, J.D.; Inal, J.M.; et al. Cannabidiol Is a Novel Modulator of Bacterial Membrane Vesicles. *Front. Cell. Infect. Microbiol.* **2019**, *9*, 324. [CrossRef]
18. Zamora-Mendoza, L.; Guamba, E.; Miño, K.; Romero, M.P.; Levoyer, A.; Alvarez-Barreto, J.F.; Machado, A.; Alexis, F. Antimicrobial Properties of Plant Fibers. *Molecules* **2022**, *27*, 7999. [CrossRef]
19. Jenul, C.; Horswill Alexander, R. Regulation of *Staphylococcus aureus* Virulence. *Microbiol. Spectr.* **2019**, *7*, 7.2.29. [CrossRef]
20. Jessberger, N.; Dietrich, R.; Granum, P.E.; Märtlbauer, E. The *Bacillus cereus* Food Infection as Multifactorial Process. *Toxins* **2020**, *12*, 701. [CrossRef]
21. Jin, Z.; Li, L.; Zheng, Y.; An, P. Diallyl disulfide, the antibacterial component of garlic essential oil, inhibits the toxicity of *Bacillus cereus* ATCC 14579 at sub-inhibitory concentrations. *Food Control* **2021**, *126*, 108090. [CrossRef]

22. Xie, Y.; Chen, J.; Xiao, A.; Liu, L. Antibacterial Activity of Polyphenols: Structure-Activity Relationship and Influence of Hyperglycemic Condition. *Molecules* **2017**, *22*, 1913. [CrossRef] [PubMed]
23. Zhang, X.; Shi, C.; Liu, Z.; Pan, F.; Meng, R.; Bu, X.; Xing, H.; Deng, Y.; Guo, N.; Yu, L. Antibacterial activity and mode of action of ε-polylysine against *Escherichia coli* O157:H7. *J. Med. Microbiol.* **2018**, *67*, 838–845. [CrossRef] [PubMed]
24. Kang, S.; Kong, F.; Shi, X.; Han, H.; Li, M.; Guan, B.; Yang, M.; Cao, X.; Tao, D.; Zheng, Y.; et al. Antibacterial activity and mechanism of lactobionic acid against *Pseudomonas fluorescens* and Methicillin-resistant *Staphylococcus aureus* and its application on whole milk. *Food Control* **2020**, *108*, 106876. [CrossRef]
25. Li, J.; Li, C.; Shi, C.; Aliakbarlu, J.; Cui, H.; Lin, L. Antibacterial mechanisms of clove essential oil against *Staphylococcus aureus* and its application in pork. *Int. J. Food Microbiol.* **2022**, *380*, 109864. [CrossRef] [PubMed]
26. Cui, H.; Zhang, C.; Li, C.; Lin, L. Antibacterial mechanism of oregano essential oil. *Ind. Crops Prod.* **2019**, *139*, 111498. [CrossRef]
27. Zhang, Y.; Wu, Y.-T.; Zheng, W.; Han, X.-X.; Jiang, Y.-H.; Hu, P.-L.; Tang, Z.-X.; Shi, L.-E. The antibacterial activity and antibacterial mechanism of a polysaccharide from *Cordyceps cicadae*. *J. Funct. Foods* **2017**, *38*, 273–279. [CrossRef]
28. Guo, F.; Liang, Q.; Zhang, M.; Chen, W.; Chen, H.; Yun, Y.; Zhong, Q.; Chen, W. Antibacterial Activity and Mechanism of Linalool against *Shewanella putrefaciens*. *Molecules* **2021**, *26*, 245. [CrossRef]
29. Alanazi, A.K.; Alqasmi, M.H.; Alrouji, M.; Kuriri, F.A.; Almuhanna, Y.; Joseph, B.; Asad, M. Antibacterial Activity of *Syzygium aromaticum* (Clove) Bud Oil and Its Interaction with Imipenem in Controlling Wound Infections in Rats Caused by Methicillin-Resistant *Staphylococcus aureus*. *Molecules* **2022**, *27*, 8551. [CrossRef]
30. Naim, N.; Bouymajane, A.; Oulad El Majdoub, Y.; Ezrari, S.; Lahlali, R.; Tahiri, A.; Ennahli, S.; Laganà Vinci, R.; Cacciola, F.; Mondello, L.; et al. Flavonoid Composition and Antibacterial Properties of *Crocus sativus* L. Petal Extracts. *Molecules* **2023**, *28*, 186. [CrossRef]
31. Shan, B.; Cai, Y.-Z.; Brooks, J.D.; Corke, H. The in vitro antibacterial activity of dietary spice and medicinal herb extracts. *Int. J. Food Microbiol.* **2007**, *117*, 112–119. [CrossRef] [PubMed]
32. Abubakar, Y.U.; Taura, D.W.; Muhammad, A.B.; Ibrahim, T.S.; Livinus, M.U. In vitro antimicrobial activity of *Cannabis sativa* (Hemp) extracts against avian pathogenic *Escherichia coli* (APEC) isolated from Broilers Chicken. *Adv. Pharm. J.* **2020**, *5*, 164–171. [CrossRef]
33. Martinenghi, L.D.; Jønsson, R.; Lund, T.; Jenssen, H. Isolation, Purification, and Antimicrobial Characterization of Cannabidiolic Acid and Cannabidiol from *Cannabis sativa* L. *Biomolecules* **2020**, *10*, 900. [CrossRef] [PubMed]
34. Nemeškalová, A.; Hájková, K.; Mikulů, L.; Sýkora, D.; Kuchař, M. Combination of UV and MS/MS detection for the LC analysis of cannabidiol-rich products. *Talanta* **2020**, *219*, 121250. [CrossRef] [PubMed]
35. Tran, J.; Elkins, A.C.; Spangenberg, G.C.; Rochfort, S.J. High-Throughput Quantitation of Cannabinoids by Liquid Chromatography Triple-Quadrupole Mass Spectrometry. *Molecules* **2022**, *27*, 742. [CrossRef] [PubMed]
36. Lin, Y.; Tang, X.; Xu, L.; Wang, S. Antibacterial properties and possible action mechanism of chelating peptides-zinc nanocomposite against *Escherichia coli*. *Food Control* **2019**, *106*, 106675. [CrossRef]
37. Funatogawa, K.; Hayashi, S.; Shimomura, H.; Yoshida, T.; Hatano, T.; Ito, H.; Hirai, Y. Antibacterial Activity of Hydrolyzable Tannins Derived from Medicinal Plants against *Helicobacter pylori*. *Microbiol. Immunol.* **2004**, *48*, 251–261. [CrossRef]
38. Liu, X.; Cai, J.; Chen, H.; Zhong, Q.; Hou, Y.; Chen, W.; Chen, W. Antibacterial activity and mechanism of linalool against *Pseudomonas aeruginosa*. *Microb. Pathog.* **2020**, *141*, 103980. [CrossRef]
39. Sheng, J.W.; Liu, D.M.; Jing, L.; Xia, G.X.; Zhang, W.F.; Jiang, J.R.; Tang, J.B. Striatisporolide A, a butenolide metabolite from *Athyrium multidentatum* (Doll.) Ching, as a potential antibacterial agent. *Mol. Med. Rep.* **2019**, *20*, 198–204. [CrossRef]
40. He, N.; Wang, P.; Wang, P.; Ma, C.; Kang, W. Antibacterial mechanism of chelerythrine isolated from root of *Toddalia asiatica* (Linn) Lam. *BMC Complement. Altern. Med.* **2018**, *18*, 261. [CrossRef]
41. Yang, Y.-J.; Lin, M.-Y.; Feng, S.-Y.; Gu, Q.; Chen, Y.-C.; Wang, Y.-D.; Song, D.-F.; Gao, M. Chemical composition, antibacterial activity, and mechanism of action of essential oil from *Litsea cubeba* against foodborne bacteria. *J. Food Process. Preserv.* **2020**, *44*, e14724. [CrossRef]
42. Song, X.; Li, R.; Zhang, Q.; He, S.; Wang, Y. Antibacterial Effect and Possible Mechanism of Salicylic Acid Microcapsules against *Escherichia coli* and *Staphylococcus aureus*. *Int. J. Environ. Res. Public Health* **2022**, *19*, 12761. [CrossRef]
43. Zhou, X.; Wang, M.; Chen, J.; Su, X. Cascade reaction biosensor based on Cu/N co-doped two-dimensional carbon-based nanozyme for the detection of lactose and β-galactosidase. *Talanta* **2022**, *245*, 123451. [CrossRef] [PubMed]
44. Ma, E.; An, Y.; Zhang, G.; Zhao, M.; Iqbal, M.W.; Zabed, H.M.; Qi, X. Enhancing the antibacterial activity of *Lactobacillus reuteri* against *Escherichia coli* by random mutagenesis and delineating its mechanism. *Food Biosci.* **2023**, *51*, 102209. [CrossRef]
45. Sun, X.-H.; Zhou, T.-T.; Wei, C.-H.; Lan, W.-Q.; Zhao, Y.; Pan, Y.-J.; Wu, V.C.H. Antibacterial effect and mechanism of anthocyanin rich Chinese wild blueberry extract on various foodborne pathogens. *Food Control* **2018**, *94*, 155–161. [CrossRef]
46. Huang, J.; Yang, L.; Zou, Y.; Luo, S.; Wang, X.; Liang, Y.; Du, Y.; Feng, R.; Wei, Q. Antibacterial activity and mechanism of three isomeric terpineols of *Cinnamomum longepaniculatum* leaf oil. *Folia Microbiol.* **2021**, *66*, 59–67. [CrossRef] [PubMed]

separations

MDPI

Article

Enantioseparation of *syn*- and *anti*-3,5-Disubstituted Hydantoins by HPLC and SFC on Immobilized Polysaccharides-Based Chiral Stationary Phases

Mladenka Jurin, Darko Kontrec *, Tonko Dražić and Marin Roje

Department of Organic Chemistry and Biochemistry, Ruđer Bošković Institute, Bijenička cesta 54, 10 000 Zagreb, Croatia; mladenka.jurin@irb.hr (M.J.); tdrazic@gmail.com (T.D.); marin.roje@irb.hr (M.R.)
* Correspondence: darko.kontrec@irb.hr

Abstract: The enantioseparation of *syn*- and *anti*-3,5-disubstituted hydantoins **5a–i** was investigated on three immobilized polysaccharide-based columns (CHIRAL ART Amylose-SA, CHIRAL ART Cellulose-SB, CHIRAL ART Cellulose-SC) by high performance liquid chromatography (HPLC) using *n*-hexane/2-PrOH (90/10, *v*/*v*) or 100% dimethyl carbonate (DMC) as mobile phases, respectively, and by supercritical fluid chromatography (SFC) using CO_2/alcohol (MeOH, EtOH, 2-PrOH; 80/20, *v*/*v*) as a mobile phase. The chromatographic parameters, such as separation and resolution factors, have indicated that Amylose-SA is more suitable for enantioseparation of the most analyzed *syn*- and *anti*-3,5-disubstituted hydantoins than Celullose-SB and Cellulose-SC in both HPLC and SFC modalities. All three tested columns showed better enantiorecognition ability toward *anti*-hydantoins compared to *syn*-hydantoins, both in HPLC and SFC modes. We have demonstrated that environmentally friendly solvent DMC can be efficiently used as the mobile phase in HPLC mode for enantioseparation of hydantoins on the immobilized polysaccharide-based chiral stationary phases.

Keywords: 3,5-disubstituted hydantoins; HPLC; SFC; enantioseparation; immobilized polysaccharide-based chiral stationary phases; DMC; green solvent

Citation: Jurin, M.; Kontrec, D.; Dražić, T.; Roje, M. Enantioseparation of *syn*- and *anti*-3,5-Disubstituted Hydantoins by HPLC and SFC on Immobilized Polysaccharides-Based Chiral Stationary Phases. *Separations* **2022**, *9*, 157. https://doi.org/10.3390/separations9070157

Academic Editor: Liangliang Liu

Received: 20 May 2022
Accepted: 17 June 2022
Published: 22 June 2022

Publisher's Note: MDPI stays neutral with regard to jurisdictional claims in published maps and institutional affiliations.

1. Introduction

Imidazolidine-2,4-dione, also well known as hydantoin, is a simple hydantoin five-membered nitrogen heterocyclic compound, with four versatile points of functionalities in its framework. Hydantoins have two nitrogens in position 1 and 3, as well as two carbonyl groups in positions 2 and 4, one of which is between the two nitrogens (Figure 1) [1–3].

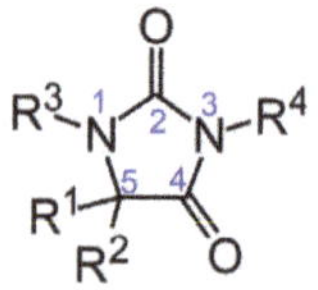

Figure 1. Chemical structure of hydantoins.

The hydantoin moiety is an important structural scaffold present in a number of drugs or drug candidates [3–5]. Phenytoin, ethotoine and norantoine are marketed as anticonvulsant drugs; nilutamide is a nonsteroidal androgen receptor antagonist for the treatment of metastatic prostate cancer [6,7], and BMS-564929 is an orally active and nonsteroidal tissue selective androgen receptor modulator [8]. In addition, hydantoins demonstrate numerous other interesting pharmacological activities, such as antibacterial [9], antiviral [10–13], antifungal [14], antiarrhythmic [15–17], antidiabetic [18,19], antitumor [20–23], antithrombotic, anti-inflammatory and antitussive [24], as well as inhibitory activity against some enzymes (human aldose reductase and leucocyte elastase) [25,26]. Finally, some herbicides [27–30], fungicides and insecticides also have a hydantoin ring in their structure [29,30].

The hydantoin ring also constitutes the core structure of various natural products, mostly isolated from different marine organisms, but also from bacteria [31]. For example, hemimycallins A and B were isolated from marine sponge *Hemimycale arabica* [26], mukanadine B was isolated from marine sponge *Agelas nakamurai* [32], midpacamide from Fidijan sponge *Agelas mauritiana* [33], and parazoanthines A-J from the Mediterranean Sea anemone *Parazoanthus axinellae* [34,35].

The enantiomers of eighteen chiral 3,5-disubstituted hydantoins were separated by Kartozia et al. using HPLC under normal phase mode on three polysaccharide columns Chiralpak AD-H, Chiralcel OD-H and Chiralcel OJ-H. In this study, a separation of seventeen out of eighteen chiral hydantoins achieved partial or baseline separation on Chiralpak AD-H. For most hydantoins, better separations were obtained on Chiralcel OD-H than on Chiralcel OJ-H [5]. More recently, the enantioseparation of eleven 3,5-disubstituted hydantoins was investigated by Yang et al. using HPLC under the normal phase mode on Chiralpak IA. In the study, the effect of polar alcoholic modifier, ethanol (EtOH), 1-propanol, 2-propanol (2-PrOH), 1-butanol and *tert*-butanol; and column temperature on retention and enantioseparation was evaluated. Additionally, two kinds of enantiomer elution order (EEO) reversals, which include solvent-induced EEO reversal for one tested chiral hydantoin and temperature-induced EEO reversals for the two hydantoins were found [3].

The chiral stationary phases (CSPs) most commonly used are based on silica coated with chiral polysaccharide derivatives, *tris*(carbamates) or *tris*(esters) of amylose or cellulose [36]. These coated CSPs are able to resolve a large variety of structurally different compounds [37,38] and are widely used in HPLC and SFC [39]. These CSPs can only be used with a limited range of solvents as mobile phases such as hydrocarbons, alcohols, acetonitrile (ACN), or hydrocarbon/alcohol and ACN/alcohols mixtures [40]. The immobilized CSPs were prepared by covalently bonding polysaccharide derivates to silica surface [36]. The immobilization allows the use of solvents that cannot be applied on the coated CSPs, such as ethers, esters, ketones and chlorinated hydrocarbons [40]. The enantiorecognition ability of polysaccharide-based CSPs depends on the interactions between the analyte enantiomers and polar carbamate moiety of the polysaccharide-based selector. Each enantiomer forms short-lived, transient diastereomeric complexes with the chiral selector through interactive forces. The complexes are formed as a result of hydrogen bonding, dipole–dipole interactions, π-π bonding, electrostatic interactions (Van der Waals forces), inclusion complexation and steric effects. The strength of these interactions depends on the structure of the analyte and the chiral selector, and on mobile phase composition [41–43].

Here, we present the use of green solvent dimethyl carbonate (DMC ($CH_3O(C{=}O)OCH_3$) as a mobile phase in HPLC mode for enantioseparation of chiral 3,5-disubstituted hydantoins **5a–i**. DMC is a nonpolar aprotic solvent, slightly soluble in water (139 g L^{-1}) [44] and miscible with alcohols, esters, ethers, and ketones [45]. It is an environmentally benign [46,47], biodegradable [48], non-corrosive [49] and non-toxic solvent [48]. It can be a potential replacement for methyl ethyl ketone, ethyl acetate, methyl isobutyl ketone, and most of other ketones [50]. Lajin and Goessler introduced DMC as a new organic solvent in HPLC-ICPMS for separation of eleven model compounds, such as theobromine, caffeine, aspirin, acetophenone, dithiodibenzoic acid, toluensulfomide, etc. [51]. They compared the elution behavior of DMC with that of other commonly used organic solvents, such as methanol (MeOH), 2-PrOH and acetonitrile. Their results showed that DMC offered stronger elution strength than MeOH and ACN for all tested compounds and stronger elution strength than 2-PrOH for most tested compounds.

In the present work, we have studied the enantioseparation of eighteen chiral 3,5-disubstituted hydantoins **5a–i** (Figure 2) by HPLC under normal standard and non-standard mobile phases. We have also studied the enantioseparation of these hydantoins by SFC using a mobile phase consisting of supercritical carbon dioxide and alcohol (80/20, *v*/*v*). Three CSPs in their immobilized form (CHIRAL ART Amylose-SA, CHIRAL ART

Cellulose-SB, CHIRAL ART Cellulose-SC) were employed to explore their enantioselectivity. Each analyzed hydantoin possessed two centers of chirality, one in the hydantoin ring and the other in the side chain (Figure 3).

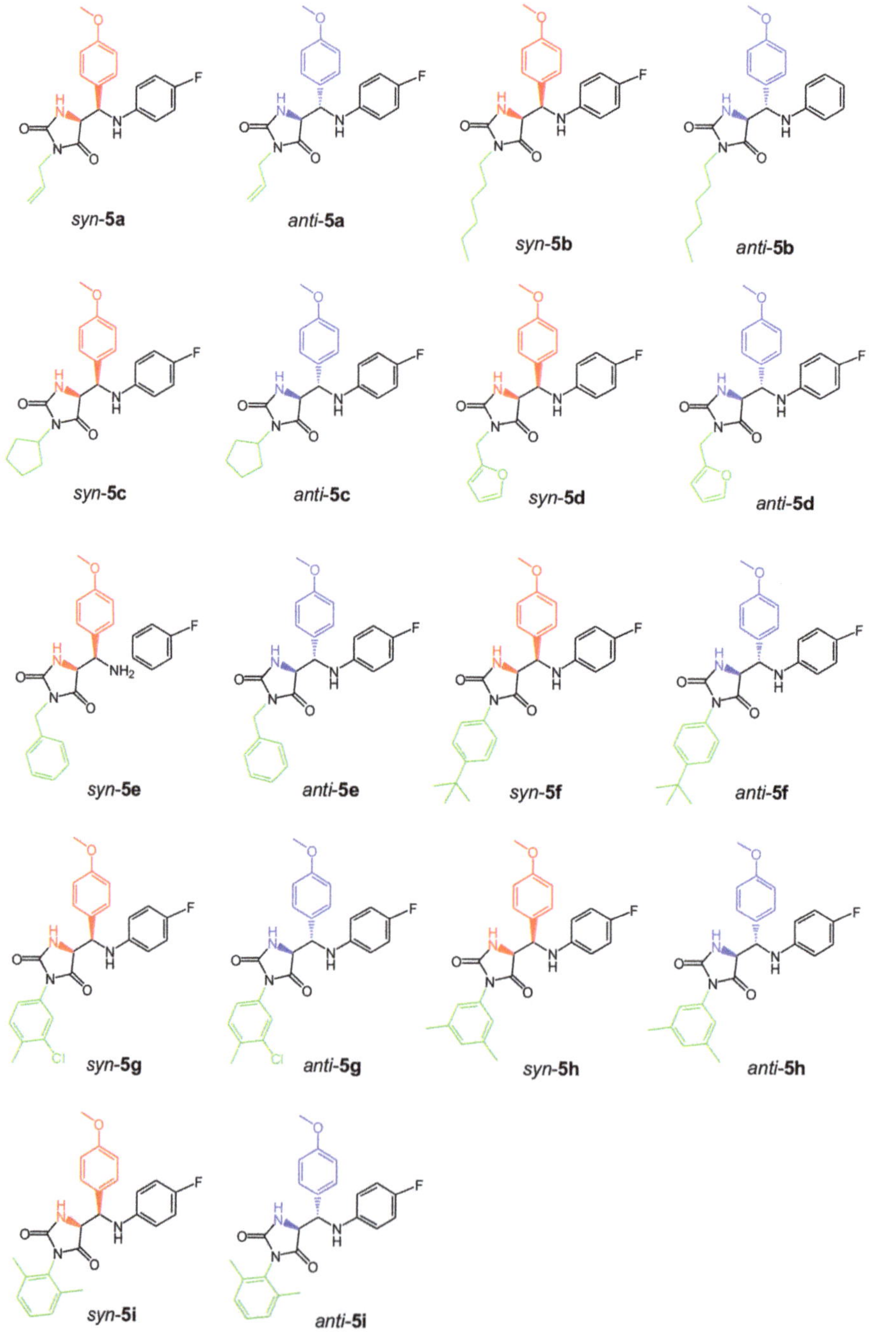

Figure 2. Chemical structures of (±)-*syn*- and (±)-*anti*-3,5-disubstituted hydantoins **5a–i**.

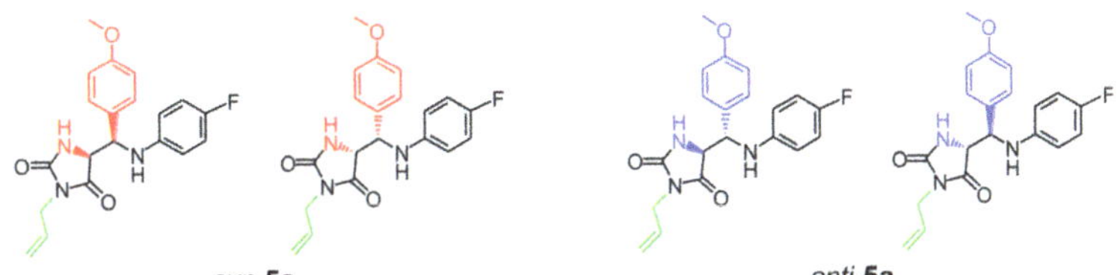

Figure 3. All four possible stereoisomers of allylhydantoin **5a**.

2. Materials and Methods

All used chemicals were purchased from commercial suppliers Sigma Aldrich (Steinheim, Germany), Merck (Darmstadt, Germany) and Fluka (Buchs, Switzerland). Dichloromethane (DCM), EtOH and ACN were dried prior to use according to standard methods [52]. EtOH, MeOH, 2-PrOH and *n*-hexane of HPLC grade were purchased from Honeywell (Seelze, Germany). DMC was purchased from Acros Organics (Geel, Belgium). Liquid CO_2 (grade 4.5) was from Messer (Zagreb, Croatia). The immobilized polysaccharide-based CSPs: CHIRAL ART Amylose-SA S-10 μm, CHIRAL ART Cellulose-SB S-10 μm and CHIRAL ART Cellulose-SC S-10 μm were purchased as bulk material from YMC (Kyoto, Japan). The empty stainless-steel HPLC columns, dimensions 250 mm × 4.6 mm ID, were purchased from Knauer GmbH (Berlin, Germany), and packed with the above mentioned CSPs.

The hydantoins used in this study were synthesized based on a procedure previously described in the literature, Scheme 1. β-Lactam ureas were prepared in four steps via Staudinger reaction [53]. Briefly, the first step included preparation of imine **1** by a condensation reaction of 4-methoxybenzaldehyde and 4-fluoroaniline in dry DCM. In a second reaction, imine **1** was treated with *N*-phthaloylglycine in the presence of triethylamine and 2-chloro-1-methylpyridinium iodide to afford a *cis/trans*-(±)-3-phthalimodo-β-lactam **2**. The *cis:trans* ratio was 1:5. In the subsequent step, deprotection of the bulky phthalimide group in the compound *trans*-**2** with ethylenediamine in dry EtOH afforded a free amine, (±)-*trans*-3-amino-β-lactam **3** [54]. The treatment of (±)-*trans*-3-amino-β-lactam **3** with various aliphatic and aromatic isocyanates in dry ACN at room temperature resulted in the isolation of (±)-*trans*-β-lactam ureas **4a–i** [55]. Diastereomeric mixtures (*syn*- and *anti*-) of racemic 3,5-disubstituted hydantoins **5a–i** were synthesized via base-promoted intramolecular amidolysis of (±)-*trans*-β-lactam ureas **4a–i** [56]. The mixtures of diastereomeric hydantoins were then separated by preparative RP-HPLC using preparative column Zorbax Extend-C18 PrepHT (250 × 9.4 mm I.D., 5-μm particle size, 300 Å pore size) from Agilent Technologies (Waldbronn, Germany) with a linear gradient AB at a flow rate of 17 mL min^{-1}, where mobile phase A was water and mobile phase B was ACN. The structures of all *syn*- and *anti*-hydantoins **5a–i** are shown in Figure 2. All compounds were characterized by NMR, IR and mass spectroscopy.

Two chromatographic systems were applied in this study. The first one was an Agilent 1200 Series HPLC System (Agilent Technologies, Waldbronn, Germany), equipped with a vacuum degasser, a quaternary pump, a thermostated column compartment, an autosampler and a variable wavelength detector. The mobile phase was *n*-hexane/2-PrOH (90/10, *v/v*) or 100% DMC. All experiments in normal-phase mode and non-standard mode were carried out under isocratic conditions at a flow rate of 1.0 mL min^{-1} and at a column temperature of 30 °C. The injection volume was 20 μL. Data analysis and processing were carried out by EZChrom Elite software version 3.1.7. (Agilent Technologies, Waldbronn, Germany).

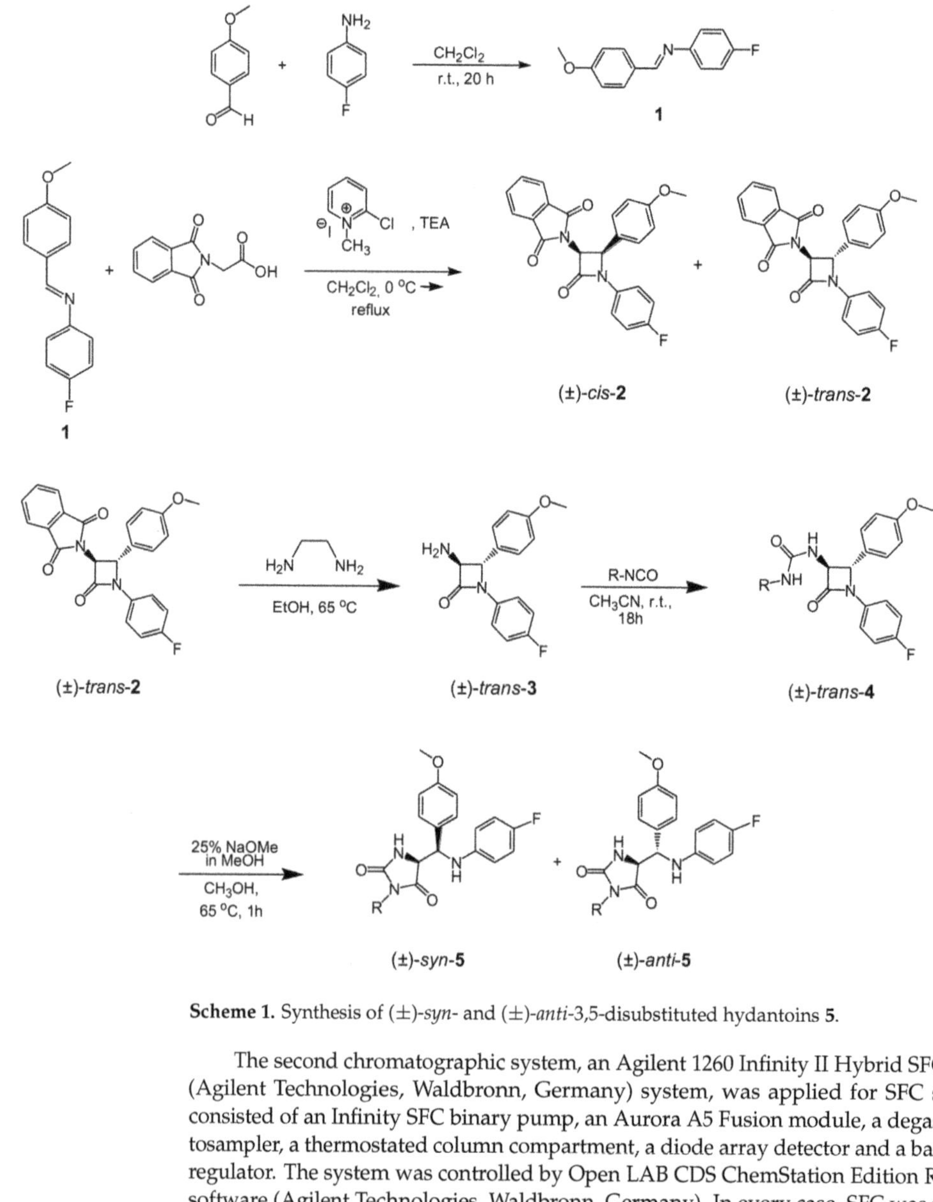

Scheme 1. Synthesis of (±)-*syn*- and (±)-*anti*-3,5-disubstituted hydantoins **5**.

The second chromatographic system, an Agilent 1260 Infinity II Hybrid SFC/UHPLC (Agilent Technologies, Waldbronn, Germany) system, was applied for SFC studies. It consisted of an Infinity SFC binary pump, an Aurora A5 Fusion module, a degasser, an autosampler, a thermostated column compartment, a diode array detector and a backpressure regulator. The system was controlled by Open LAB CDS ChemStation Edition Rev. C01.08 software (Agilent Technologies, Waldbronn, Germany). In every case, SFC was performed in isocratic mode at a flow rate of 4.0 mL min^{-1} and a column temperature of 35 °C. The injection volume was 20 μL and the outlet pressure was set at 15 MPa. The mobile phases applied in SFC consisted of CO_2 and MeOH, EtOH or 2-PrOH, each in the ratio 80/20, *v*/*v*. Detection was performed at a wavelength of 254 nm using a diode-array detector.

Sample solutions of the analytes were prepared by dissolving hydantoin compounds in *n*-hexane/2-PrOH (90/10, *v*/*v*), DMC or MeOH in 0.5 mg mL^{-1} concentration and filtered through RC-45/25 Chromafil® Xtra 0.45 μm syringe filter (Macherey-Nagel GmbH & Co. KG, Düren, Germany). The HPLC columns were packed by using the typical slurry method, where *n*-hexane/2-PrOH (90:10, *v*/*v*) was used as slurring solvent to prepare *n*-hexane/2-PrOH-YMC bulk materials (CHIRAL ART Amylose-SA, CHIRAL ART Cellulose-SB and CHIRAL ART Cellulose-SC) suspensions with sonication, respectively.

The suspensions were packed into stainless-steel columns (250 × 4.6 mm I.D.) by the conventional high pressure downward slurry technique using a Knauer pneumatic HPLC pump (Knauer Gmbh, Berlin, Germany). In the following text, these columns are marked as Amylose-SA, Cellulose-SB and Cellulose-SC. The chiral selectors in Amylose-SA, Cellulose-SB and Cellulose-SC are amylose *tris*-(3,5-dimethylphenylcarbamate), cellulose *tris*-(3,5-dimethylphenylcarbamate) and cellulose *tris*-(3,5-dichlorophenylcarbamate), respectively; all three are shown in Figure 4.

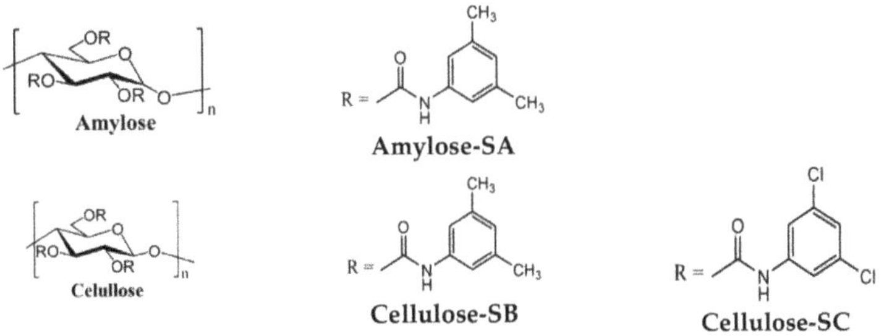

Figure 4. Chemical structures of chiral selectors.

The retention factor of the first and the second eluted enantiomer (k_1 and k_2), the separation factor (α), and the resolution (R_s) are calculated according to the usual formulae:

$$k_1 = (t_{r1} - t_0)/t_0 \quad (1)$$

$$k_2 = (t_{r2} - t_0)/t_0 \quad (2)$$

$$\alpha = k_2/k_1 \quad (3)$$

$$R_s = 2 \times (t_{r2} - t_{r1})/(w_1 + w_2) \quad (4)$$

where t_0 is the dead time, t_{r1} and t_{r2} are the retention times of the first and second eluted enantiomers, respectively, and w_1 and w_2 are the corresponding base peak widths. In HPLC mode, the dead time, which is the retention time of a nonadsorbing component, was determined by injection of 1,3,5-tri-*tert*-butylbenzene, while in SFC mode the first negative signal by injecting MeOH was used.

3. Results and Discussion

The use of DMC in our enantioseparation studies also requested the use of immobilized CSPs. DMC and the other organic solvents of medium polarity (non-standard HPLC solvents), such as acetone, dichloromethane, chloroform, ethyl acetate, methyl *tert*-butyl ether and tetrahydrofuran can be used on immobilized chiral selectors. Contrary, these solvents cannot be used with coated polysaccharide-based chiral selectors, because they can dissolve or swell the polysaccharide derivative [57–59]. In this study, our intention was to demonstrate an efficient replacement of hydrocarbon-based mobile phase with DMC and/or supercritical CO_2.

3.1. Enantioseparation on Amylose-SA

The results of the HPLC and SFC enantioseparation using Amylose-SA column are summarized in Table 1. The Amylose-SA allowed the enantioseparation of all eighteen analyzed 3,5-disubstituted hydantoins under the *n*-hexane/2-PrOH (90/10, *v*/*v*) mobile phase system. Among them, for seventeen pairs of enantiomers the baseline separation was achieved, while partial separation was observed only for the enantiomers of the compound *syn*-**5i**. In the normal phase mode, the retention factors (k_1) of the first eluting enantiomers of the *syn*- and *anti*-compounds **5h** and **5i** were higher than that of other hydantoins, which implied that the interactions between these analytes and CSP were the strongest. The possible reason may be the presence of two methyl groups at either *ortho*- or *meta*-position of the N3 phenyl ring in **5h** and **5i**, respectively. On the contrary, the lower retention of compounds *syn*-**5b**, *anti*-**5b**, *syn*-**5c** and *anti*-**5c** was the result of their weaker interaction with CSP. The possible reason may be the presence of alkyl or cycloalkyl substituent at the N3 position of the hydantoin ring, which, unlike other tested hydantoins, cannot provide additional π-π interactions with the CSP. It is obvious that the longer retained compounds did not always accomplish higher separation factors and resolution. Moreover, this column exhibited better chiral recognition toward *anti*-hydantoins **5a–i** compared to *syn*-hydantoins **5a–i**. When DMC was used as the mobile phase, among eighteen hydantoins, a baseline separation was achieved for eight hydantoins while six hydantoins were partially separated. The enantiomers of the hydantoins *syn*-**5a**, *syn*-**5d**, *anti*-**5c**, and *anti*-**5f** did not separate on this column under the same condition. As seen from Table 1, better enantioselectivity of compounds *syn*-**5b**, *syn*-**5f**, *syn*-**5g**, *syn*-**5i** was achieved with DMC as the mobile phase, while the resolution was always higher with *n*-hexane/2-PrOH (90/10, *v*/*v*) as the mobile phase. Moreover, under DMC conditions, *anti*-hydantoins **5a**, **5b**, **5d**, **5e**, **5h** and **5i** showed higher α and R_s values compared to α and R_s of their *syn*-isomers. *syn*-Hydantoins **5c** and **5f** showed better results (in terms of α and R_s) than *anti*-**5c** and *anti*-**5f**; the *syn*-isomers were partially separated, while *anti*-isomers were not resolved under DMC. Furthermore, the enantiomers of compound *anti*-**5g** achieved greater resolution and lower value of separation factor compared to the enantiomers of compound *syn*-**5g** under non-standard mobile phase. DMC as the hydrogen bond acceptor is capable of interacting with the polarized hydrogen atom of the carbamate N-H group (hydrogen bond donor) of amylose-based selector, competing with hydantoin compounds for hydrogen bonding sites, therefore accelerating the elution rate. As shown in Figure 5, *syn*-allyl hydantoin **5a** expressed a superior R_s of 3.23 and α of 1.61 on Amylose-SA under *n*-hexane/2-PrOH (90/10, *v*/*v*) compared to R_s 0 and α of 1.44 under 100% DMC. *anti*-Allyl hydantoin **5a** showed R_s of 6.36 and α of 2.48 under *n*-hexane/2-PrOH (90/10, *v*/*v*) compared to R_s of 1.57 and α of 2.25 under 100% DMC. In particular, Amylose-SA showed a significantly higher resolution and enantioselectivity values for the *anti*-allyl hydantoin **5a** under the normal and non-standard HPLC conditions.

Table 1. Chromatographic parameters for the enantioselective separations of racemic *syn*-and *anti*-3,5-disubstituted hydantoins on Amylose-SA.

Compound	Condition *	k_1	k_2	α	R_s	Compound	Condition *	k_1	k_2	α	R_s
syn-**5a**	A	6.57	10.58	1.61	3.23	*anti*-**5a**	A	5.33	13.30	2.48	6.36
	B	0.09	0.09	1.00	-		B	0.16	0.36	2.25	1.57
	C	1.45	1.85	1.28	0.80		C	1.23	2.32	1.89	2.32
	D	1.40	1.78	1.27	0.76		D	1.20	2.15	1.80	2.07
	E	1.34	1.90	1.42	1.21		E	1.34	2.43	1.81	2.23
syn-**5b**	A	2.47	4.51	1.83	3.74	*anti*-**5b**	A	2.04	5.74	2.81	6.27
	B	0.10	0.22	2.20	0.41		B	0.22	0.50	2.27	1.81
	C	1.54	2.19	1.42	1.36		C	1.56	2.76	1.76	2.29
	D	1.39	1.96	1.41	1.21		D	1.32	2.42	1.83	2.28
	E	1.31	2.01	1.53	1.49		E	1.36	2.70	1.99	2.55
syn-**5c**	A	2.85	5.30	1.86	3.98	*anti*-**5c**	A	2.85	8.03	2.82	6.43
	B	0.18	0.28	1.56	0.87		B	0.60	0.60	1.00	-
	C	1.34	3.19	2.38	2.90		C	1.70	4.45	2.62	4.60
	D	1.47	2.63	1.79	2.33		D	1.61	3.76	2.34	4.52
	E	1.59	2.63	1.65	1.99		E	1.59	3.93	2.47	3.58

Table 1. *Cont.*

Compound	Condition *	k_1	k_2	α	R_s	Compound	Condition *	k_1	k_2	α	R_s
syn-**5d**	A	6.05	11.63	1.92	4.42	*anti*-**5d**	A	4.95	17.52	3.54	8.63
	B	0.19	0.19	1.00	-		B	0.14	0.30	2.14	1.11
	C	2.08	2.75	1.32	1.12		C	1.78	3.37	1.89	2.72
	D	1.84	2.52	1.37	1.30		D	1.64	3.07	1.87	2.68
	E	1.78	2.82	1.58	1.88		E	1.84	3.99	2.17	3.58
syn-**5e**	A	7.97	16.34	2.05	4.12	*anti*-**5e**	A	4.49	16.01	3.57	8.31
	B	0.17	0.24	1.41	0.54		B	0.28	0.47	1.68	1.38
	C	2.87	3.77	1.31	1.27		C	2.68	4.88	1.82	2.77
	D	2.44	3.57	1.46	1.86		D	2.51	4.33	1.73	3.05
	E	2.39	4.16	1.74	2.63		E	2.64	5.76	2.18	3.99
syn-**5f**	A	4.76	8.54	1.79	3.17	*anti*-**5f**	A	3.70	10.04	2.71	4.59
	B	0.14	0.45	3.21	1.97		B	0.32	0.32	1.00	-
	C	3.08	8.34	2.71	4.20		C	3.04	15.51	5.10	7.65
	D	2.91	7.39	2.54	4.03		D	2.82	12.39	4.39	6.56
	E	2.66	6.01	2.26	3.65		E	2.59	9.15	3.53	5.64
syn-**5g**	A	6.85	14.14	2.06	4.30	*anti*-**5g**	A	3.92	10.39	2.65	4.94
	B	0.17	0.61	10.11	3.02		B	0.37	1.72	4.65	5.23
	C	4.63	11.91	2.57	4.46		C	4.25	17.62	4.14	6.20
	D	4.35	10.87	2.50	4.70		D	4.00	15.40	3.85	6.57
	E	3.93	9.39	2.39	4.25		E	3.78	13.00	3.44	5.91
syn-**5h**	A	10.43	21.05	2.02	4.30	*anti*-**5h**	A	6.17	15.89	2.58	4.94
	B	0.15	0.45	3.00	2.19		B	0.25	0.96	3.84	4.55
	C	2.67	5.95	2.23	3.55		C	2.40	7.22	3.01	4.65
	D	2.62	5.90	2.25	3.68		D	2.31	6.86	2.97	4.93
	E	1.91	6.05	3.17	3.02		E	2.39	6.57	2.75	4.57
syn-**5i**	A	14.34	16.13	1.12	0.75	*anti*-**5i**	A	16.81	22.33	1.33	1.77
	B	0.10	0.17	1.70	0.42		B	0.26	0.50	1.92	1.54
	C	2.43	3.30	1.33	1.23		C	2.77	4.25	1.53	1.76
	D	2.81	3.74	1.33	1.34		D	3.41	5.10	1.50	1.83
	E	3.52	5.11	1.45	1.88		E	4.74	8.26	1.74	2.90

* Chromatographic conditions: mobile phase, A, *n*-hexane/2-PrOH (90/10, *v*/*v*), flow rate 1 mL min^{-1}; B, 100% DMC, flow rate 1 mL min^{-1}; C, CO_2/MeOH (80/20, *v*/*v*), flow rate 4 mL min^{-1}, backpressure 15 MPa; D, CO_2/EtOH (80/20, *v*/*v*), 4 mL min^{-1}, backpressure 15 MPa; E, CO_2/2-PrOH (80/20, *v*/*v*), flow rate 4 mL min^{-1}, backpressure 15 MPa. Detection wavelength for each condition was 254 nm. Column temperature of conditions A and B is 30 °C, of C, D and E is 35 °C. The chromatographic parameters k_1, k_2, α and R_s are defined in Section 2.

Under SFC conditions, we investigated the effect of three modifiers (MeOH, EtOH, 2-PrOH) on enantioseparation of hydantoin derivatives **5a–i**. It is interesting to note that 2-PrOH yields more baseline enantioseparations than MeOH and EtOH, Table 1. By comparing the data obtained with the modifiers MeOH, EtOH or 2-PrOH the retention factors of the first eluted enantiomers of compounds *syn*-**5c**, *syn*-**5i**, *anti*-**5a**, *anti*-**5d** and *anti*-**5i** were the highest with 2-PrOH as the modifier. The retention factors of *syn*-**5c**, *syn*-**5i** and *anti*-**5i** hydantoins increased as the mobile phase modifier changed from MeOH to EtOH and then to 2-PrOH, which should be due to the decrease in solvent polarity and the increase in bulkiness of alcoholic modifiers (due to the chain length and branching). The compounds *syn*-**5a**, *syn*-**5b**, *syn*-**5d**, *syn*-**5e**, *syn*-**5f**, *syn*-**5g**, *syn*-**5g**, *anti*-**5f**, *anti*-**5g** and *anti*-**5i** followed the opposite trend; the retention factor of the compounds increasing in the order 2-PrOH > EtOH > MeOH. MeOH, EtOH and 2-PrOH are all protic solvents, and they are capable of interacting with amylose *tris*-(3,5-dimethylphenylcarbamate) through hydrogen bonding, and thus compete with the hydantoin compounds for the hydrogen bonding site, thus accelerating the elution rate. When applying a branched-chain alcohol 2-PrOH, it forms weaker hydrogen bonds with CSP than MeOH and EtOH, so the retention factor is expected to be longer. However, it was observed that 2-PrOH gave shorter retention times than straight-chain alcohols (MeOH and EtOH) for ten analyzed compounds *syn*-**5a**, *syn*-**5b**, *syn*-**5d**, *syn*-**5e**, *syn*-**5f**, *syn*-**5g**, *syn*-**5h**, *anti*-**5c**, *anti*-**5f** and *anti*-**5f**. The molecular structure type and the steric effects of the modifiers influence the enantioselectivity and retention of the analyte. The chiral recognition mechanisms of the analytes with Amylose-SA are very complex, and largely depend on the structure of hydantoin derivatives, i.e., the nature of the functional group at the N3 position of the hydantoin ring.

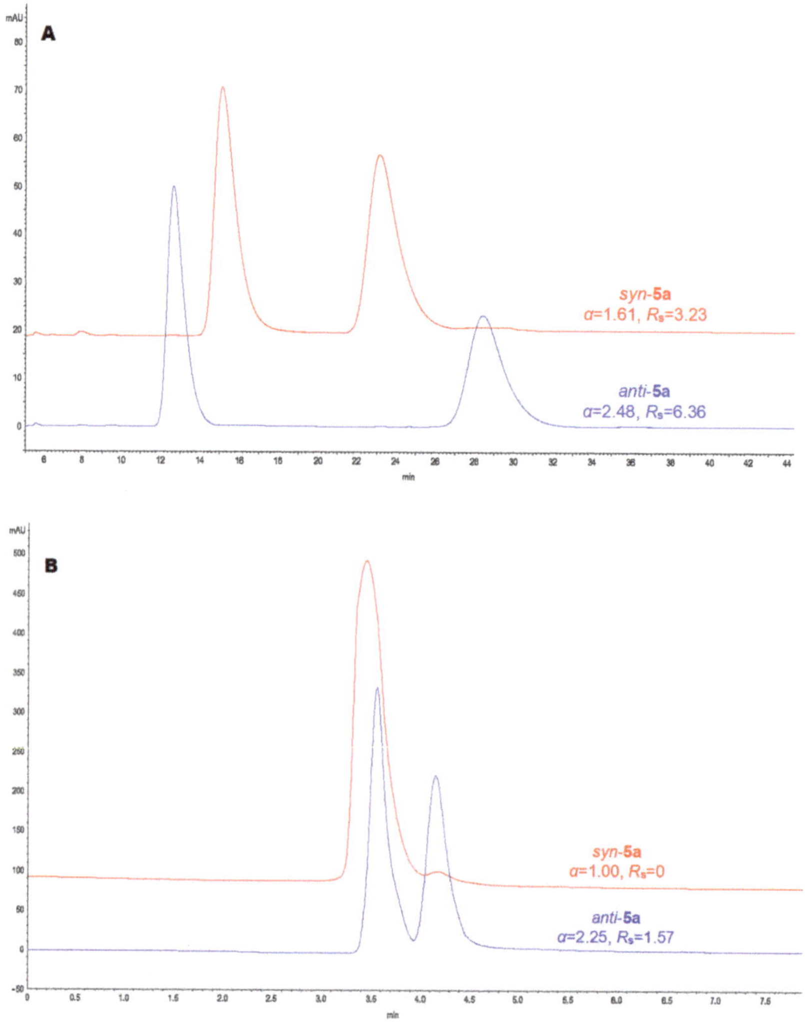

Figure 5. HPLC overlay chromatograms of (±)-*syn*-**5a** and (±)-*anti*-**5a** on Amylose-SA column with: (**A**) *n*-hexane/2-PrOH (90/10, *v*/*v*) and (**B**) 100% DMC as mobile phases.

The Amylose-SA column under SFC mode exhibits better chiral recognition ability toward *anti*-hydantoins compared to *syn*-hydantoins. As shown in Figure 6, under the mobile phase of CO_2/EtOH (80/20, *v*/*v*), both *syn*- and *anti*-allyl hydantoins **5a** have the lowest R_s and α and the largest values for MeOH as the modifier.

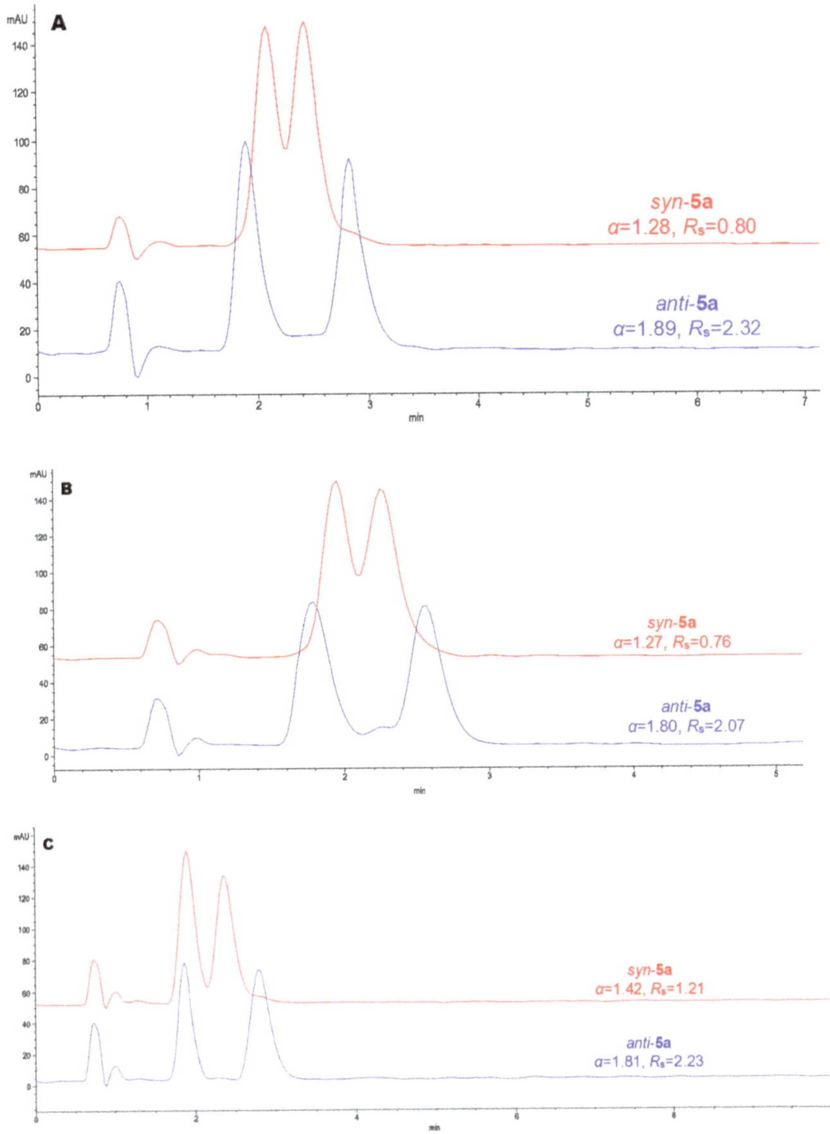

Figure 6. SFC overlay chromatograms of (±)-*syn*-**5a** and (±)-*anti*-**5a** on Amylose-SA column with: (**A**) CO_2/MeOH (80/20, *v*/*v*), (**B**) CO_2/EtOH (80/20, *v*/*v*) and (**C**) CO_2/2-PrOH(80/20, *v*/*v*) as mobile phases.

3.2. Enantioseparation on Cellulose-SB

This CSP contains the same 3,5-dimethylphenylcarbamate substituent as Amylose-SA; however, they differ only in the nature of polysaccharide backbone, i.e., cellulose and amylose. When operating in the normal phase HPLC mode, the Cellulose-SB column provided a better separation for enantiomers of *anti*-hydantoins **5a–i** compared to *syn*-isomers. It is obvious that the enantiomers of all analyzed *anti*-hydantoins were well separated (R_S > 2.14) on Cellulose-SB along with good enantioselectivity. In contrast to *anti*-hydantoins, seven *syn*-hydantoins were baseline separated on this column while two hydantoins *syn*-**5a** and *syn*-**5c** showed only partial enantioseparation. In general, lower retention and higher α and R_S values were obtained for *anti*-hydantoins **5a–i**. In the normal phase mode, the retention factor of the first eluting enantiomers of the *syn*- and *anti*-compounds **5a** and **5d–i** were always higher than that of other two hydantoins **5b** and **5c**. A possible reason may be the presence of a allyl group (compound **5a**), a furan ring (compound **5d**) or a phenyl ring (compounds **5e–i**) at the N3 position of these hydantoins, which could provide additional π-π interactions between these analytes (donor) and CSP (acceptor). On the other hand, compounds **5b** and **5c** showed the lowest retention under the normal phase mode. This indicates that the interaction between compounds **5b** and **5c** with the stationary phase is weak, possibly due to a hexyl or cycloalkyl substituent at the N3 position of the hydantoin ring, which cannot provide additional interactions with the CSP like other analyzed hydantoins.

When DMC was used as the mobile phase, compounds *anti*-**5a**, *anti*-**5c**, *anti*-**5e** and *anti*-**5i** achieved baseline separation. Among them, compound *anti*-**5i** containing the 2,6-dimethylphenyl group on N3 showed the best separation (R_S = 2.21 and α = 1.90), followed by *anti*-**5c** containing the cyclopentyl moiety on N3 with a *Rs* value of 2.15 and α of 2.46. Other *anti*-hydantoins **5b**, **5d**, **5f**, **5g** and **5h** were partially separated. In addition, all nine *syn*-hydantoins **5a–i** were not enantioseparated on this column under the same conditions. It is possible that DMC altered the supramolecular structure of the cellulose chiral selector by modifying the size of the interaction cavity between the polysaccharide chains and made them more suitable for interactions with the *anti*-hydantoins **5a–i**. As seen from Table 2, all analyzed hydantoins were always longer retained in the normal phase than in non-standard phase mode, which indicates that the lower retention of hydantoins was the result of their weaker interaction with CSP. Typical chromatograms of the resolution of the enantiomers of allyl hydantoins *syn*-**5a** and *anti*-**5a** under *n*-hexane/2-PrOH and dimethyl carbamate are shown in Figure 7. It can be seen that the higher retention, and higher separation factor and resolution was obtained for *anti*-**5a** compared to *syn*-**5a** under both HPLC modes.

Table 2. Chromatographic parameters for the enantioselective separations of racemic *syn*-and *anti*-3,5-disubstituted hydantoins on Cellulose-SB.

Compound	Condition *	k_1	k_2	*A*	R_s	Compound	Condition *	k_1	k_2	α	R_s
syn-**5a**	A	13.95	14.93	1.07	0.84	*anti*-**5a**	A	7.15	10.38	1.45	4.58
	B	0.12	0.12	1.00	-		B	0.17	0.30	1.76	1.68
	C	2.91	3.25	1.12	0.71		C	2.08	3.20	1.54	2.83
	D	2.77	3.05	1.10	0.46		D	1.91	2.62	1.37	1.75
	E	3.71	3.71	1.00	-		E	2.42	3.37	1.39	2.13
syn-**5b**	A	8.04	10.33	1.28	3.09	*anti*-**5b**	A	4.64	6.36	1.37	3.63
	B	0.12	0.12	1.00	-		B	0.20	0.27	1.35	0.89
	C	3.07	3.07	1.00	-		C	2.18	3.20	1.47	2.45
	D	2.82	2.82	1.00	-		D	1.94	2.53	1.30	1.45
	E	3.40	3.80	1.12	0.78		E	2.46	3.18	1.29	1.60

Table 2. *Cont.*

Compound	Condition *	k_1	k_2	A	R_s	Compound	Condition *	k_1	k_2	α	R_s
	A	7.69	8.63	1.12	1.42		A	3.47	4.19	1.21	2.14
	B	0.14	0.14	1.00	-		B	0.13	0.32	2.46	2.15
syn-**5c**	C	3.46	3.46	1.00	-	*anti*-**5c**	C	2.26	2.99	1.32	1.73
	D	3.06	3.06	1.00	-		D	1.93	2.29	1.19	0.98
	E	3.74	3.74	1.00	-		E	2.22	2.66	1.20	1.09
	A	19.82	23.62	1.19	2.28		A	11.63	18.08	1.55	5.49
	B	0.09	0.09	1.00	-		B	0.18	0.26	1.44	1.01
syn-**5d**	C	4.10	4.60	1.12	0.99	*anti*-**5d**	C	3.16	4.77	1.51	3.27
	D	3.90	4.30	1.10	0.73		D	2.86	3.93	1.37	2.33
	E	5.24	5.24	1.00	-		E	3.63	5.22	1.44	2.87
	A	17.03	22.08	1.30	3.34		A	11.37	17.44	1.53	5.05
	B	0.11	0.11	1.00	-		B	0.19	0.32	1.68	1.57
syn-**5e**	C	5.83	7.02	1.20	1.90	*anti*-**5e**	C	5.09	7.94	1.56	4.45
	D	5.27	6.27	1.19	1.65		D	4.32	6.09	1.41	2.98
	E	7.06	7.06	1.00	-		E	5.40	7.67	1.42	3.39
	A	19.70	34.70	1.76	6.63		A	10.32	27.59	2.67	10.44
	B	0.11	0.11	1.00	-		B	0.14	0.23	1.64	0.85
syn-**5f**	C	5.85	8.78	1.50	3.95	*anti*-**5f**	C	4.66	11.98	2.57	5.74
	D	5.43	7.72	1.42	3.37		D	4.43	7.00	1.58	4.23
	E	7.74	9.96	1.29	2.48		E	5.86	9.57	1.63	4.45
	A	25.61	32.02	1.25	2.97		A	11.70	23.86	2.04	8.12
	B	0.15	0.15	1.00	-		B	0.22	0.30	1.36	1.08
syn-**5g**	C	8.39	10.84	1.29	2.81	*anti*-**5g**	C	6.03	10.75	1.78	6.02
	D	7.58	9.39	1.24	2.22		D	5.42	8.12	1.50	3.90
	E	10.73	11.66	1.11	1.09		E	6.98	10.70	1.53	4.34
	A	21.82	30.92	1.42	4.16		A	13.66	27.91	2.04	8.32
	B	0.15	0.15	1.00	-		B	0.23	0.33	1.43	0.88
syn-**5h**	C	5.96	8.85	1.48	4.01	*anti*-**5h**	C	4.17	7.29	1.75	5.15
	D	3.86	5.71	1.48	3.29		D	3.86	5.71	1.48	3.29
	E	7.95	10.00	1.26	2.34		E	4.93	7.99	1.62	4.24
	A	22.09	31.27	1.42	4.16		A	8.39	36.14	4.31	13.06
	B	0.13	0.13	1.00	-		B	0.20	0.38	1.90	2.21
syn-**5i**	C	6.25	8.43	1.35	3.05	*anti*-**5i**	C	4.82	8.51	1.77	5.48
	D	5.87	8.20	1.40	3.26		D	3.91	7.13	1.82	5.35
	E	7.89	11.91	1.51	4.21		E	4.52	10.55	2.33	7.86

* Chromatographic conditions: mobile phase, A, *n*-hexane/2-PrOH (90/10, *v*/*v*), flow rate 1 mL min^{-1}; B, 100% DMC, flow rate 1 mL min^{-1}; C, CO_2/MeOH (80/20, *v*/*v*), flow rate 4 mL min^{-1}, backpressure 15 MPa; D, CO_2/EtOH (80/20, *v*/*v*), 4 mL min^{-1}, backpressure 15 MPa; E, CO_2/2-PrOH (80/20, *v*/*v*), flow rate 4 mL min^{-1}, backpressure 15 MPa. Detection wavelength for each condition was 254 nm. Column temperature of condition A and B is 30 °C, of C, D and E is 35 °C. The chromatographic parameters k_1, k_2, α and R_s are defined in Section 2.

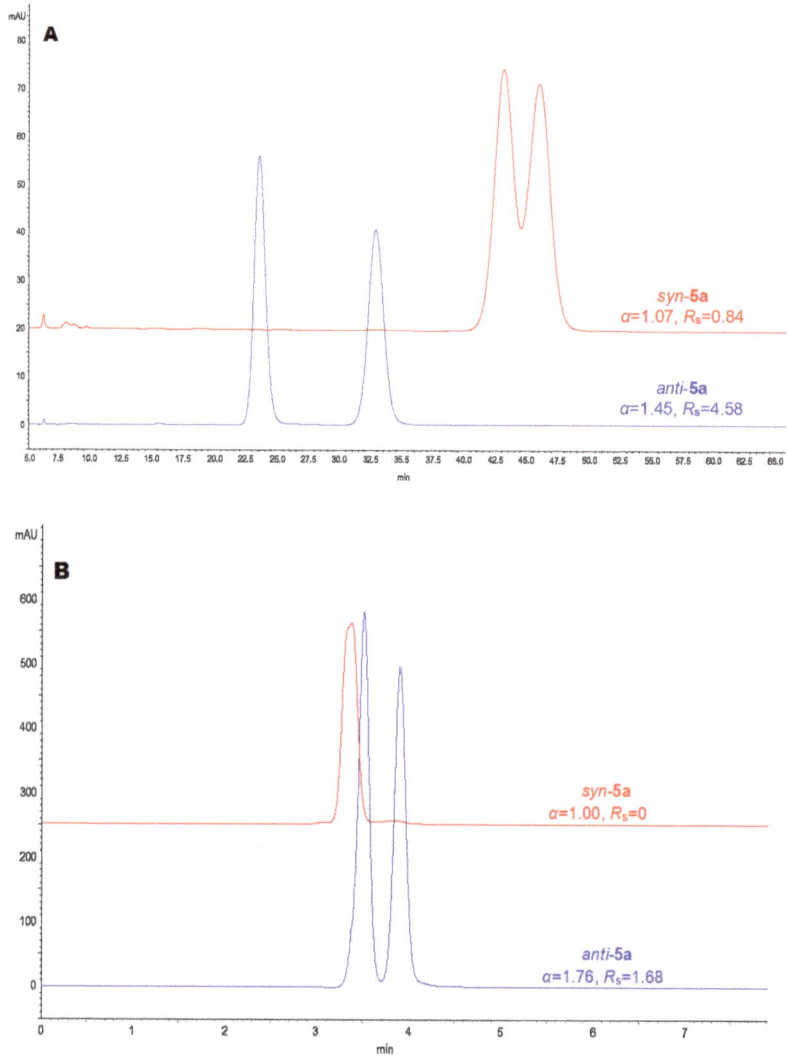

Figure 7. HPLC overlay chromatograms of (±)-*syn*-**5a** and (±)-*anti*-**5a** on Cellulose-SB column with: (**A**) *n*-hexane/2-PrOH (90/10, *v*/*v*) and (**B**) 100% DMC as mobile phases.

The effects of supercritical carbon dioxide and alcohol modifiers (MeOH, EtOH and 2-PrOH) on the enantioseparation of *syn*- and *anti*-hydantoins are listed in Table 2. From the obtained results, we can notice that compounds *syn*-**5f**, *syn*-**5h**, *syn*-**5i**, *anti*-**5a**, *anti*-**5d**, *anti*-**5e**, *anti*-**5f**, *anti*-**5g**, *anti*-**5h** and *anti*-**5i** were completely separated using all three alcohol modifiers, along with a good resolution. Baseline separation of *syn*-**5e** with benzyl substituent at the N3 position of the hydantoin ring and *syn*-**5g** with 3-chloro-4-methylphenyl substituent at the same position was achieved when MeOH or EtOH were selected as the alcohol modifier; among them, a better separation was obtained with MeOH. Baseline separation of *anti*-**5b** was achieved with MeOH and 2-PrOH, while the baseline separation of *anti*-**5c** was achieved using MeOH, and EtOH compounds *syn*-**5a** and *syn*-**5d** were partially separated using MeOH and EtOH as the polar modifiers. The enantiomers of compound *syn*-**5b** were partially separated on the Cellulose-SB with mobile phase CO_2/2-PrOH (80/20, *v*/*v*), while the enantiomers of *syn*-**5a** and *syn*-**5d** were not resolved with the same mobile phase. No chiral resolution of hydantoin *syn*-**5c** was observed on this column with either MeOH, EtOH or 2-PrOH as modifiers. As seen from Table 2, compounds *syn*- and *anti*-**5g** with 3-chloro-4-methylphenyl substituent at the N3 position of the hydantoin ring were always longer retained than other analytes, which implied that the interactions between these two analytes and CSP were the strongest. On the contrary, lower retention of compounds *syn*- and *anti*-**5a**, *syn*- and *anti*-**5b**, and *syn*- and *anti*-**5c** was the result of their weaker interaction with CSP. As can be seen by comparing the results in Table 2, the retention factors of all *syn*-hydantoins were higher than *anti*-hydantoins with all three modifiers. Obviously, when branched alcohol, 2-PrOH, was used as the alcohol modifier, the retention time of most compounds was longer than that of using linear alcohols, which indicated that the steric effect of the modifier likely contributed to the decreased strengths of the interactions between the mobile phase and the CSP, resulting in the reduced elution ability of the mobile phase. The Cellulose-SB column in SFC mode exhibited better enantioseparation toward *anti*-hydantoins compared to *syn*-hydantoins. As shown in Figure 8, higher separation and resolution of the compound *anti*-**5a** with allyl substituent at the N3 position of the hydantoin ring was achieved using the mobile phase CO_2/alcohol (80/20, *v*/*v*).

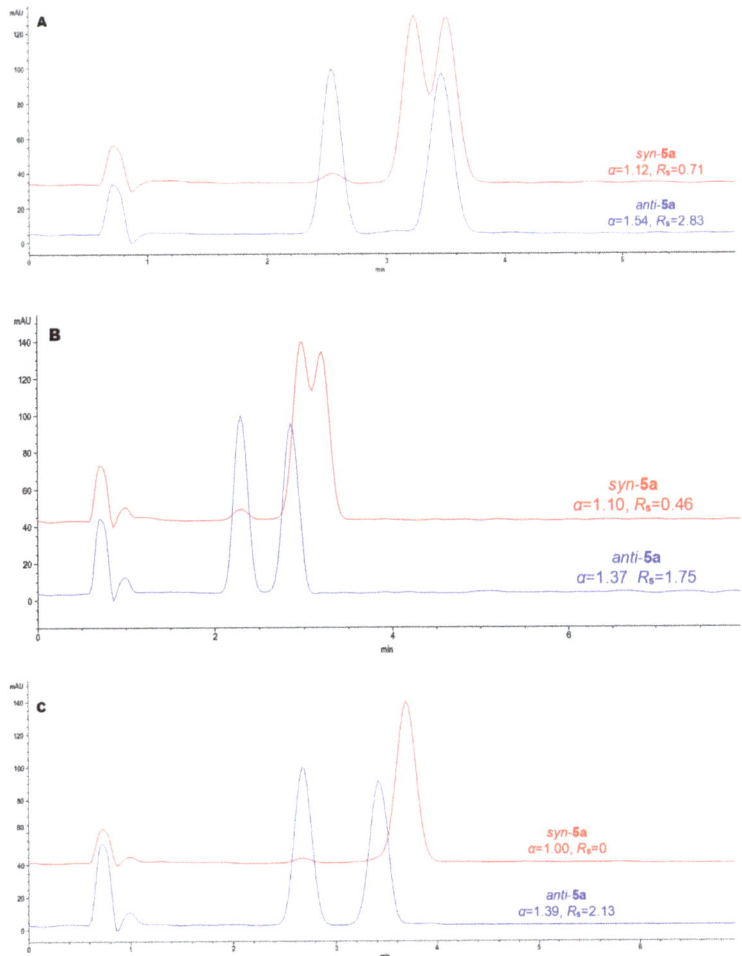

Figure 8. SFC overlay chromatograms of (±)-*syn*-**5a** and (±)-*anti*-**5a** on Cellulose-SB column with: (**A**) CO_2/MeOH (80/20, *v*/*v*), (**B**) CO_2/EtOH (80/20, *v*/*v*) and (**C**) CO_2/2-PrOH(80/20, *v*/*v*) as mobile phases.

3.3. Enantioseparation on Cellulose-SC

The results of the enantioseparations using the Cellulose-SC column in the normal and non-standard phase HPLC mode and SFC mode are summarized in Table 3.

Table 3. Chromatographic parameters for the enantioselective separations of racemic *syn*-and *anti*-3,5-disubstituted hydantoins on Cellulose-SC.

Compound	Condition *	k_1	k_2	α	R_s	Compound	Condition *	k_1	k_2	α	R_s
	A	5.22	5.76	1.10	0.47		A	3.07	6.97	2.27	4.44
	B	0.06	0.06	1.00	-		B	0.11	0.11	1.00	-
syn-**5a**	C	0.96	0.96	1.00	-	*anti*-**5a**	C	0.79	0.79	1.00	-
	D	0.76	0.76	1.00	-		D	0.68	0.68	1.00	-
	E	1.13	1.37	1.21	0.27		E	0.87	2.05	2.36	2.26
	A	3.39	3.96	1.17	0.80		A	2.03	4.70	2.32	4.38
	B	0.07	0.07	1.00	-		B	0.12	0.12	1.00	-
syn-**5b**	C	1.03	1.03	1.00	-	*anti*-**5b**	C	0.83	0.83	1.0	-
	D	0.80	0.80	1.00	-		D	0.70	0.70	1.00	-
	E	1.20	1.61	1.34	0.80		E	0.97	2.37	2.44	2.47
	A	2.54	3.79	1.49	2.10		A	1.53	2.24	1.46	-
	B	0.07	0.07	1.00	-		B	0.14	0.14	1.00	-
syn-**5c**	C	1.02	1.49	1.46	1.14	*anti*-**5c**	C	0.80	1.03	1.29	0.52
	D	0.77	1.07	1.39	0.66		D	0.62	0.79	1.27	0.52
	E	1.24	1.53	1.23	0.48		E	0.85	1.22	1.44	0.77
	A	8.51	9.61	1.13	0.68		A	5.09	13.57	2.67	5.27
	B	0.05	0.05	1.00	-		B	0.10	0.10	1.00	-
syn-**5d**	C	1.38	1.38	1.00	-	*anti*-**5d**	C	1.20	1.20	1.00	-
	D	1.06	1.06	1.00	-		D	0.93	1.08	1.16	-
	E	1.62	2.02	1.25	0.70		E	1.35	3.51	2.60	3.41
	A	6.19	6.19	1.00	-		A	3.80	6.22	1.64	2.61
	B	0.06	0.06	1.00	-		B	0.11	0.11	1.00	-
syn-**5e**	C	1.79	1.79	1.00	-	*anti*-**5e**	C	1.50	1.50	1.00	-
	D	1.28	1.28	1.00	-		D	1.11	1.11	1.00	-
	E	2.03	2.03	1.00	-		E	1.63	2.73	1.67	1.92
	A	8.91	8.91	1.00	-		A	6.52	13.81	2.12	3.29
	B	0.05	0.05	1.00	-		B	0.09	0.09	1.00	-
syn-**5f**	C	2.14	2.14	1.00	-	*anti*-**5f**	C	1.74	1.74	1.00	-
	D	1.77	2.37	1.34	1.14		D	1.55	1.55	1.00	-
	E	3.13	9.66	3.08	5.38		E	3.00	6.65	2.22	3.55
	A	13.49	33.60	2.49	5.00		A	11.02	23.05	2.09	3.99
	B	0.07	0.07	1.00	-		B	0.13	0.13	1.00	-
syn-**5g**	C	2.91	2.91	1.00	-	*anti*-**5g**	C	2.01	2.54	1.26	0.98
	D	2.41	3.15	1.31	1.22		D	2.15	2.15	1.00	-
	E	4.77	12.01	2.52	4.86		E	5.01	9.30	1.86	3.32

Table 3. *Cont.*

Compound	Condition *	k_1	k_2	α	R_s	Compound	Condition *	k_1	k_2	α	R_s
syn-**5h**	A	12.55	12.55	1.00	-	*anti*-**5h**	A	13.80	13.80	1.00	-
	B	0.09	0.09	1.00	-		B	0.13	0.16	1.23	-
	C	2.30	3.40	1.48	1.87		C	2.22	2.22	1.00	-
	D	1.98	4.66	2.35	3.96		D	2.03	2.86	1.41	1.24
	E	4.77	12.01	2.52	4.86		E	5.18	17.36	3.35	6.58
syn-**5i**	A	4.66	7.23	1.55	2.17	*anti*-**5i**	A	3.88	3.88	1.00	-
	B	0.04	0.04	1.00	-		B	0.09	0.09	1.00	-
	C	1.34	1.92	1.43	1.27		C	1.26	1.26	1.00	-
	D	0.94	1.48	1.57	1.27		D	0.89	0.89	1.00	-
	E	1.53	2.34	1.53	1.53		E	1.44	1.44	1.00	-

* Chromatographic conditions: mobile phase, A, *n*-hexane/2-PrOH (90/10, *v*/*v*), flow rate 1 mL min^{-1}; B, 100% DMC, flow rate 1 mL min^{-1}; C, CO_2/MeOH (80/20, *v*/*v*), flow rate 4 mL min^{-1}, backpressure 15 MPa; D, CO_2/EtOH (80/20, *v*/*v*), 4 mL min^{-1}, backpressure 15 MPa; **E**, CO_2/2-PrOH (80/20, *v*/*v*), flow rate 4 mL min^{-1}, backpressure 15 MPa. Detection wavelength in each condition was 254 nm. Column temperature of condition A and B is 30 °C, of C, D and E is 35 °C. The chromatographic parameters k_1, k_2, α and R_s are defined in Section 2.

When *n*-hexane/2-PrOH (90/10, *v*/*v*) was used as the mobile phase, the retention factors of the first-eluting enantiomers of compounds *syn*- and *anti*-**5g** and **5h** were higher than of other hydantoins. This implies that the interactions between these analytes and CSP were the strongest. A possible reason may be the presence of two groups at the *meta*-position of the phenyl ring attached to the N3 position of these hydantoins. On the contrary, the lower retention of compounds *anti*-**5b**, *syn*-**5c** and *anti*-**5c** was the result of their weaker interaction with CSP, possibly due to an alkyl or cycloalkyl substituent at the N3 position of the hydantoin ring. The longer retention was not evidently always accompanied with better enantioseparation. Among all eighteen analytes, baseline separations of nine pairs of enantiomers were achieved using this mobile phase system. It is interesting to notice that the Cellulose-SC column did not show chiral recognition ability toward any of the *syn*- and *anti*-hydantoin racemates with DMC as the mobile phase. An explanation for this possible supramolecular effect could be the same as for the Cellulose-SB column. The HPLC mobile phase composition was found to influence the retention time and resolution of the analyzed hydantoin enantiomers. The effects of *n*-hexane/2-PrOH (90/10, *v*/*v*) versus DMC on enantioselectivity of allyl hydantoins *syn*-**5a** and *anti*-**5a** using the Cellulose-SC column are shown in Figure 9. The enantiomers of *anti*-**5a** were well separated (R_s = 4.44), while the enantiomers of its diastereoisomer *syn*-**5a** were only partially separated (R_s = 0.47) under the normal phase HPLC mode. No chiral recognition was observed for both allyl hydantoins *syn*-**5a** and *anti*-**5a** under DMC mobile phase.

The Cellulose-SC column did not exhibit an enantiorecognition ability for most of the tested racemates using CO_2/MeOH (80/20, *v*/*v*). For compounds *syn*-**5c**, *anti*-**5c**, *anti*-**5g** and *syn*-**5i**, only a partial enantioseparation was achieved. However, baseline separation was achieved for only one hydantoin, *syn*-**5h**, with 3,5-dimethylphenyl group at the N3 position of the hydantoin ring when MeOH was selected as a modifier. When 2-PrOH was used as the alcoholic modifier, eleven hydantoin racemates were baseline separated. The compounds *syn*-**5a-d** and *anti*-**5c** attained partial separation, while enantiomers of compounds *syn*-**5e** and *anti*-**5i** were not separated. When EtOH was used, six compounds attained partial separation and only compound *syn*-**5h** was separated to baseline. Moreover, when using 2-PrOH α, values were larger than those obtained with EtOH as the modifier.

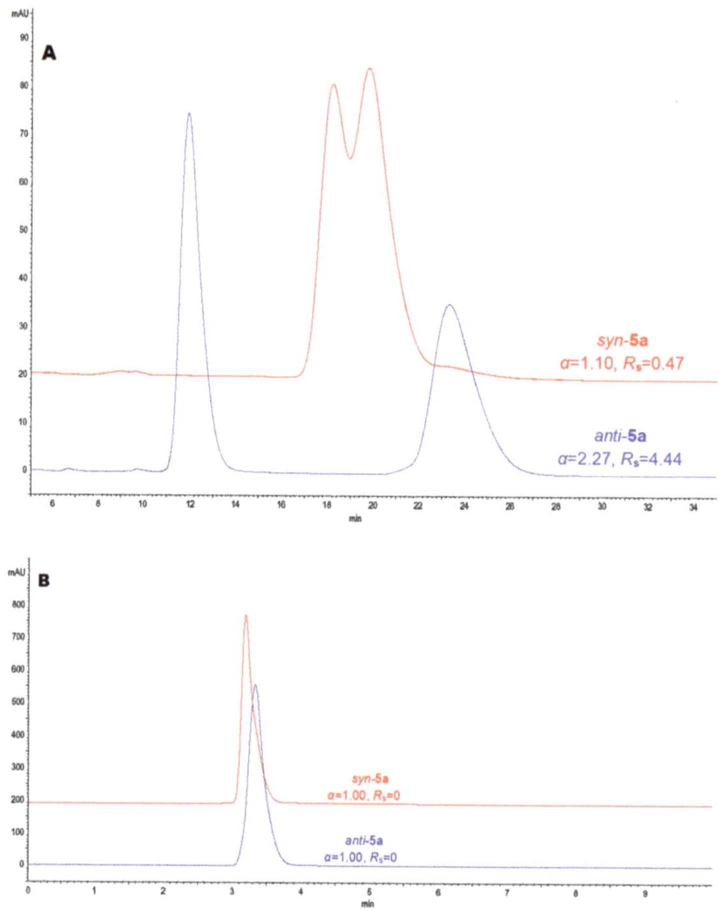

Figure 9. HPLC overlay chromatograms of (±)-*syn*-**5a** and (±)-*anti*-**5a** on Cellulose-SC column with: (**A**) *n*-hexane/2-PrOH (90/10, *v*/*v*) and (**B**) 100% DMC as mobile phases.

3.4. Recognition Complementarities of Three Tested Immobilized CSPs

As described in Sections 3.1–3.3, a number of baseline-separated hydantoins using three different immobilized CSPs is seventeen, sixteen, and nine for columns Amylose-SA, Cellulose-SB and Cellulose-SC, respectively, when using the *n*-hexane/2-PrOH (90/10, *v*/*v*) mobile phase system. Under non-standard HPLC conditions (DMC), the enantiomers of eight hydantoins were separated at baseline on the Amylose-SA column, while baseline separation of only four hydantoins was achieved on the column Cellulose-SB. No chiral resolution was observed on the Cellulose-SC with DMC as the mobile phase. Among eighteen hydantoins, baseline separations of fourteen pairs of enantiomers were achieved on the immobilized Cellulose-SB column by SFC, using the mobile phase CO_2/MeOH (80/20, *v*/*v*) followed by the Amylose-SA column (thirteen enantiomers of eighteen tested compounds) and the Cellulose-SC (one enantioseparation of eighteen). Operating in SFC mode, the Cellulose-SC column provided poor enantioseparations for these kinds of compounds when MeOH and EtOH were used as polar modifiers. However, when switched to 2-PrOH as the modifier, baseline separation of the eleven hydantoins was obtained on this column. Baseline separation of a great number of 3,5-disubstituted hydantoin enantiomers was achieved on the Amylose-SA and Cellulose-SB columns with CO_2/alcohol (MeOH, EtOH, 2-PrOH) as the mobile phase.

The rate of baseline separation (r.b.s.) is defined as the ratio of baseline-separated analytes to the total samples [60]. The r.b.s. values for the immobilized column Amylose-SA, Cellulose-SB and Cellulos-SC under normal phase and non-standard HPLC conditions and under SFC conditions were always higher for *anti*-hydantoins **5a–i** than for *syn*-hydantoins **5a–i**. The Cellulose-SC column exhibits quite poor performances for the series of hydantoins under investigation. Neither *syn*-hydantoins **5a–i** nor *anti*-hydantoins **5a–i** were separated on Cellulose-SC under the mobile phase DMC. Furthermore, the *anti*-hydantoins **5a–i** were not separated on this column using the mobile phase of supercritical CO_2 and the alcoholic modifiers (MeOH and EtOH). With regard to the number of baseline separations, the enantioseparation ability of the three columns decreased in the order Amylose-SA > Cellulose-SB > Cellulose-SC. Amylose-SA provided greater enantioresolution toward the majority of the tested analytes. The amylose-based CSP is considered to be more helical in nature than cellulose-derived CSP [52]. Consequently, the difference in helical structures between amylose and cellulose resulted in different enantiorecognition behaviors [61,62]. It is obvious from Table 4 that the Amylose-SA column has better chiral recognition capacities than Cellulose-SB. Furthermore, it can be seen that the Cellulose-SB column is more efficient than the Cellulose-SC column due to the presence of chlorine atoms in later chiral selector. The electronegative nature of chlorine atoms makes phenyl ring electrons deficient, and this consequently leads to poor π-π interactions and low chiral recognition capabilities.

Table 4. Enantioseparation efficiencies of the three tested immobilized CSPs.

Column/CPS	Mobile Phase Condition (*v*/*v*)	r.b.s. *		
		syn	*anti*	*Syn + anti*
Amylose-SA	Hex/2-PrOH = 90/10	0.89	1.00	0.94
	DMC	0.33	0.56	0.44
	CO_2/MeOH = 80/20	0.44	1.00	0.72
	CO_2/EtOH = 80/20	0.56	1.00	0.78
	CO_2/2-PrOH = 80/20	0.78	1.00	0.89
Cellulose-SB	Hex/2-PrOH = 90/10	0.78	1.00	0.89
	DMC	0	0.44	0.22
	CO_2/MeOH = 80/20	0.56	1.00	0.78
	CO_2/EtOH = 80/20	0.56	0.78	0.67
	CO_2/2-PrOH = 80/20	0.33	0.89	0.61
Cellulose SC	Hex/2-PrOH = 90/10	0.33	0.67	0.50
	DMC	0	0	0
	CO_2/MeOH = 80/20	0.11	0	0.06
	CO_2/EtOH = 80/20	0.11	0	0.06
	CO_2/2-PrOH = 80/20	0.44	0.78	0.61

* r.b.s was defined as the rate of baseline separation.

4. Conclusions

In this comprehensive study, the chiral separation of eighteen 3,5-disubstituted hydantoins **5a–i** were conducted on three immobilized polysaccharide-based CSPs (Amylose-SA, Cellulose-SB and Cellulose-SC) by HPLC under normal and non-standard mobile phase and by SFC, using carbon dioxide and different alcohol modifiers (MeOH, EtOH and 2-PrOH). The column Amylose-SA turned out to be the best in both HPLC and SFC modalities. All three CSPs showed better chiral recognition toward *anti*-3,5-disubstituted hydantoins compared to *syn*-isomers, both in HPLC and SFC modes. In the HPLC, the results were better when *n*-hexane/2-PrOH (90/10, *v*/*v*) was used, in terms of higher separation and resolution, but with longer analysis times. We have shown that DMC can be efficiently used as a mobile phase in chiral separation of 3,5-disubstituted hydantoins on the immobilized polysaccharide-based CSPs, especially on the Amylose-SA column. Using DMC, no chiral recognition of any *syn*-hydantoins was observed on Cellulose-SB, and all of *syn*- and *anti*-hydantoins on Cellulose-SC. The columns Amylose-SA and Cellulose-SB provided fine or excellent separations for these types of compounds.

Author Contributions: M.J. performed synthesis, chromatographic analyses and writing. T.D. performed literature search and overview. D.K. and M.R. performed the study design, data analysis, revising, final approval, and handled the accountability of all aspects of the work. All authors contributed to the article and approved the submitted version. All authors have read and agreed to the published version of the manuscript.

Funding: This study was supported by the Croatian Government and the European Union through the European Regional Development Fund-the Competitiveness and Cohesion Operational Programme (KK.01.1.1.01) through the project Bioprospecting of the Adriatic Sea (KK.01.1.1.01.0002) granted to The Scientific Centre of Excellence for Marine Bioprospecting-BioProCro.

Institutional Review Board Statement: Not applicable.

Informed Consent Statement: Not applicable.

Data Availability Statement: Data are contained within the article.

Acknowledgments: We would like to thank the Croatian Government and the European Union through the European Regional Development Fund-the Competitiveness and Cohesion Operational Programme (KK.01.1.1.01) for funding this research through the project Bioprospecting of the Adriatic Sea (KK.01.1.1.01.0002) granted to The Scientific Centre of Excellence for Marine Bioprospecting-BioProCro. We also thank the Center for NMR for recording spectra.

Conflicts of Interest: The authors declare no conflict of interest.

References

1. Cho, S.; Kim, S.-H.; Shin, D. Recent applications of hydantoin and thiohydantoin in medicinal chemistry. *Eur. J. Med. Chem.* **2019**, *164*, 517–545. [CrossRef] [PubMed]
2. Gawas, P.P.; Buthanapalli, R.; Veeraiah, N.; Nutalapati, V. Multifunctional hydantoins: Recent advances in optoelectronics and medicinal drugs from Academia to the chemical industry. *J. Mater. Chem. C* **2021**, *9*, 16341–16377. [CrossRef]
3. Yang, X.Y.; Su, L.; Hou, X.B.; Ding, S.Y.; Xu, W.F.; Wang, B.H.; Fang, H. High-performance liquid chromatographic enantioseparation of 3,5-disubstituted hydantoins analogs and temperature-induced reversals of elution orders on a polysaccharide-based chiral stationary phase. *J. Chrom. A* **2014**, *1355*, 291–295. [CrossRef]
4. Konnert, L.; Lamaty, F.; Martinez, J.; Colacino, E. Recent Advances in the Synthesis of Hydantoins: The State of the Art of a Valuable Scaffold. *Chem. Rev.* **2017**, *117*, 13757–13809. [CrossRef] [PubMed]
5. Kartozia, I.; Kanyonyo, M.; Happaerts, T.; Lambert, D.M.; Scriba, G.K.E.; Chankvetadze, B. Comparative HPLC enantioseparation of new chiral hydantoin derivatives on three different polysaccharide type chiral stationary phases. *J. Pharm. Biomed. Anal.* **2002**, *27*, 457–465. [CrossRef]
6. Velázquez-Macías, R.F.; Aguilar-Patiño, S.; Cortez-Betancourt, R.; Rojas-Esquivel, I.; Fonseca-Reyes, G.; Contreras-González, N. Evaluation of efficacy of buserelin plus nilutamide in Mexican Male patients with advanced prostate cancer. *Rev. Mex. Urol.* **2016**, *76*, 346–351. [CrossRef]
7. Ito, Y.; Sadar, M.D. Enzalutamide and blocking androgen receptor in advanced prostate cancer: Lessons learnt from the history of drug development of antiandrogens. *Res. Rep. Urol.* **2018**, *10*, 23–32. [CrossRef]
8. Ostrowski, J.; Kuhns, J.-E.; Lupisella, J.A.; Manfredi, M.C.; Beehler, B.C.; Krystek, S.R., Jr.; Bi, Y.; Sun, C.; Seethala, R.; Golla, R.; et al. Pharmacological and X-Ray structural characterization of a novel selective androgen receptor modulator: Potent hyperanabolic stimulation of skeletal muscle with hypostimulation of prostate in rats. *Endocrinology* **2007**, *148*, 4–12. [CrossRef]
9. Cherukuvada, S.; Babu, N.J.; Nangia, A. Nitrofurantoin–*p*-aminobenzoic acid cocrystal: Hydration stability and dissolution rate studies. *J. Pharm. Sci.* **2011**, *100*, 3233–3244. [CrossRef]
10. Kim, D.; Wang, L.; Caldwell, C.G.; Chen, P.; Finke, P.E.; Oates, B.; MacCoss, M.; Mills, S.G.; Malkowitz, L.; Gould, S.L.; et al. Discovery of human CCR5 antagonists containing hydantoins for the treatment of HIV-1 infection. *Bioorg. Med. Chem. Lett.* **2001**, *11*, 3099–3102. [CrossRef]
11. El-Barbary, A.A.; Khodair, A.I.; Pedersen, E.B.; Nielsen, C. *S*-Glucosylated hydantoins as new antiviral agents. *J. Med. Chem.* **1994**, *37*, 73–77. [CrossRef] [PubMed]
12. Verlinden, Y.; Cuconati, A.; Wimmer, E.; Rombaut, B. The antiviral compound 5-(3,4-dichlorophenyl) methylhydantoin inhibits the post-synthetic cleavages and the assembly of poliovirus in a cell-free system. *Antivir. Res.* **2000**, *48*, 61–69. [CrossRef]
13. Rajic, Z.; Zorc, B.; Raic-Malic, S.; Ester, K.; Kralj, M.; Pavelic, K.; Balzarini, J.; Clercq, E.D.; Mintas, M. Hydantoin Derivatives of L- and D-amino acids: Synthesis and evaluation of their antiviral and antitumoral activity. *Molecules* **2006**, *11*, 837–848. [CrossRef] [PubMed]

14. Marton, J.; Enisz, J.; Hosztafi, S.; Timar, T. Preparation and fungicidal activity of 5-substituted hydantoins and their 2-thio analog. *J. Agric. Food Chem.* **1993**, *41*, 148–152. [CrossRef]
15. Knabe, J.; Baldauf, J.; Ahlhem, A. Racemates and enantiomers of basic, substituted 5-phenylhydantoins, synthesis and antiarrhythmic action. *Die Pharm.* **1997**, *52*, 912–919.
16. Matsukura, M.; Daiku, Y.; Ucda, K.; Tanaka, S.; Igarashi, T.; Minami, N. Synthesis and antiarrhythmic activity of 2,2-dialkyl-1′-(*N*-substituted aminoalkyl)-spiro-[chroman-4,4′-imidazolidine]-2′,5′-diones. *Chem. Pharm. Bull.* **1992**, *40*, 1823–1827. [CrossRef]
17. Anger, T.; Madge, D.J.; Mulla, M.; Riddall, D. Medicinal chemistry of neuronal voltage-gated sodium channel blockers. *J. Med. Chem.* **2001**, *44*, 115–137. [CrossRef]
18. Somsák, L.; Kovács, L.; Tóth, M.; Ösz, E.; Szilágyi, L.; Györgydeák, Z.; Dinya, Z.; Docsa, T.; Tóth, B.; Gergely, P. Synthesis of and a comparative study on the inhibition of muscle and liver glycogen phosphorylases by epimeric pairs of D-gluco- and D-xylopyranosylidene-spiro-(thio)hydantoins and *N*-(D-Glucopyranosyl) amides. *J. Med. Chem.* **2001**, *44*, 2843–2848. [CrossRef] [PubMed]
19. Oka, M.; Matsumoto, Y.; Sugiyama, S.; Tsuruta, N.; Matsushima, M. A potent aldose reductase inhibitor, (2*S*,4*S*)-6-Fluoro-2′,5′-dioxospiro[chroman-4,4′-imidazolidine]-2-carboxamide (Fidarestat): Its absolute configuration and interactions with the aldose reductase by X-ray crystallography. *J. Med. Chem.* **2000**, *43*, 2479–2483. [CrossRef]
20. Nakabayashi, M.; Regan, M.M.; Lifsey, D.; Kantoff, P.W.; Taplin, M.-E.; Sartor, O.; Oh, W.K. Efficacy of nilutamide as secondary hormonal therapy in androgen-independent prostate cancer. *BJU Int.* **2005**, *96*, 783–786. [CrossRef]
21. Kassouf, W.; Tanguay, S.; Aprikian, A.G. Nilutamide as second line hormone therapy for prostate cancer after androgen ablation fails. *J. Urol.* **2003**, *169*, 1742–1744. [CrossRef] [PubMed]
22. Struck, R.F.; Kirk, M.C.; Rice, L.S.; Suling, W.J. Isolation, synthesis and antitumor evaluation of spirohydantoin aziridine, a mutagenic metabolite of spirohydantoin mustard. *J. Med. Chem.* **1986**, *29*, 1319–1321. [CrossRef] [PubMed]
23. Foulds, G.; O'Brien, M.M.; Bianchine, J.R.; Gabbay, K.H. Kinetics of an orally absorbed aldose reductase inhibitor, sorbinil. *Clin. Pharmacol. Ther.* **1981**, *30*, 693–700. [CrossRef] [PubMed]
24. Lu, H.; Kong, D.; Wu, B.; Wang, S.; Wang, Y. Synthesis and evaluation of anti-inflammatory and antitussive activity of hydantion derivatives. *Lett. Drug Des. Discov.* **2012**, *9*, 638–642. [CrossRef]
25. Fiallo, M.M.L.; Kozlowski, H.; Garnier-Suillerot, A. Mitomycin antitumor compounds: Part 1. CD studies on their molecular structure. *Eur. J. Pharm. Sci.* **2001**, *12*, 487–494. [CrossRef]
26. Youssef, D.T.A.; Shaala, L.A.; Alshali, K.Z. Bioactive hydantoin alkaloids from the red sea marine sponge *Hemimycale arabica*. *Mar. Drugs* **2015**, *13*, 6609–6619. [CrossRef]
27. Mio, S.; Ichinose, R.; Goto, K.; Sugaai, S.; Sato, S. Synthetic studies on (+)-hydantocidin (1): A total synthesis of (+)-hydantocidin, a new herbicidal metabolite from microorganism. *Tetrahedron* **1991**, *47*, 2111–2120. [CrossRef]
28. Mio, S.; Kumagawa, Y.; Sugaai, S. Three-step synthetic pathway to fused bicyclic hydantoins involving a selenocyclization step. *Tetrahedron Lett.* **1993**, *34*, 7391–7394. [CrossRef]
29. Gregoriou, M.; Noble, M.; Watson, K.; Garman, E.; Krulle, T.; Delafuente, C.; Fleet, G.; Oikonomakos, N.; Johnson, L. The structure of a glycogen phosphorylase glucopyranose spirohydantoin complex at 1.8 Å resolution and 100 K: The role of the water structure and its contribution to binding. *Protein Sci.* **1998**, *7*, 915–927. [CrossRef]
30. Shiozaki, M. Syntheses of hydantocidin and C-2-thioxohydantocidin. *Carbohyd. Res.* **2002**, *337*, 2077–2088. [CrossRef]
31. Kalník, M.; Gabko, P.; Bella, M.; Koóš, M. The Bucherer–Bergs multicomponent synthesis of hydantoins—excellence in simplicity. *Molecules* **2021**, *26*, 4024. [CrossRef] [PubMed]
32. Uemoto, H.; Tsuda, M.; Kobayashi, J. Mukanadins A—C, New bromopyrrole alkaloids from marine sponge *Agelas nakamurai*. *J. Nat. Prod.* **1999**, *62*, 1581–1583. [CrossRef] [PubMed]
33. Jimenez, C.; Crews, P. Mauritamide A and accompanying oroidin alkaloids from the sponge *Agelas mauritiana*. *Tetrahedron Lett.* **1994**, *35*, 1375–1378. [CrossRef]
34. Audion, C.; Cocandeau, V.; Thomas, O.P.; Bruschini, A.; Holderith, S.; Genta-Jouve, G. Metabolome consistency: Additional Parazoanthines from the Mediterranean zoanthid *Parazoanthus axinellae*. *Metabolites* **2014**, *4*, 421–432. [CrossRef] [PubMed]
35. Cachet, N.; Genta-Jouve, G.; Regalado, E.L.; Mokrini, R.; Amade, P.; Culioli, G.; Thomas, O.P. Parazoanthines A—E, hydantoin alkaloids from the Mediterranean sea anemone *Parazoanthus axinellae*. *J. Nat. Prod.* **2009**, *72*, 1612–1615. [CrossRef]
36. Chankvetadze, B. Recent trends in preparation, investigation and application of polysaccharide-based chiral stationary phases for separation of enantiomers in high-performance liquid chromatography. *TrAC* **2020**, *122*, 115709. [CrossRef]
37. Horváth, S.; Eke, Z.; Németh, G. A protocol to replace dedication to either normal phase or polar organic mode for chiral stationary phases containing amylose tris (3,5-dimethylphenylcarbamate). *J. Chrom. A* **2022**, *1673*, 463052. [CrossRef]
38. Yamamoto, C.; Okamoto, Y. Optically Active Polymers for Chiral Separation. *Bull. Chem. Soc. Jpn.* **2004**, *77*, 227–257. [CrossRef]
39. Foroughbakhshfasaei, M.; Dobó, M.; Boda, F.; Szabó, Z.-I.; Tóth, G. Comparative Chiral Separation of Thalidomide Class of Drugs Using Polysaccharide-Type Stationary Phases with Emphasis on Elution Order and Hysteresis in Polar Organic Mode. *Molecules* **2022**, *27*, 111. [CrossRef]
40. Cirilli, R.; Ferretti, R.; Gallinella, B.; De Santis, E.; Zanitti, L.; La Torre, F. High-performance liquid chromatography enantioseparation of proton pump inhibitors using the immobilized amylose-based Chiralpak IA chiral stationary phase in normal-phase, polar organic and reversed-phase conditions. *J. Chrom. A* **2008**, *1177*, 105–113. [CrossRef]

41. Peluso, P.; Mamane, V.; Dallocchio, R.; Dessi, A.; Cossu, S. Noncovalent interactions in high-performance liquid chromatography enantioseparations on polysaccharide-based chiral selectors. *J. Chrom. A.* **2020**, *1623*, 461202. [CrossRef] [PubMed]
42. De Klerck, K.; Mangelings, D.; Heyden, Y.V. Supercritical fluid chromatography for the enantioseparation of pharmaceuticals. *J. Pharm. Biomed. Anal.* **2012**, *69*, 77–92. [CrossRef] [PubMed]
43. Bajtai, A.; Ilisz, I.; Berkecz, R.; Fülöp, F.; Lindner, W. Polysaccharide-based chiral stationary phases as efficient tools for diastereo- and enantioseparation of natural and synthetic *Cinchona* alkaloid analogs. *J. Pharm. Biomed. Anal.* **2021**, *193*, 113724. [CrossRef] [PubMed]
44. Tundo, P.; Selva, M. The Chemistry of Dimethyl Carbonate. *Acc. Chem. Res.* **2002**, *35*, 706–716. [CrossRef] [PubMed]
45. Ono, Y. Catalysis in the production and reactions of dimethyl carbonate, an environmentally benign building block. *Appl. Catal. A Gen.* **1997**, *155*, 133–166. [CrossRef]
46. Kim, K.H.; Lee, E.Y. Environmentally-benign dimethyl carbonate-mediated production of chemicals and biofuels from renewable bio-oil. *Energies* **2017**, *10*, 1790. [CrossRef]
47. Arico, F.; Tundo, P. Dimethyl carbonate: A modern green reagent and solvent. *Russ. Chem. Rev.* **2010**, *79*, 479–489. [CrossRef]
48. Tundo, P. New developments in dimethyl carbonate chemistry. *Pure Appl. Chem.* **2001**, *73*, 1117–1124. [CrossRef]
49. Abdalla, A.O.G.; Liu, D. Dimethyl carbonate as a promising oxygenated fuel for combustion: A review. *Energies* **2018**, *11*, 1552. [CrossRef]
50. Nomanbhay, S.; Ong, M.Y.; Chew, K.; Show, P.-L.; Lam, M.K.; Chen, W.-H. Organic carbonate production utilizing crude glycerol derived as by-product of biodiesel production: A Review. *Energies* **2020**, *13*, 1483. [CrossRef]
51. Lajin, B.; Goessler, W. Introducing dimethyl carbonate as a new eluent in HPLC-ICPMS: Stronger elution with less carbon. *J. Anal. At. Spectrom.* **2021**, *36*, 1272–1279. [CrossRef]
52. Armarego, W.L.F. *Purification of Laboratory Chemicals*, 8th ed.; Butterworth Heinemann: Oxford, UK, 2017; p. 31.
53. Jurin, M.; Kontrec, D.; Dražić, T.; Roje, M. Enantioseparation of (±)-trans-β-lactam Ureas by Supercritical Fluid Chromatography. *Croat. Chem. Acta* **2020**, *93*, 203–213. [CrossRef]
54. Bandyopadhyay, D.; Cruz, J.; Banik, B.K. Novel synthesis of 3-pyrrole substituted β-lactams via microwave-induced bismuth nitrate-catalyzed reaction. *Tetrahedron* **2012**, *68*, 10686–10695. [CrossRef]
55. Radolović, K.; Habuš, I.; Kralj, B. New thiazolidinone and triazinethione conjugates derived from amino-β-lactams. *Heterocycles* **2009**, *78*, 1729–1759. [CrossRef]
56. Mehra, V.; Kumar, V. Facile diastereoselective synthesis of functionally enriched hydantoins via base-promoted intramolecular amidolysis of C-3 functionalized azetidin-2-ones. *Tetrahedron Lett.* **2013**, *54*, 6041–6044. [CrossRef]
57. Ghanem, A.; Wang, C. Enantioselective separation of racemates using CHIRALPAK IG amylose-based chiral stationary phase under normal standard, non-standard and reversed phase high performance liquid chromatography. *J. Chrom. A* **2018**, *1532*, 89–97. [CrossRef]
58. Ikai, T.; Yamamoto, C.; Kamigaito, M.; Okamoto, Y. Immobilized polysaccharide derivatives: Chiral packing materials for efficient HPLC resolution. *Chem. Rec.* **2007**, *7*, 91–103. [CrossRef]
59. Zhu, B.; Zhao, F.; Yu, J.; Wang, Z.; Song, Y.; Li, Q. Chiral separation and a molecular modeling study of eight azole antifungals on the cellulose *tris* (3,5-dichlorophenylcarbamate) chiral stationary phase. *New J. Chem.* **2018**, *42*, 13421–13429. [CrossRef]
60. Pelusoa, P.; Mashiko, V.; Aubert, E.; Cossu, S. High-performance liquid chromatography enantioseparation of atropisomeric 4,4′-bipyridines on polysaccharide-type chiral stationary phases: Impact of substituents and electronic properties. *J. Chrom. A* **2012**, *1251*, 91–100. [CrossRef]
61. Okamoto, Y. Chiral polymers for resolution of enantiomers. *J. Polym. Sci. Polym. Chem.* **2009**, *47*, 1731–1739. [CrossRef]
62. Yang, G.S.; Zhan, C.Y.; Fu, G.H.; Vazquez, P.P.; Frenich, A.G.; Vidal, J.L.M.; Aboul-Enein, H.Y. Chiral separation of organic phosphonate compounds on polysaccharide-based chiral stationary phases. *Chromatographia* **2004**, *59*, 631–635. [CrossRef]

separations

MDPI

Article

Validation, Optimization and Hepatoprotective Effects of Boeravinone B and Caffeic Acid Compounds from *Boerhavia diffusa* Linn

Kamal Y. Thajudeen [1], Abdulrhman Alsayari [1], Shehla Nasar Mir Najib Ullah [1], Shahana Salam [2], Muhammed Elayadeth-Meethal [3] and Ilyas Uoorakkottil [4,*]

1 Department of Pharmacognosy, College of Pharmacy, King Khalid University, Abha 62529, Saudi Arabia; kthajudeen@kku.edu.sa (K.Y.T.); alsayarikku@gmail.com (A.A.); sajibullah@kku.edu.sa (S.N.M.N.U.)

2 Department of Pharmaceutical Chemistry, College of Pharmacy, Prince Sattam Bin Abdulaziz University, Al-kharj 11942, Saudi Arabia; s.salam@psau.edu.sa

3 Department of Animal Breeding and Genetics, College of Veterinary and Animal Sciences, Kerala Veterinary and Animal Sciences University, Wayanad 673576, Kerala, India; dremmuhammed@gmail.com

4 Department of Pharmacognosy and Phytochemistry, Moulana College of Pharmacy, Perinthalmanna 679321, Kerala, India

* Correspondence: dr.ilyasmcp@gmail.com

Abstract: Boerhavia diffusa, also known as Punarnava, is a plant of the Nyctaginaceae family that has been utilized in traditional medicine to cure a variety of ailments. The goal of this study was to use response surface methodology (RSM) to optimize the maximum percentage yield of boeravinone B and caffeic acid from *Boerhavia diffusa* roots, and simultaneous determination of boeravinone B and caffeic acid in newly developed single solvent system and demonstrate the hepatoprotective benefits of boeravinone B and caffeic acid. The extraction process examined extraction time, extraction temperature and solvent concentration, which were optimized via Box–Behnken experimental design. The proposed HPTLC method for the quantification of boeravinone B and caffeic acid were successfully validated and developed. The method was validated in term of linearity and detection limit, quantification limit, range, precision, specificity and accuracy. The separation of boeravinone B and caffeic acid bands was achieved on HPTLC plate using formic acid: ethyl acetate: toluene (1:3:5 *v*/*v*) as developing system. Densitometric analyses of boeravinone B and caffeic acid was carried out in the absorbance mode at 254 nm. The maximum percentage yield of caffeic acid and boeravinone B from *Boerhavia diffusa* require appropriate extraction parameters such as temperature, time, organic solvents and water content, which can be achieved using the Box-Behnken statistical design provide time: temperature: solvent ratio (30:45:40 *v*/*v*) for extraction of caffeic acid and 60:60:40 *v*/*v* for extraction of boeravinone B. The boeravinone B (200 μg/mL) and caffeic acid (200 μg/mL) showed the most significant hepatoprotective activity compared with standard sylimarin in HepG2 cell induced with galactosamine 40 mM toxicity. The findings supported B. diffusa's traditional use as a functional food forhuman health benefits.

Keywords: *Boerhavia diffusa*; boeravinone B; caffeic acid; optimization; hepatoprotective activity; HPTLC

Citation: Thajudeen, K.Y.; Alsayari, A.; Najib Ullah, S.N.M.; Salam, S.; Elayadeth-Meethal, M.; Uoorakkottil, I. Validation, Optimization and Hepatoprotective Effects of Boeravinone B and Caffeic Acid Compounds from *Boerhavia diffusa* Linn. *Separations* **2022**, *9*, 177. https://doi.org/10.3390/separations9070177

Academic Editor: Liangliang Liu

Received: 24 May 2022
Accepted: 26 June 2022
Published: 18 July 2022

1. Introduction

Boerhavia diffusa has been traditionally used by Indian and Brazilian indigenous and tribal people. Ayurvedic medicine traditionally uses the plant's root and leaves to treat a wide range of ailments, such as viral jaundice and liver dysfunction. This plant isreported to have behavioral and neuroendocrine effect and anti-depressant activity [1], anti-angiogenic activity [2], anti-convulsant and anti-epileptic activity [3], anti-inflammatory activity [4], inhibitory effect of prostatic hyperplasia [5], anti-genotoxic effect [6], thrombolytic, cytotoxic and anti-microbial activities [7], arsenic trioxide-induced cardiotoxicity [8], anti-urolithic activity [9], anti-hyperglycaemic and reno-protective effect [10], acetaminophen-induced liver toxicity [11], anti-proliferative and anti-estrogenic effects [12], immunomodulatory and anti-metastatic activity [13,14], radio-protective activity [15], potent anti-breast cancer activity [16], intestinal activity [17], immunosuppressive activity [18], anti-diabetic activity [19], Ca^{2+} channel antagonistic activity [20], chemo-preventive action [21], glycol-induced urolithiasis [22], and inhibition of human cervical cancer cells [23]. Main Rotenoids area type isoflavone compound (known as boeravinones) that are isolated from the roots of this plant; they showed potent anti-cancer [24,25], anti-inflammtory [26], cardioprotective [27] powerful anti-oxidant and geno-protective properties [28], intestinal motility and spasmolytic activity [29]. Caffeic acid (3,4-dihydroxycinnamic acid) is one of the major polyphenolic compounds found in many plant species, showing anti-viral activity [30], immunostimulatory activity [31], cardioprotective activity [32], hepatoprotective activity [33], anti-cancer activity [34] and anti-hepatocellular carcinoma activity [35]. For detecting phenolic compounds in medicinal plants, several analytical techniques such as liquid chromatography–mass spectrometry (LCMS), high-performance liquid chromatography (HPLC), high-performance thin layer chromatography (HPTLC)and ultra-performance liquid chromatography (UPLC). Boeravinone B and caffeic acid are rarely applied by use of their similar analytical techniques. Ilyas et al. in 2013 reported only identification and quantification of caffeic acid and boeravinone in single-solvent system toluene: ethyl acetate: formic acid: methanol (3:3:0.8:0.2 *v/v*) but tailing peak and separation and resolution of band of caffeic acid and boeravinone was very poor [36]. Here, a validated high-performance thin layer chromatography (HPTLC) method has been newly developed based on single solvent system (toluene: ethyl acetate: formic acid (5:4:1 *v/v*) for simultaneous validation, quantification and optimization of boeravinone B and caffeic acid in the plants. The selected HPTLC method is accurate, precise, simple, specific, less time-consuming, cost-effective and can separate boeravinone B and caffeic acid compounds from their other constituents. Because of advantages such as ease of sample preparation, optimization of specific chemicals and comparison of several samples on a single plate, comparable chromatographic techniques are frequently used to evaluate retinoid and phenolic acids. The comparison of the entire chromatograms allows for the detection of minor similarities and differences between the plants under investigation. The objective of the study was to simultaneously quantify, validate, optimize and to evaluate hepatoprotective activity of boeravinone B and caffeic acid in hydroalcoholic extracts of *B. diffusa*.

2. Materials and Methods

2.1. Chemicals and Instruments

We purchased standard boeravinone B and caffeic acid from NRPL, Bangalore, India. Ethyl acetate, toluene, methanol and formic acid (CDH Labs, Mumbai, India) were used as developing systems for HPTLC analysis. All samples and standards used for the analysis were filtered through a membrane filter, 0.22 μm pore size (HIMEDIA, Mumbai, India). Extraction method using a Soxhlet apparatus (Omega, Mumbai, India) and Rotary evaporator (Buchi R-114, Switzerland). CAMAG HPTLC system (Muttenz, Switzerland) equipped with a Linomat IV sample applicator was used for quantification and optimization. HepG2 cell line (National Centre for Cell Science (NCCS), Pune, India), Dulbecco's modified eagle medium (DMEM), fetal bovine serum (FBS) and MTT assay kit, trypsin EDTA, penicillin and streptomycin and DMSO (Sigma-Aldrich Co. LLC, St. Louis, MO, USA), Trypan blue solution and, galactosamine and absolute ethanol (HimediaLab Pvt. Ltd., Mumbai, India).Tissue culture flasks, 96- and 24-well microculture plates, eppendorf tube, inverted microscope, serological pipette, heamocytometer (HimediaLab Pvt. Ltd., Mumbai, India), laminar flow hoods (Khera instrument, New Delhi, India), CO_2 incubator (NuAire, Plymouth, MN, USA), water bath, Deepfreezer (−20 °C).

2.2. Collection and Extraction of Plant Materials

In February 2011, we collected the roots of *B. diffusa in* Maruthmallai, Kanyakumari district, Tamilnadu, India. It was identified and authenticated by Dr. V. Chelladurai, Research Officer, Central Council for Research in Ayurveda and Siddha (Govt. of India), Tirunelveli, Tamil Nadu. The dried roots were grinded into a coarse powder and subjected to extraction by using 95% methanol and 50% (v/v) hydro-alcohol for 6 h at 37 °C, respectively. The methanol and hydro-alcoholic extracts were evaporated and dried at 45 °C with help of rotary evaporator under reduced pressure. The yield of the extract prepared with 95% methanol was 5.29% weight-for-weight, while the extract prepared with hydro-alcoholic alcohol yielded 9.9% weight-for-weight.

2.3. HPTLC Instrumentation

2.3.1. Preparation of Sample Solution

A total of 100 mg of extracts were dissolved in 10 mL of methanol and sonicated the solution for 15 min and then makeup with 10 mL HPLC grade methanol, filtered through a membrane filter, 0.22 µm pore size before injecting into the HPTLC system.

2.3.2. Preparation of Plate

Prior to usage, pre-coated silica gel 60 F254 aluminum TLC plates (Merck, Germany) were washed in methanol and dried. The samples were applied in the form of bands of 4 mm width using Linomat V applicator (Muttenz, Switzerland) with a 100 µL syringe. The application rate was kept constant at 200 nL s^{-1}, and the space between the two bands was 9 mm.

2.3.3. Slit and Scanning Speed

The slit size was fixed at 5 0.45 mm, and the scanning speed was set to 5 mm s^{-1}.

2.3.4. Chromatographic Conditions

The mobile phase was 1:3:5 $v/v/v$ formic acid, ethyl acetate, and toluene and the development volume was20 mL. In a 10 × 10 cm twin-trough glass chamber saturated with the mobile phase, linear ascending development was performed. At room temperature, the optimal chamber saturation time for the mobile phase was 20 min. The chromatogram run was 8.5 cm long, and the TLC plates were dried in an oven at 50 °C for 5 min after development. CAMAG TLC Scanner 3 in absorbance-reflectance mode at 254 nm, operated by Win-CATS software, was used for densitometric scanning. The deuterium lamp was used as a source of radiation. The peak area response was compared to the amount of medication in a linear regression.

2.3.5. Preparation of Standards

The standard solution of bioactive compounds was prepared by dissolving 1.0 mg of boeravinone B and caffeic acid in 10 mL of methanol as stock solution and stored at 2 to 8 °C. These boeravinone B and caffeic acid were further diluted as per the requirement to plot a final concentration for quantification.

2.3.6. Analytical Method Evaluation

The lowest diluted solutions of the two reference compounds in the calibration curves were further diluted to a series of concentrations with HPLC grade methanol for the determinations of limits of detection (LOD) and quantification (LOQ). The LOD and LOQ under the present chromatographic conditions were determined at a signal-to-noise (S/N) ratio of 3 and 10, respectively. Calibration curves, regression equation along with LOD and LOQ for each compound have been validated as per the ICH guidelines Q2 (R1) (2005). Linearity was assessed with the aid of serially diluted calibration solutions of all standards. Calibration graphs were plotted on the basis of triplicate analysis of each calibration solutions by using peak area against concentration. To evaluate the accuracy, the preanalyzed samples were spiked with standard at three different known concentration levels i.e., 50, 100 and 150% and the mixtures were re-analyzed by the proposed method. Precision of the method was determined by carrying out the intra-day and inter-day variation tests. The inter-day and intra-day variations for the determination of boeravinone B and caffeic acid were carried out at different levels of concentrations 100, 200, 400 ng/band. The identification of bands was carried out in triplicate. The R.S.D. was taken as a measure of precision. The resolution between the peaks of caffeic acid and Boeravinone B found satisfactory (2.1 ± 0.2). The developed method resolution found to be good enough to avoid any possible merging between the peaks. Satisfactory recovery values suggest that there was no interference happening between the standard peaks.

2.4. Optimization of Boeravinone B and Caffeic Acidin Hydroalcoholic Extracts

Using the Box-Behnken statistical design, three factors and levels were considered. There were seventeen runs of it. To optimize the design, the Stat-Ease V6 software (Minneapolis, MN, USA) was used (Minneapolis, Stat-Ease Inc., MN, and USA). This style is suitable for investigating constructing second-order polynomial models and quadratic response surfaces. The designed style contains the replicated center point of the four-dimensional cube and a group of purposes lying at the center of every edge that confirmed the region of interest. The dependent and independent variables are recorded in Table 1.

Table 1. Independent and dependent variables selected in Box–Behnken design.

Factors Independent Variables	Levels Used Low (−1)	Medium	High (+1)
B_1 = Time in min	30	60	90
B_2 = Temperature (°C)	30	45	60
B_3 = Solvent concentration (*v*/*v*)	40	60	80
Dependent variables			
C_1 = Boeravinone B	Maximized percentage yield		
C_2 = Caffeic acid	Maximized percentage yield		

The quadratic equation created by the Box–Behnken Design-Expert equation

$$R = A_0 + A_1 B_1 + A_2 B_2 + A_3 B_3 + A_4 B_1 B_2 + A_5 B_1 B_3 + A_6 B_2 B_3 + A_7 B_1^2 + A_8 B_2^2 + A_9 B_3^2 \quad (1)$$

where A_0 is the intercept, A_1 to A_9 are the regression coefficients, R is the dependent variable and B_1, B_2 and B_3 are the independent variables.

The data of the maximum percentage yield of boeravinone B and caffeic acid in the hydro-alcoholic extract of plant materials were recorded through arithmetical improvement by using Design-Expert software system. The Box–Behnken Design style spreadsheet is shown in Table 2.

Table 2. Mentioned responses in Box–ehnken design experiment for 17 analytical trails.

Run	Factor-1 (A_1): Time (Min)	Factor-2(A_2): Temperature (°C)	Factor-3 (A_3): Solvent Concentration Methanol (%)	Water Content (%)	% Yield Boeravinone B (C_1)	% Yield Caffeic Acid(C_2)
01	30	30	60	40	0.044	0.011
02	90	30	60	40	0.048	0.018
03	30	60	60	40	0.045	0.017
04	90	60	60	40	0.039	0.017
05	30	45	40	60	0.050	0.036
06	90	45	40	60	0.027	0.048
07	30	45	80	20	0.011	0.046
08	90	45	80	20	0.032	0.0384
09	60	30	40	60	0.036	0.041

Table 2. *Cont.*

Run	Factor-1 (A_1): Time (Min)	Factor-2(A_2): Temperature (°C)	Factor-3 (A_3): Solvent Concentration		% Yield Boeravinone B (C_1)	% Yield Caffeic Acid(C_2)
			Methanol (%)	Water Content (%)		
10	60	60	40	60	0.047	0.068
11	60	30	60	40	0.035	0.067
12	60	60	80	20	0.015	0.043
13	60	45	60	40	0.0086	0.066
14	60	45	60	40	0.0085	0.065
15	60	45	60	40	0.0086	0.066
16	60	45	60	40	0.0085	0.065
17	60	45	60	40	0.0085	0.066

2.5. *The Hepatoprotective Activity of Caffeic Acid and Boeravinone B*

Cells were seeded at a density of 1 × 10^5 cells/ in each well of the 24-well plates and incubated overnight. After 24 h, the media was flicked off, the cells were treated with boeravinone B and caffeic acid (100 and 200 µg/mL) in separate wells of a 24-well plate and incubated for 2 h. Silymarin (250, 500 µg/mL) was used as a reference standard. After incubation, the cell was treated with D-galactosamine at a concentration of 40 mM and was allowed to incubate for 2 h. After incubation, the cells were washed and incubated for 1 h with MTT (20 µL of 5 mg/mL of MTT in PBS) was added to each well. Formation of formazan crystal was observed under a microscope after 2 h. In case crystal, formation was not proper, incubation was continued for an hour more. Media was removed and the remaining formazan crystals were dissolved in 200 µLof DMSO in each well. The cell culture plate was kept on a shaker for 15 min; the absorbance was recorded by ELISA reader at 540 nm.

The % hepatoprotection of boeravinone B, caffeic acid and silymarin was obtained by the following formula:

$$\% \text{ Hepatoprotection} = (\text{Optical Density of Test sample} / \text{Optical Density of Control}) \times 100$$

3. Results

3.1. *Development of Mobile Phase*

Separation of standard markers studies were carried out on the working standard solution of boeravinone B and caffeic acid in HPLC-grade methanol. Originally, many trials were established with different developing systems such as ethyl acetate: chloroform (5:5), ethyl acetate: chloroform (7:3), ethyl acetate: toluene (7:3), ethyl acetate: toluene (5:5). Finally, toluene: ethyl acetate: formic acid (5:4:1 *v*/*v*) for simultaneous quantification and optimization of boeravinone B and caffeic acid gave a well-defined and sharp peak. We obtained sharp bands upon saturating the twin trough plate development chamber with the developing system at room temperature for 30 min. The band of boeravinone B and caffeic acid was recorded on HPTLC plate scanned at 254 nm (Figure 1).

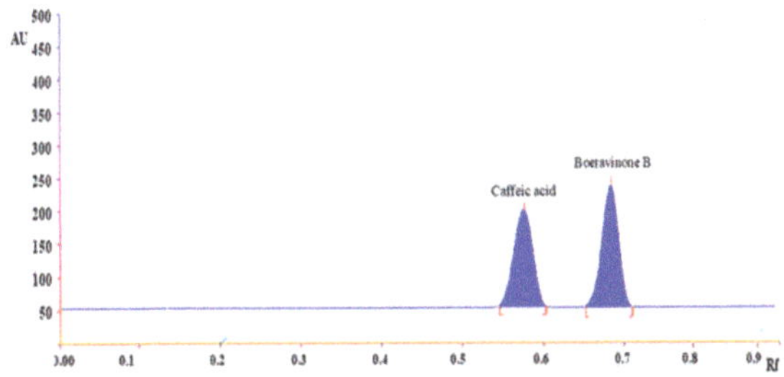

Figure 1. HPTLC densitogram of caffeic acid (R_f = 0.58), boeravinone B (R_f = 0.71) in formic acid: ethyl acetate: toluene (1:3:5 *v*/*v*) presented on HPTLC plate scaned at 254 nm.

3.2. Method of Validation

Figure 1 shows a representative chromatogram of boeravinone b and caffeic acid in the established HPTLC technique. The HPTLC chromatogram in Figure 1 shows a retention factor of caffeic acid (R_f = 0.58), boeravinone B (R_f = 0.71), respectively. The calibration curve's linear regression data revealed a good linear relationship throughout a concentration range of 60 to 240 μg mL^{-1} with a correlation coefficient (R2) value of 0.992 indicated acceptable linearity for caffeic acid and 50 to 250 μg mL^{-1} with a correlation coefficient (R2) value of 0.98.91 indicated acceptable linearity for boeravinone B (Table 2). Correlation coefficient is a statistical tool used to measure the strength or degree of this relationship, and, here, a high correlation coefficient value (a value extremely close to 1.0) indicates a high level of linear relationship between the peak area and concentration of caffeic acid and boeravinone B. The LOD and LOQ were calculated using the ICH Guidelines Q2 (R1) (2005) and were 50 ng, 200 ng for boeravinone B and 40 ng, 200 ng for caffeic acid, respectively (Table 3). The intraday and interday assay was used to determine the repeatability of the proposed HPTLC method, and the results was expressed in the term of %RSD (Table 3) showed that the repeatability of method. The developed HPTLC method's outstanding precision was suggested by the low-percent RSD for repeatability and intermediate precision. The specificity of the developed HPTLC method for the analysis of boeravinone B and caffeic acid in the hydroalcoholic extract was recorded by comparing the spectra obtained in the standards and sample analyses. These spectra's peak start, peak apex, and peak end positions were identical. The recovery trials were done to see how sensitive the method for estimating caffeic acid and boeravinone B. The standard addition approach was used to increase the concentration of caffeic acid and boeravinone B in the sample extract by 50%, 100%, and 150%. The percentage recoveries of the three concentrations ranged from 98.91% to 101.11% for boeravinone B and 99.91% to 100.71% for caffeic acid, indicating a high level of precision. The system suitability parameters of HPLC to establish that the total system operated successfully, system suitability criteria such as peak symmetry, resolution (Rs), capacity factor (K), and selectivity was tested. As demonstrated in Table 3, the obtained values were within acceptable ranges.

Table 3. Method validation parameters for boeravinon b and caffeic acid in *B. diffusa*.

Parameters	Boeravinone B	Caffeic Acid
LOD (ng)	50	40
LOQ (ng)	200	200
Resolution (Rs)	2.11 ± 0.74	1.93 ± 1.2
Capacity factor (k′)	0.31	0.24
Selectivity (α)	1.1	1.17
Tailing factor	1	1
Specificity	Specific	Specific
Interdayprecision (%RSD)	0.93	0.76
Intraday precision (%RSD)	0.99	0.79
Regression equation	Y = −204.4 + 8.6X	Y = −484.7X + 11.2
Average recovery %	98.91 ± 1.2	99.91 ± 0.81
R_f values	0.76	0.63

3.3. Box–Behnken Design-Experiment

A three-level factorial, Box–Behnken Design-Expert applied statistical modeling experimental style was performed by applying 17 investigational trial runs, as recorded in Table 2. Standard boeravinone B and caffeic acid determined that the preferred model was the squared model and also the equivalent values of standard deviation and percentage coefficient of variation for the various planned models is listed in Table 4 together with the regression equations typically for ultimately elite responses. Only statically significant ($p < 0.0001$) for boeravinone B and caffeic acid respectively coefficient is enclosed in the equations. A positive value indicated the optimization of boeravinone B and caffeic acid in hydroalcoholic extracts, while a negative value represents an inverse relationship between the response and the factor. It is clear from the equations that the factor time (B_1), as well as temperature (B_2) has an uncooperative effect and 40:60:80 v/v solvent ratios (B_3) have a productive on the responses (C_1 and $C_{2,}$). It also indicates that the correlation between factors and response is not always linear. When more than one factor is replaced at the same time, a factor can represent a various degree of response. Relationships of B1 and B2, as well as B1 and B3, induce uncooperative impacts on the response. However, the result in the case of the square root of various factors does not repeat history, as shown by its performance. In the case of the square root of various factors, $B^2{}_2$ (extraction temperature) and $B^2{}_3$ (solvent concentration) produced positive results, and $B^2{}_1$ (time of extraction) showed the negative impact on the responses in Table 4. The last mixture ratios of the extractions were established based on percentage yield of boeravinone B and caffeic acid.

Table 4. Summary of results of regression analysis for models and responses (Y_1-Y_2) regression equations for Quadratic models.

Models F Value	R^2	Adjusted R_2	Predicted R_2	SD	C.V.%
		Boeravinone B (Y_1)			
Linear	0.1218	−0.0808	−0.4134	0.018	-
$2F_1$	0.2745	−0.1609	−0.9616	0.018	-
Cubic	0.9899	0.9598	-	0.0034	-
Quadratic	0.9899	0.9769	0.9836	0.0026	9.48
Caffeic acid (Y_2)					
Linear	0.0033	−0.2267	−0.8001	0.023	-
$2F_1$	0.1187	−0.4101	−2.3379	0.024	-
Cubic	0.9999	0.9995	-	0.00045	-
Quadratic	0.9995	0.9989	0.9938	0.00069	1.52

3.4. *Response of Independent Factors on Boeravinone B (C_1)*

Regression equation of the fitted quadratic models:

$$\text{Boeravinone B} = 0.00733 + 0.000075\,B_1 - 0.0023\,B_2\ 0.00812\,B_3 - 0.0029\,B_1\,B_2 + 0.011\,B_1\,B_3 + 0.0078\,B_2\,B_3 + 0.017\,B_1{}^2 + 0.021\,B_2{}^2 + 0.0054\,B_3{}^2 \ldots \quad (2)$$

Factor B_1, which is the time (min) for extraction of plant materials used indicated the percentage yield of boeravinone B in productive direction to that produced with factor B_2. The productive coefficient composition of factor A produced the increase in percentage yield of boeravinone B, which improves in factor B_1. It was confirmed that, when an improved time of extraction was applied, a higher concentration of the percentage of boeravinone B was obtained (Figure 2A) shows the results of time in min of extraction on boeravinone B percentage yield.

Factor B_2 (extraction temperature) was found to have an uncooperative result on the boeravinone B percentage yield compared to factor B_3 (solvent concentration). As the factor B_2 (temperature of extraction) level improved, demonstrated a positive impact on the percentage yield of boeravinone B (Figure 2B), showing that the extraction temperature should be increased to 60 degrees Celsius to get the highest percentage yield of boeravinone B.

In contrast to Factor B3 (extraction temperature), which indicated the maximum percentage yield of boeravinone B, factor B3 (solvent concentration (water: methanol)) produced a favourable impact on the percentage yield of boeravinone B. Boeravinone B concentration increased when a higher solvent concentration ratio was used (Figure 2C), demonstrating the influence of solvent ratio levels on boeravinone B percentage yield.

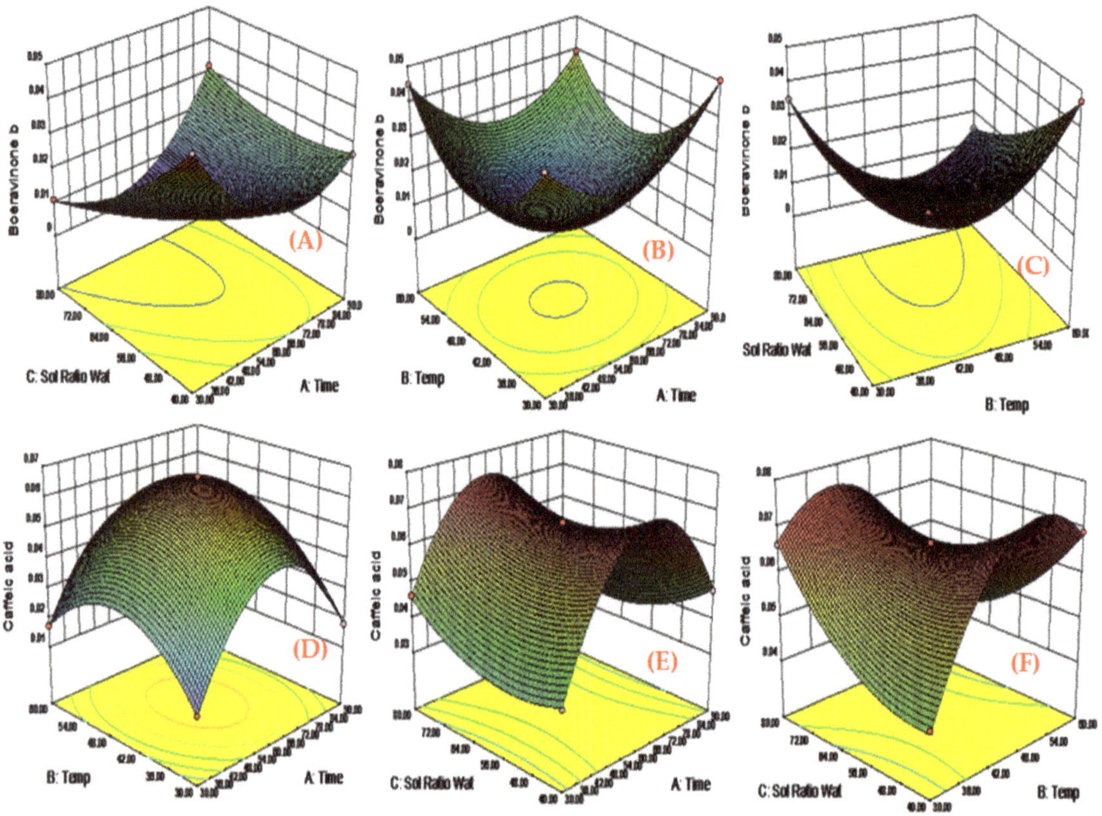

Figure 2. Response surface plots of factor B_2 vs. B_1 against boeravinone B (C_1) and caffeic acid (C_2) respectively. (**A**) when an improved time of extraction was applied, higher concentrations of percentage of boeravinone B were obtained; (**B**) as the factor temperature level improved, the boeravinone B percentage yield also showed an uncooperative automatic impact; (**C**) when an elevated level of solvent ratio was applied, boeravinone B concentration also produced uncooperative impact; (**D**) when an increase time of extraction was applied, an elevated level of percentage yield of caffeic acid was investigated due to elevated level of penetration of solvent into plant extracts; (**E**) when the level of temperature increases, the sharp percentage yield of caffeic acid conjointly inflated slowly as a result of thermostable compound; (**F**) once an elevated level of solvent magnitude relation was applied, the percentage yield of caffeic acid was minimized as a result of the medium polar compound.

3.5. Responseof Independent Factors on Caffeic Acid (C_2)

As the equation shows that B_1 has a productive impact on the caffeic acid percentage yield, When factor B (time of extraction) was increased, a productive level of caffeic acid percentage yield was obtained due to solvent concentration penetrating into the plant materials, as shown in figure. This is because the equation shows that B1 has a productive impact on the caffeic acid percentage yield (Figure 2D).

$$\text{Caffeic acid} = 0.066 + 0.0014\, B_1 + 0.00093\, B_2 + 0.000013\, B_3 - 0.0016\, B_1 B_2 - 0.005\, B_1 B_3 - 0.013\, B_2 B_3 - 0.031\, B_1^2 - 0.019\, B_2^2 + 0.0074\, B_3^2 \quad (3)$$

Factor B_2 is perceived to have an increased profound impact on the proportion of caffeic acid percentage yield compared to issue B_1 and B_2. Because the factor B (temperature level) inflated, the sharp percentage yield of caffeic acid conjointly inflated slowly as a result of thermostable compound. Figure 2E shows the impact of temperature levels on caffeic acid proportion yield.

Factor B_3 (40:60:80 v/v solvent ratio) indicated a distinguished lower caffeic acid percentage yield as compared to that discovered on the issue B_2, which showed a broader negative impact. It had been discovered that, once an elevated level of solvent magnitude relation was applied, the percentage yield of caffeic acid was minimized as a result of the medium polar compound. Figure 2F indicates the impact of solvent magnitude affinity levels on percentage yield of caffeic acid.

3.6. Hepatoprotective Activity of Bioactive Compounds on HpeG$_2$ Cell Induced with Galactosamine

In the study, the above-mentioned fractions were screened for hepatoprotective activity against GalN-induced cytotoxicity, by pre-incubating the cells with or without the extracts or sylimarin. A significant decrease in cell viability was observed upon treatment of HepG2 cells with GalN (40 Mm). Results are expressed as mean ± standard error mean (S.E.M). The percentage protection is represented in Figure 3 and Table 5.

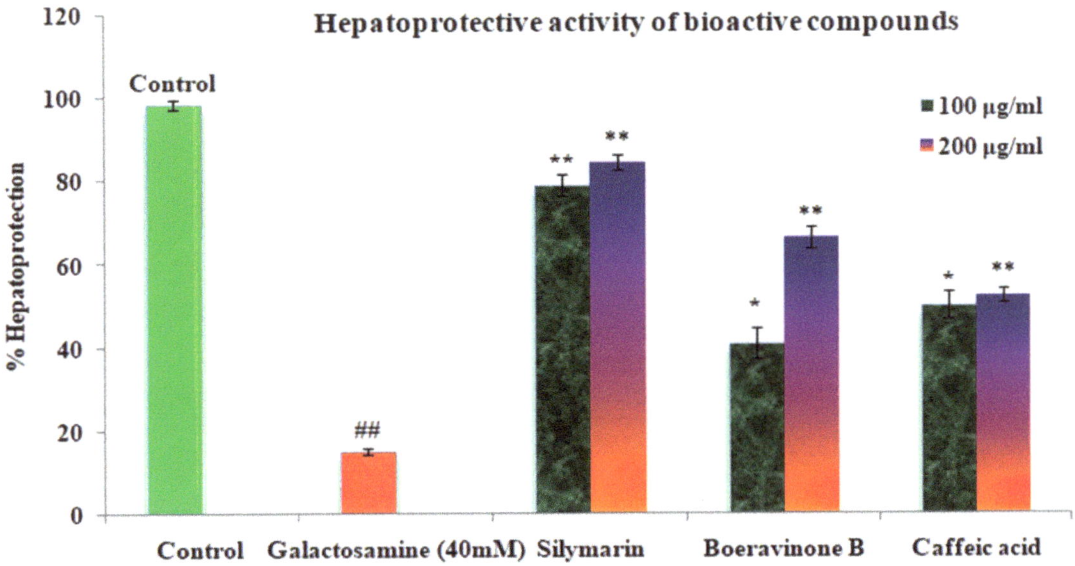

Figure 3. Effect of the caffeic acid and boeravinone B on GlaN-induced toxicity in HepG2 cells. HepG2 cells were incubated in the presence/ absence of various compounds for 2h, prior to treatment with GlaN (40mm) for 2h. Thereafter, the cells were processed for the MTT assay. The results are expressed as mean ± SEM. ## Galactosamine toxic group significant, ns $p > 0.05$, ** $p < 0.01$, * $p < 0.05$ compared to the GalN (40 mM) control.

Table 5. In-vitro hepatoprotective activity of potent compounds present in *B. diffusa*.

Dose Treatment	Control	GLN(40 mM)	Silymarin	Boeravinone B	Caffeic Acid
100 µg/mL	98.3 ± 1.1	15 ± 0.6 ##	78.7 ± 2.6 **	40.89 ± 4.7 *	46.17 ± 4.4 *
Absorbance at 540 nm	3.5	0.459	2.757	1.431	1.616
200 µg/mL			84.35 ± 1.9 **	66.21 ± 2.6 **	52.34 ± 1.7 **
Absorbance at 540 nm			2.952	2.318	1.832

ns: non-significant, $n = 3$, data ± S.E.M. Groups I-4 were compared against group II using Dunnett's posthoc test. ## Galactosamine toxic group significant, ns $p > 0.05$, ** $p < 0.01$, * $p < 0.05$.

4. Discussion

A significant number of medicinal plants are being employed for healing and for the development of new pharmaceuticals all around the world. Understanding the therapeutic significance of natural extracts or their constituents, as well as the standardization of crude pharmaceuticals, necessitate phytochemical study [37]. The literature review supports the fact that many natural products exhibiting hepatoprotective activity need to be standardized and optimized due to degradation of polyphenolic compounds during extraction time.

Several plants that are traditionally used are characterized for the presence of active ingredients using diverse methods [38,39]. Recently, *B. diffusa* has been increasingly propagated as a medicinal plant. The current study aims to develop and validate high performance thin layer chromatography (HPTLC) method for the simultaneous validation, quantification of boeravinone B and caffeic acid in the hydro alcoholic extract of *B. diffusa* in single solvent system. The HPTLC method is accurate, precise, simple, specific, less time consuming, cost-effective, and can separate boeravinone B and caffeic acid compounds from their other constituents such as boeravinone A, C, D, E, F, G and flavonoid glycosides, etc.

In this method, the hydroalcoholic extract was found to contain a comparatively elevated amount of boeravinone B and caffeic acid compared to alcoholic and aqueous extracts. This may be because medium polar retinoid and flavonoid glycosides present in hydroalcoholic extract. The percentage yield of extracts depends on the temperature, time, and solvents of the extraction as well as the chemical moiety of secondary metabolites. Under the same time and temperature conditions, the solvent used and the chemical property of the sample are the most important factors. In the present investigation, the application of the Box–Behnken Design for the optimization of boeravinone B and caffeic acid based on temperature of extraction (°C), time of extraction (min) and solvent concentration ratio (%v/v).

Shimma Ibrahim et al., in 2017, reported that natural plants have garnered great interest in the last few decades as therapeutic agents, since they produce inexpensive and less toxic products than synthetic drugs. To date, there is only one protective natural drug (Silymarin) that too is not curative and also has limitations in protecting the liver from viral attacks. Karimi, G. et al., in 2011, established that silymarin is useful forthe treatment of liver cirrhosis and hepatitis, and displaysimmumodulatory activity [40]. However, silymarin's structure containsa flavonoid moiety. Phenolic compounds (coumarins, flavonoids, xanthones, retinoid and tannins) are widely distributed in plants with potent anti-oxidant and anti-inflammatoryeffects, regulating immunity by interfering with immune cell regulation, proflammatory cytokines, and synthesis and gene expression and inhibitingphsophatidylinostide -3- kinase/ protein kinase [41].In the present study, isolated constituents caffeic acid (100 microgram/mL) and boeravinone B (200 µg/mL) showed significant hepatoprotective activity y compared with standard sylimarin in HepG2 cell induced with galactosamine 40 mM toxicity.

Caffeic acid and boeravinone B have been of interest for theirhealth benefits, and the present HPTLC analytical study could be a future application to purify, identify, quantify, and optimize boeravinone B and caffeic acid in *Boerhaviadiffusa*.

5. Conclusions

The proposed method is innovative, since it is the first analytical method to report on the validation, quantification, and optimization of boeravinone B and caffeic acid in *Boerhavia diffusa* (Linn).The proposed HPTLC method was performed according to the guidelines of the International Conference on Harmonization (ICH) for the quantification of boeravinone B and caffeic acid, which were successfully validated and developed. The method was validated in term of linearity and detection limit, quantification limit, range, precision, specificity, and accuracy. The maximum percentage yield of caffeic acid and boeravinone B from *Boerhavia diffusa* require appropriate extraction parameters such as temperature, time, organic solvents, and water content, which can be achieved using the Box–Behnken statistical design, providing a time: temperature: solvent ratio

(30:45:40 v/v) for the extraction of caffeic acid and 60:60:40 v/v for the extraction of boeravinone B. The hepatoprotection of standard silymarin showed ranging from 78.7 percent at 100 μg/mL to 84.34 percent 200 μg/mL, caffeic acid varied from 46.17 percent at 100 μg/mL to 52.34 percent at 200 μg/mL and boeravinone B ranged from 40.89 percent at 100 μg/mL to 62.21 percent at 200 μg/mL. The boeravinone B and caffeic acid showed the most significant hepatoprotective activity compared with standard silymarin in HepG2 cell induced with galactosamine 40 mM toxicity. The findings supported *Boerhavia diffusa*'s traditional use as a functional food for human health benefits.

Author Contributions: The contribution of the authors is as follows: conceptualization, I.U. and K.Y.T.; methodology, I.U.; investigation, I.U.; data curation, I.U., S.S., M.E.-M., S.N.M.N.U. and A.A.; writing—original draft preparation, I.U., K.Y.T. and M.E.-M.; writing—review and editing, I.U. and M.E.-M.; visualization, I.U.; supervision, K.Y.T.; funding acquisition, K.Y.T., S.N.M.N.U., A.A. and S.S. All authors have read and agreed to the published version of the manuscript.

Funding: This research received no external funding.

Institutional Review Board Statement: Not applicable.

Informed Consent Statement: All participants in the study provided their informed permission.

Data Availability Statement: The data that supports the finding of this study are also available from the corresponding author upon request.

Acknowledgments: The whole authors extend their appreciation to the Deanship of Scientific Research at King Khalid University for funding this work through General Research Project under Grant No: GRP/ 105/43.

Conflicts of Interest: The authors declare no conflict of interest.

References

1. Dhingra, D.; Valecha, R. Behavioural and neuroendocrine effects of aqueous extract of *Boerhaavia diffusa* Linn. in mice using tail suspension and forced swim tests—A preliminary study. *Indian J. Exp. Biol.* **2014**, *52*, 53–59. [PubMed]
2. Saraswati, S.; Alhaider, A.A.; Agrawal, S.S. Punarnavine, an alkaloid from *Boerhaavia diffusa* exhibits anti-angiogenic activity via downregulation of VEGF in vitro and in vivo. *Chem. Biol. Interact.* **2013**, *206*, 204–213. [CrossRef] [PubMed]
3. Sharma, J.; Gairola, S.; Gaur, R.D.; Painuli, R.M.; Siddiqi, T.O. Ethnomedicinal plants used for treating epilepsy by indigenous communities of sub-Himalayan region of Uttarakhand, India. *J. Ethnopharmacol.* **2013**, *150*, 353–370. [CrossRef] [PubMed]
4. Bairwa, K.; Singh, I.N.; Roy, S.K.; Grover, J.; Srivastava, A.; Jachak, S.M. Rotenoids from *Boerhaavia diffusa* as potential anti-inflammatory agents. *J. Nat. Prod.* **2013**, *76*, 1393–1398. [CrossRef]
5. Vyas, B.A.; Desai, N.Y.; Patel, P.K.; Joshi, S.V.; Shah, D.R. Effect of *Boerhaavia diffusa* in experimental prostatic hyperplasia in rats. *Indian J. Pharmacol.* **2013**, *45*, 264–269. [CrossRef]
6. Aher, V.; Chattopadhyay, P.; Goyary, D.; Veer, V. Evaluation of the genotoxic and antigenotoxic potential of the alkaloid punarnavine from *Boerhaavia diffusa*. *Planta Med.* **2013**, *79*, 939–945. [CrossRef]
7. Apu, A.S.; Liza, M.S.; Jamaluddin, A.T.; Howlader, M.A.; Saha, R.K.; Rizwan, F.; Nasrin, N. Phytochemical screening and in vitro bioactivities of the extracts of aerial part of *Boerhavia diffusa* Linn. *Asian Pacific J. Trop. Biomed.* **2012**, *2*, 673–678. [CrossRef]
8. Vineetha, V.P.; Prathapan, A.; Soumya, R.S.; Raghu, K.G. Arsenic trioxide toxicity in H9c2 myoblasts–damage to cell organelles and possible amelioration with *Boerhavia diffusa*. *Cardiovasc. Toxicol.* **2013**, *13*, 123–137. [CrossRef]
9. Yasir, F.; Waqar, M.A. Effect of indigenous plant extracts on calcium oxalate crystallization having a role in urolithiasis. *Urological. Res.* **2011**, *39*, 345–350. [CrossRef]
10. Singh, P.K.; Baxi, D.; Doshi, A.; Ramachandran, A.V. Antihyperglycaemic and renoprotective effect of *Boerhaavia diffusa* L. in experimental diabetic rats. *J. Comp. Int. Med.* **2011**, *8*, 1–20. [CrossRef]
11. Olaleye, M.T.; Akinmoladun, A.C.; Ogunboye, A.A.; Akindahunsi, A.A. Antioxidant activity and hepatoprotective property of leaf extracts of *Boerhaavia diffusa* Linn against acetaminophen-induced liver damage in rats. *Food Chem. Toxicol.* **2010**, *48*, 2200–2205. [CrossRef] [PubMed]
12. Sreeja, S.; Sreeja, S. An in vitro study on antiproliferative and antiestrogenic effects of *Boerhaavia diffusa* L. extracts. *J. Ethnopharmacol.* **2009**, *126*, 221–225. [CrossRef] [PubMed]
13. Manu, K.A.; Kuttan, G. *Boerhaavia diffusa* stimulates cell-mediated immune response by upregulating IL-2 and downregulating the pro-inflammatory cytokines and GM-CSF in B16F-10 metastatic melanoma bearing mice. *J. Exp. Ther. Oncol.* **2008**, *7*, 17–29. [PubMed]
14. Manu, K.A.; Kuttan, G. Anti-metastatic potential of Punarnavine, an alkaloid from *Boerhaavia diffusa* Linn. *Immunobiology* **2009**, *214*, 245–255. [CrossRef] [PubMed]

15. Manu, K.A.; Leyon, P.V.; Kuttan, G. Studies on the protective effects of *Boerhaavia diffusa* L. against gamma radiation induced damage in mice. *Integr. Cancer Ther.* **2007**, *6*, 381–388. [CrossRef] [PubMed]
16. Ahmed-Belkacem, A.; Macalou, S.; Borrelli, F.; Capasso, R.; Fattorusso, E.; Taglialatela-Scafati, O.; Di Pietro, A. Nonprenylated rotenoids, a new class of potent breast cancer resistance protein inhibitors. *J. Med. Chem.* **2007**, *50*, 1933–1938. [CrossRef]
17. Borrelli, F.; Milic, N.; Ascione, V.; Capasso, R.; Izzo, A.A.; Capasso, F.; Petrucci, F.; Valente, R.; Fattorusso, E.; Taglialatela-Scafati, O. Isolation of new rotenoids from *Boerhaavia diffusa* and evaluation of their effect on intestinal motility. *Planta Med.* **2005**, *71*, 928–932. [CrossRef]
18. Pandey, R.; Maurya, R.; Singh, G.; Sathiamoorthy, B.; Naik, S. Immunosuppressive properties of flavonoids isolated from *Boerhaavia diffusa* Linn. *Int. Immunopharmacol.* **2005**, *5*, 541–553. [CrossRef]
19. Satheesh, M.A.; Pari, L. Antioxidant effect of *Boerhavia diffusa* L. in tissues of alloxan induced diabetic rats. *Indian J. Exp. Biol.* **2004**, *42*, 989–992.
20. Lami, N.; Kadota, S.; Kikuchi, T.; Momose, Y. Constituents of the roots of *Boerhaavia diffusa* L. III. Identification of Ca2+ channel antagonistic compound from the methanol extract. *Chem. Pharmaceut. Bulletin.* **1991**, *39*, 1551–1555. [CrossRef]
21. Bharali, R.; Azad, M.R.; Tabassum, J. Chemopreventive action of Boerhaavia diffusa on DMBA-induced skin carcinogenesis in mice. *Indain J. Physiol.Pharmacol.* **2003**, *47*, 459–464.
22. Jarald, E.E.; Kushwah, P.; Edwin, S.; Asghar, S.; Patni, S.A. Effect of Unex on ethylene glycol-induced urolithiasis in rats. *Indian J. Pharmacol.* **2011**, *43*, 466–468. [CrossRef] [PubMed]
23. Srivastava, R.; Saluja, D.; Dwarakanath, B.S.; Chopra, M. Inhibition of Human Cervical Cancer Cell Growth by Ethanolic Extract of *Boerhaavia diffusa* Linn. (Punarnava) Root. *Evid. -Based Complementary Altern. Med. Ecam* **2011**, *2011*, 427031. [CrossRef] [PubMed]
24. Wang, Z.; Xiao, S.; Huang, J.; Liu, S.; Xue, M.; Lu, F. Chemoprotective Effect of Boeravinone B against DMBA/Croton Oil Induced Skin Cancer via Reduction of Inflammation. *J. Oleo. Sci.* **2021**, *70*, 955–964. [CrossRef]
25. Huang, Y.; Sun, Y.; Wang, W.W.; Zhang, L. Boeravinone B a natural rotenoid exerts anticancer activity via inducing internalization and degradation of inactivated EGFR and ErbB2 in human colon cancer cells. *Am. J. Transl. Res.* **2018**, *10*, 4183–4192.
26. Bairwa, K.; Jachak, S.M. Anti-inflammatory potential of a lipid-based formulation of a rotenoid-rich fraction prepared from Boerhavia diffusa. *Pharm. Biol.* **2015**, *53*, 1231–1238. [CrossRef]
27. Chen, Y.; Peng, L.; Shi, S.; Guo, G.; Wen, H. Boeravinone B alleviates gut dysbiosis during myocardial infarction-induced cardiotoxicity in rats. *J. Cell Mol. Med.* **2021**, *25*, 6403–6416. [CrossRef]
28. Aviello, G.; Canadanovic-Brunet, J.M.; Milic, N.; Capasso, R.; Fattorusso, E.; Taglialatela-Scafati, O.; Fasolino, I.; Izzo, A.A.; Borrelli, F. Potent antioxidant and genoprotective effects of boeravinone G, a rotenoid isolated from Boerhaavia diffusa. *PLoS ONE* **2011**, *6*, 9628. [CrossRef]
29. Mishra, S.; Aeri, V.; Gaur, P.K.; Jachak, S.M. Phytochemical, Therapeutic, and Ethnopharmacological Overview for a Traditionally Important Herb: *Boerhavia diffusa* Linn. *Biomed. Res. Int.* **2014**, *2014*, 808302. [CrossRef]
30. Erdemli, H.K.; Akyol, S.; Armutcu, F.; Akyol, O. Antiviral properties of caffeic acid phenethyl ester and its potential application. *J. Intercult Ethnopharmacol.* **2015**, *4*, 344–347. [CrossRef]
31. Al-Hatamleh, M.A.I.; Hatmal, M.M.; Sattar, K.; Ahmad, S.; Mustafa, M.Z.; Bittencourt, M.C.; Mohamud, R. Antiviral and Immunomodulatory Effects of Phytochemicals from Honey against COVID-19: Potential Mechanisms of Action and Future Directions. *Molecules* **2020**, *25*, 5017. [CrossRef] [PubMed]
32. Bıçakç, N.; Karaboğa, I.; Dökmeci, A.H.; Güzel, S.; Fidanol Erboğa, Z. Cardioprotective effect of caffeic acid phenethyl ester on cardiac contusion following blunt chest trauma in rats. *Biotech. Histochem.* **2019**, *94*, 442–448. [CrossRef] [PubMed]
33. Mirzaei, S.; Gholami, M.H.; Zabolian, A.; Saleki, H.; Farahani, M.V.; Hamzehlou, S.; Far, F.B.; Sharifzadeh, S.O.; Samarghandian, S.; Khan, H.; et al. Caffeic acid and its derivatives as potential modulators of oncogenic molecular pathways: New hope in the fight against cancer. *Pharmacol. Res.* **2021**, *171*, 105759. [CrossRef] [PubMed]
34. Tolba, M.F.; Azab, S.S.; Khalifa, A.E.; Abdel-Rahman, S.Z.; Abdel-Naim, A.B. Caffeic acid phenethyl ester, a promising component of propolis with a plethora of biological activities: A review on its anti-inflammatory, neuroprotective, hepatoprotective, and cardioprotective effects. *IUBMB Life* **2013**, *65*, 699–709. [CrossRef] [PubMed]
35. Brautigan, D.L.; Gielata, M.; Heo, J.; Kubicka, E.; Wilkins, L.R. Selective toxicity of caffeic acid in hepatocellular carcinoma cells. *Biochem. Biophys. Res. Commun.* **2018**, *505*, 612–617. [CrossRef]
36. Ilyas, U.K.; Katare, D.P.; Ambardar, N.; Aeri, V. HPTLC densitometric quantification of caffeic acid and boeravinone B in Boerhavia diffusa Linn. *Int. J. Phytopharm.* **2013**, *4*, 184–189.
37. Ekor, M. The growing use of herbal medicines: Issues relating to adverse reactions and challenges in monitoring safety. *Front Pharmacol.* **2014**, *4*, 177. [CrossRef]
38. Sasidharan, S.; Chen, Y.; Saravanan, D.; Sundram, K.M.; Yoga Latha, L. Extraction, isolation and characterization of bioactive compounds from plants' extracts. *Afr. J. Tradit. Complement. Altern. Med.* **2011**, *8*, 1–10. [CrossRef]
39. Wink, M. Modes of Action of Herbal Medicines and Plant Secondary Metabolites. *Medicines* **2015**, *2*, 251–286. [CrossRef]
40. Karimi, G.; Vahabzadeh, M.; Lari, P.; Rashedinia, M.; Moshiri, M. "Silymarin", a promising pharmacological agent for treatment of diseases. *Iran. J. Basic Med. Sci.* **2011**, *14*, 308–317.
41. Yahfoufi, N.; Alsadi, N.; Jambi, M.; Matar, C. The Immunomodulatory and Anti-Inflammatory Role of Polyphenols. *Nutrients* **2018**, *10*, 1618. [CrossRef] [PubMed]

separations

MDPI

Article

Green Synthesis of Zinc Oxide Nanoparticles Using Aqueous Extracts of *Hibiscus cannabinus* L.: Wastewater Purification and Antibacterial Activity

Xitao Yang [1], Xuan Cao [2], Chenxiao Chen [1], Liping Liao [1], Sitian Yuan [1] and Siqi Huang [1,]*

[1] Institute of Bast Fiber Crops, Chinese Academy of Agricultural Sciences, Changsha 410205, China; 82101225155@caas.cn (X.Y.); chenchenxiao@caas.cn (C.C.); lipingliaocaas@yahoo.com (L.L.); sitianyuan2022@126.com (S.Y.)
[2] College of Life Sciences, Shangrao Normal University, Shangrao 334001, China; 15607926335@163.com
* Correspondence: huangsiqi@caas.cn; Tel.: +86-731-88998505

Abstract: The green preparation of metal oxide nanoparticles is an environmentally friendly method, which could reduce the use of toxic solvents and their impact on the environment. The purpose of this study is to investigate the green synthesis of zinc oxide (ZnO) nanoparticles using extracts of *Hibiscus cannabinus* leaves and to evaluate their potential applications in environmental remediation. In this work, ZnO nanoparticles were successfully prepared and thoroughly characterized using UV–vis, Fourier transform infrared analysis (FTIR), X-ray diffraction (XRD), transmission electron microscope (TEM) analysis, and scanning electron microscope (SEM) with energy dispersive x-ray analysis (EDAX). As a result, the synthesized ZnO nanoparticles showed a good adsorption capacity for Congo red (CR), and satisfactory antioxidant and antibacterial activities. They exhibited good adsorption and removal abilities for CR in aqueous solutions. With the conditions optimized, the adsorption kinetics and isotherms were fitted to the pseudo-second-order model and the Langmuir model. The ZnO nanoparticles could also effectively scavenge 2-2-diphenyl-1-picrylhydrazyl (DPPH) and 2,2′-azino-di(3-ethylbenzthiazoline sulphonate) (ABTS) radicals, and appeared to inhibit the growth of *Escherichia coli* and *Staphylococcus aureus* bacteria. Based on the identified adsorption capacity, the green synthesized ZnO nanoparticles demonstrated their potential to be used in the removal of dyeing wastewater and in the further purification of water due to their antioxidant activity and antibacterial activity.

Keywords: adsorption; antibacterial; antioxidant; Congo red; *Hibiscus cannabinus* L.; zinc oxide

Citation: Yang, X.; Cao, X.; Chen, C.; Liao, L.; Yuan, S.; Huang, S. Green Synthesis of Zinc Oxide Nanoparticles Using Aqueous Extracts of *Hibiscus cannabinus* L.: Wastewater Purification and Antibacterial Activity. *Separations* **2023**, *10*, 466. https://doi.org/10.3390/separations10090466

Academic Editor: Dimosthenis Giokas

Received: 26 July 2023
Revised: 11 August 2023
Accepted: 14 August 2023
Published: 24 August 2023

1. Introduction

With the rapid advancement of printing and dyeing technologies, synthetic dyes have been extensively utilized in various industries over the past few decades. Consequently, a significant amount of printing and dyeing wastewater is discharged into water bodies, constituting a substantial portion of industrial wastewater discharge [1]. This dyeing wastewater poses a severe threat in terms of water pollution due to its intense color, high toxicity, elevated organic pollutant content, and poor biodegradability of organic compounds [2]. Multiple approaches have been introduced for the treatment of dye wastewater, including physical methods (such as adsorption and membrane technology), chemical methods (including electrochemistry and advanced oxidation), and biological methods. Among these, physical methods are commonly used to treat water-soluble dyes due to their simplicity, cost-effectiveness, and notable decolorization efficacy [3].

Nanomaterials possess remarkable physicochemical properties and biocompatibility, rendering them highly valuable in the fields of biomedical, environmental, and energy research [4]. Zinc oxide (ZnO) nanoparticles are a new type of inorganic nanomaterial with various special properties, such as photocatalytic, antioxidant, antibacterial, and antifungal activities, which gives them significant value to various research fields [5]. Traditional synthesis methods for ZnO nanoparticles include the hydrothermal, sol–gel, chemical precipitation, and spray pyrolysis techniques [6]. Meanwhile, in recent years, green synthesis technology using plant extracts has become increasingly popular due to its simplicity, low costs, safety, non-toxicity, abundant raw materials, and environmentally friendly characteristics [7]. Plant extracts are readily available, inexpensive, and often contain significant amounts of reducing and capping agents, ensuring their suitability for large-scale production [8]. Dangana et al. prepared ZnO nanoparticles using extracts obtained from the leaves of Chaya Cnidoscolus aconitifolius and used them as nano fertilizers on *Sorghum bicolour* plants [9]. Elshafie et al. carried out the green synthesis of ZnO nanoparticles using *Moringa oleifera* leaves [10]. Rini et al. reported a green method of synthesizing ZnO nanoparticles using pineapple-peel extract and evaluated their antibacterial activity [11]. Moreover, almond shells, sugar cane bagasse, eggshells, Viscum album leaves, peppermint tea dregs, marine brown algae, and many plant extracts have been used for the synthesis of ZnO nanoparticles [12–15]. Hence, the green synthesis approach to obtaining ZnO nanoparticles has significant potential for further exploration and development.

Hibiscus cannabinus L., commonly referred to as kenaf, is an annual herbaceous plant belonging to the Malvaceae family [16]. It has a wide distribution across regions such as India, China, and Thailand. *Hibiscus cannabinus* serves as an important industrial fiber crop, finding applications in various fields such as textiles, animal feed, papermaking, and construction. This plant contains numerous bioactive components including polyphenols, alkaloids, tannins, and saponins [17]. Consequently, *Hibiscus cannabinus* exhibits medicinal properties with potential anti-tumor, antioxidant, anti-inflammatory, anti-hypertensive, and anti-proliferative activities [18]. Apart from their primary use as a bast fiber, the leaves of *Hibiscus cannabinus* could offer an additional source for the production of nanomaterials.

This article focused on the synthesis of ZnO nanoparticles using extracts of *Hibiscus cannabinus* L. leaves and investigated their ability to adsorb organic dyes (Figure 1). The crystalline properties, elemental composition, and structure of the synthesized nanoparticles were analyzed using various characterization techniques, including X-ray diffraction (XRD), UV–Vis spectroscopy, Fourier transform infrared (FTIR) spectroscopy, transmission electron microscopy (TEM), scanning electron microscopy (SEM), and energy-dispersive X-ray analysis (EDAX). The presence in dyeing wastewater of azo dye compounds, which are extensively utilized synthetic dyes, poses a significant threat to human health and aquatic organisms due to these compounds' carcinogenic properties [19]. Moreover, the aromatic structure of azo dyes presents challenges in terms of their degradation, leading to prolonged environmental pollution [20]. Exposure to azo dyes can result in various ailments such as edema, allergies, skin irritation, and respiratory issues [21]. To assess the efficacy of removing contaminants, the adsorption of Congo red (CR) dye was evaluated using synthesized ZnO nanoparticles. To enhance our understanding of their efficacy in environmental remediation, the antioxidant potential of the ZnO nanoparticles synthesized using the green method was assessed using the 2-2-diphenyl-1-picrylhydrazyl (DPPH) and 2,2′-azino-di(3-ethylbenzthiazoline sulpho-nate) (ABTS) radical scavenging assays. Additionally, the antibacterial properties of these nanoparticles were further examined against *Escherichia coli* (*E. coli*) and *Staphylococcus aureus* (*S. aureus*) bacteria.

Figure 1. Diagram of the ZnO nanoparticles synthesized using green methods with *Hibiscus cannabinus* leaf extracts.

2. Materials and Methods

2.1. Materials

All the chemicals and reagents used in this study were of analytical grade, purchased from Macklin Inc. (Shanghai, China), and used without treatment. *Hibiscus cannabinus* leaves were collected from the Innovation experimental base of the Institute of Bast Fiber Crops, Chinese Academy of Agricultural Sciences, Changsha, China, in 2023. The samples were dried, smashed, and stored at 4 °C for further use.

2.2. Green Preparation of ZnO Nanoparticles

First, 10 g of *Hibiscus cannabinus* leaves was weighed and transferred in a 250 mL round-bottom flask containing 100 mL of water at a ratio of 1:10 (g/mL). The mixture was then extracted under heating at 90 °C for 4 h. After extraction and cooling to room temperature, the extraction solution was centrifuged at 3000 rpm for 3 min. The supernatant was stored at 4 °C.

For the green preparation of the ZnO nanoparticles, 20 mL of *Hibiscus cannabinus* leaf extract was added to 50 mL of 0.02 mol/L zinc sulfate solution in a conical flask and incubated under magnetic stirring for 1.5 h. Then, some 0.2 mol/L sodium hydroxide solution was slowly added to the mixture to adjust the pH to 8. The mixture was magnetically stirred for 3.5 h. After the reaction, the products were centrifuged at 4000 rpm for 5 min. The supernatant was discarded, and the precipitates were washed twice with water and dried at 70 °C.

For the gas chromatography–mass spectrometer (GC–MS) analysis, 10 g of *Hibiscus cannabinus* leaves was weighed and transferred to a 250 mL round-bottom flask containing 100 mL of ethanol at a ratio of 1:10 (g/mL). The mixture was then extracted under heating at 80 °C for 4 h. After extraction and cooling to room temperature, the solution was centrifuged at 3000 rpm for 3 min. The supernatant was concentrated via reduced pressure distillation, filtrated using a 0.45 μm membrane, and subjected to GC–MS analysis.

2.3. Characterizations

The green-synthesized ZnO nanoparticles were characterized using a SEM (Apreo 2 Thermo Scientific, Waltham, MA, USA) equipped with an X-ray fluorescence spectrometer Oxford INCA X-max 80 for EDAX, TEM (JEM-2100F, JEOL, Tokyo, Japan), and XRD (Rigaku SmartLab 9, Tokyo, Japan) using Cu-Kα radiation at 40 kV and 40 mA, and UV spectroscopy was undertaken using a Shimadzu UV-2700 absorption spectrophotometer (Kyoto, Japan) and FT-IR (IRSpirit, Shimadzu, Kyoto, Japan) in the range of 400–4000 cm^{-1}.

The GC–MS analytical method was used to detect the components in the ethanol extract of *Hibiscus cannabinus* leaves. The analysis was carried out on a 7890B GC equipped with a 5977A MSD (Agilent Technologies, Santa Clara, CA, USA) under the following conditions: 1 μL of the sample was injected in the HP-5MS (5% phenyl methyl siloxane) capillary column (30 m × 250 μm × 0.25 μm); helium was used as the carrier gas at a constant flow of 1.2 mL/min; and the temperatures of the injector and MSD were set to 250 °C and 280 °C, respectively. The column was first maintained at 60 °C for 2 min; then, the temperature increased to 280 °C at the rate of 5 °C/min, and it was finally maintained for 9 min. The MS parameters were operated at 70 eV and scanned from 40 to 600 amu. The chemical information was identified via their mass spectral data using the standard NIST library 2011 during the GC–MS analysis [22].

2.4. Adsorption of Dyes

First, 0.5 g of the ZnO nanoparticles was added to 50 mL of the aqueous solution of the CR dye at different initial concentrations in a conical flask. The solution was shaken at 200 rpm on an oscillator at 25 °C to reach adsorption equilibrium. A portion of the solution was taken at various time intervals and filtered using a 0.45 μm membrane. The solution was then monitored using a UV spectrophotometer (UV2700 UV–Vis Spectrophotometer (Shimadzu, Kyoto, Japan)) at 510 nm. The adsorption capacity was calculated according to the following formula [23]:

$$q_t = \frac{(C_0 - C_t) \times V}{w} \quad (1)$$

where q_t is the adsorbed amount of the sample at a certain time, C_0 is the initial concentration of the sample, C_t is the concentration of the sample at a certain time, V is the volume of the solution, and w is the weight of the used adsorbent.

2.5. Antioxidant Activity

The antioxidant activity of ZnO nanoparticles was analyzed via DPPH and ABTS free radical scavenging assays. The procedures were slightly changed according to the report by Hu et al. [24]. Different weights of ZnO nanoparticles and 3.0 mL of 0.1 mmol/L DPPH radical solution were mixed and shaken on an oscillator for 20 min in the dark. After that, the solution was filtrated with a 0.45 μm membrane and the absorbance was measured at 517 nm using a UV spectrophotometer. The scavenging rate of each sample was calculated using the following formula:

$$\text{Scavenging rate } (\%) = \left(1 - \frac{A_s}{A_0}\right) \times 100\% \quad (2)$$

where A_0 is the absorbance of the control group and A_s is the absorbance of the sample.

For the ABTS assay, the same volume of 2.45 mmol/L potassium persulfate solution and 7 mmol/L ABTS solution was mixed and incubated in the dark for 18 h [25]. The ABTS stock solution was diluted to obtain an absorbance between 0.7 and 0.72 at 734 nm before use. Different amounts of ZnO nanoparticles and 3.0 mL of the ABTS radical solution were mixed and shaken on an oscillator for 20 min in the dark. The solution was filtrated using a 0.45 μm membrane after the reaction and the absorbance was measured at 734 nm. The scavenging rate of each sample was calculated according to Formula (1).

2.6. Antibacterial Activity

The antibacterial activity of the ZnO nanoparticles was evaluated using the disc diffusion method [26]. *Escherichia coli* and *Staphylococcus aureus* bacteria were grown in an LB broth containing 5 g/L of beef extract, 10 g/L of peptones, 20 g/L of agar, and 5 g/L of NaCl, and they were diluted to 1×10^{-6} CFU prior to use. Then, 0.2g of the ZnO nanoparticles was weighed and pressed into a tablet using a tablet press machine. The sterilized filter paper was immersed in different concentrations of the extract for further use. Then, 100 μL of the bacterial suspensions was evenly spread on an agar plate, and the ZnO nanoparticle tablet (Φ = 13 mm) or the sterilized filter paper (Φ = 6 mm) was carefully placed on the plate. The plates were incubated at 37 °C for 24 h, and the diameter of the inhibition zone was measured.

The minimal inhibitory concentration of the prepared ZnO nanoparticles was determined according to the micro-dilution broth method [27]. In brief, 180 μL of the nutrient broth and different weights of the ZnO nanoparticles were mixed in each well of a sterile 96-well plate, followed by the addition of 20 μL of the bacterial solution (1×10^{-6} CFU). The 96-well plate was sealed and placed on a shaker at 37 °C for cultivation. The absorbance at 600 nm (OD_{600}) of each well of the plate was measured using a microplate reader (Epoch, BioTek Instruments Inc., Winooski, VT, USA) at predetermined time intervals, the growth curve was plotted based on the obtained data. The nutrient broth without a sample was used as a negative control. The lowest concentration of ZnO nanoparticles inhibiting the growth of bacteria was defined as the minimal inhibitory concentration value.

3. Results and Discussion

3.1. *Characterizations of ZnO Nanoparticles*

3.1.1. UV–Vis Spectroscopic Analysis

After preparation, the ZnO nanoparticles were thoroughly characterized (Figure 1). The UV–Vis absorption spectra of the ZnO nanoparticles and the *Hibiscus cannabinus* leaf extracts were observed in the range of 190 to 500 nm (Figure 2a). For the ZnO nanoparticles, a weak absorption peak was observed around 270 nm. Compared to the reported data, many of the prepared ZnO nanoparticles showed adsorption at a wavelength around 370 nm; there is a blue (short wavelength) shift of the peak in this study [28,29]. The spectrum of the *Hibiscus cannabinus* leaf extract showed adsorptions at 265 nm and 325 nm. This might be because of the existence of extract molecules, since many compounds in extract show adsorption around 254 nm. The highest peak at 265 nm could explain the shift in the spectrum of the ZnO nanoparticles.

3.1.2. FTIR Analysis

The FTIR spectrum of the ZnO nanoparticles is shown in the range 400–4000 cm^{-1} in Figure 2b. The prepared ZnO nanoparticles showed the band at 3433 cm^{-1} as O-H stretching and N-H stretching, the band at 2360 cm^{-1} as strong (O=C=O) stretching carbon dioxide, the band at 1600 cm^{-1} as the chemisorbed and/or physisorbed water on the particle surface, the band at 1401 cm^{-1} as the angular deformation of C-H, and the band at 1078 cm^{-1} as being ascribed to the C-O ether bond of the glucose ring of starch [30]. Moreover, the intense peak at 486 cm^{-1} was attributed to the ZnO stretching frequency of Zn-O bonds [31]. These peaks showed the typical groups of ZnO nanoparticles and many groups from the extract.

3.1.3. XRD Analysis

The XRD pattern of the synthesized ZnO nanoparticles was recorded in the range of 20° to 80° of 2θ. As shown in Figure 2c, the ZnO nanoparticles showed peaks assigned to the (100), (002), (101), (102), (110), and (112) planes [32]. Unlike many of the reported data, these peaks were hidden by several wide peaks, which might have come from the amorphous materials [33]. This might be due to the presence of natural compounds during the preparation procedures.

3.1.4. Morphology Analysis

SEM and TEM were used to analyze the morphology of the ZnO nanoparticles. As shown in Figure 2d, the SEM image shows that there was an aggregation of ZnO nanoparticles, and most of the ZnO nanoparticles demonstrated flake shapes at the nanoscale. This kind of aggregation can also be observed in some other reports, and it does not affect the related applications of ZnO nanoparticles [31,34]. Like the SEM image, the shapes of the ZnO nanoparticles observed in TEM were not uniform (Figure 2e). Based on our calculations, the average size of the products was around 70 nm (Supplementary File, Figure S1).

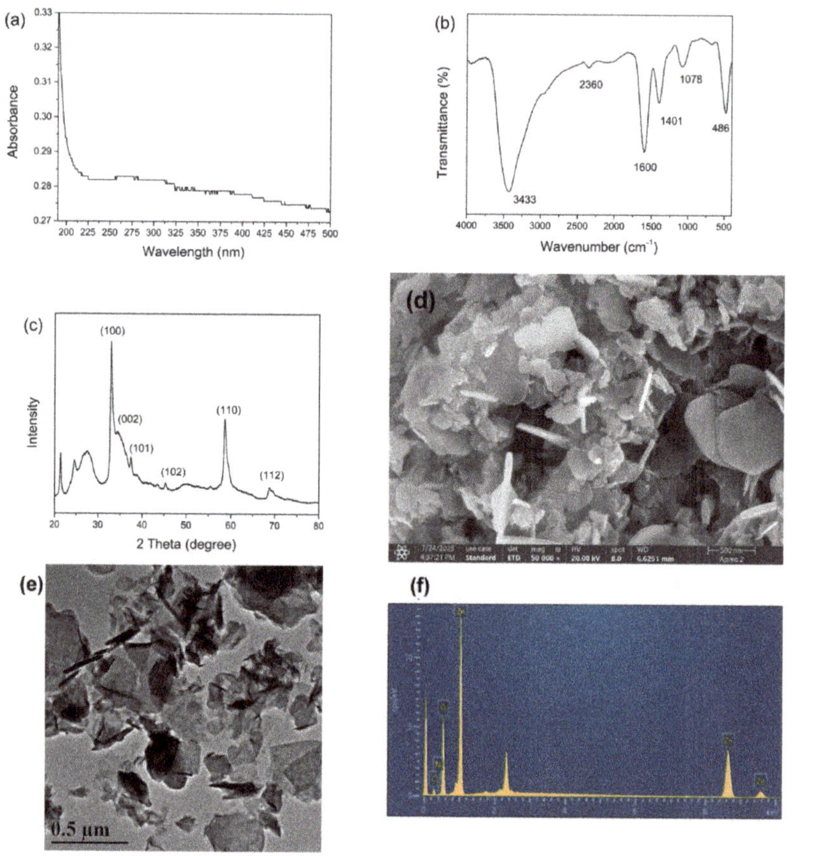

Figure 2. (**a**) UV–Vis spectrum, (**b**) FTIR spectrum, (**c**) XRD pattern, (**d**) SEM image, (**e**) TEM image, and (**f**) EDAX spectra of ZnO nanoparticles.

3.1.5. Elemental Composition Analysis

The elemental composition of the ZnO nanoparticles (C, N, O, and Zn) was analyzed via EDAX analysis (Figure 2e). As a result, 22.54 At% of C, 46.30 At% of O, 30.88 At% of Zn, and 0.29 At% of N were detected. Except for the composition of ZnO (Zn and O), the high contents of carbon and oxygen exhibited the existence of organic natural active compounds from the natural extract. The ZnO sample was made up of only Zn and O elements, with no other impurities, which was consistent with the XRD pattern.

3.2. GC–MS Analysis of Hibiscus cannabinus Leaf Extracts

The GC–MS analysis of *Hibiscus cannabinus* leaves extract was performed, and the results were depicted in Figure 3. The chemical composition of the extract plays a crucial role in the green synthesis of nanoparticles, as various components can act as stabilizing, functional, and capping agents [35]. As a result, about thirteen phytochemicals were identified in *Hibiscus cannabinus* leaf aqueous extracts, as presented in Table 1. Among them, methyl linolenate (33.3%), methyl hexadecanoate (13.1%), phytol (14.6%), and β-sitosterol (11.5%) were main components in the aqueous extract; they have been reported to exhibit diverse biological properties [36–38]. These phytochemicals were the active molecules in the formation of the ZnO nanoparticles [39]. They were also found to exist in the characterizations.

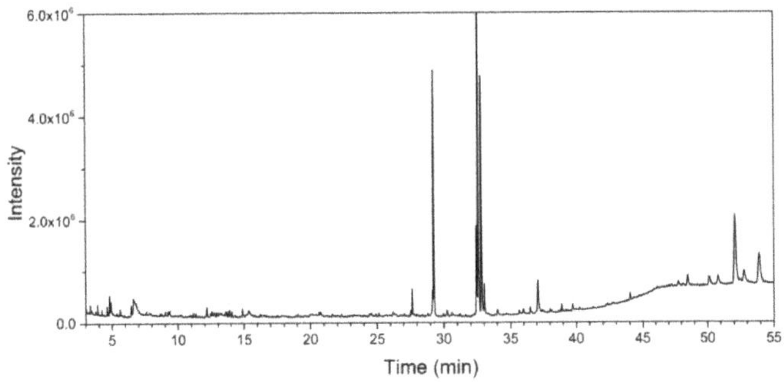

Figure 3. GC–MS spectrum of the *Hibiscus cannabinus* leaf extract.

Table 1. The phytochemicals identified in the *Hibiscus cannabinus* leaf extract.

Rt (min)	Mw	Molecular Formula	Molecular Name
4.795	84.0	$C_4H_4O_2$	Butenolide
4.954	88.1	$C_4H_8O_2$	Acetaldol
6.799	92.0	$C_3H_8O_3$	Glycerin
12.156	198.2	$C_{14}H_{30}$	Tetradecane
27.660	268.3	$C_{18}H_{36}O$	Hexahydrofarnesyl acetone
29.289	270.3	$C_{17}H_{34}O_2$	Methyl hexadecanoate
32.438	294.3	$C_{19}H_{34}O_2$	Methyl linoleate
32.597	292.2	$C_{19}H_{32}O_2$	Methyl linolenate
32.820	296.3	$C_{20}H_{40}O$	Phytol
33.068	438.8	$C_{29}H_{58}O_2$	Methyl montanate
37.121	281.3	$C_{18}H_{35}NO$	Oleamide
52.122	414.4	$C_{29}H_{50}O$	β-Sitosterol
53.961	426.4	$C_{30}H_{50}O$	Amyrin

3.3. Impact of Different Parameters on the Adsorption of the CR Dye

3.3.1. Effect of the Initial Dye Concentration

Figure 4a illustrates the adsorption efficiency of various dye concentrations (ranging from 0.2 to 0.5 g/L). It can be observed that, as the initial concentration of the dye solution increases, the adsorption efficiency of CR gradually diminishes. However, when the initial concentration of CR is lower than 0.3 g/L, the adsorption efficiency is maintained at the highest level. As the concentration of CR continually increases, the limited adsorption sites eventually reach saturation [40].

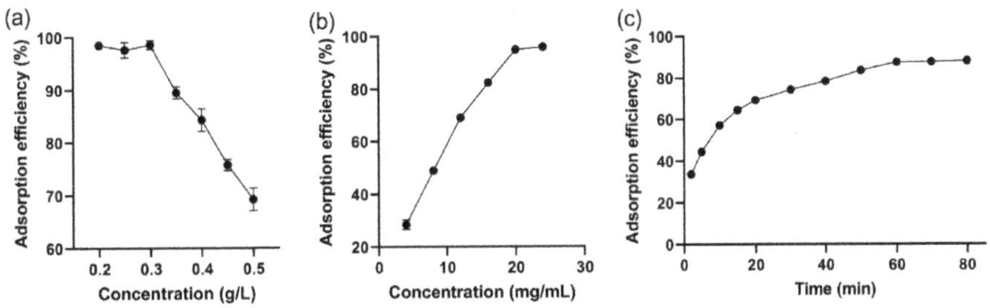

Figure 4. Effects of (**a**) the initial dye concentration, (**b**) the adsorbent dosage, and (**c**) the contact time on the adsorption efficiency of CR.

3.3.2. Effect of Adsorbent Dosage

The impact of the adsorbent dosage on the CR adsorption efficiency was investigated by varying the ZnO nanoparticle dosages from 4 mg/mL to 24 mg/mL, as depicted in

Figure 4b. It was observed that, as the adsorbent dosage increased, the adsorption efficiency of CR also increased, and it eventually reached a plateau when the adsorbent dosage exceeded 20 mg/mL. This is mainly because more active sites for CR are available with the increase in the adsorbent dosage [23]. In this study, an adsorbent dosage of 20 mg/mL was selected as the optimum concentration.

3.3.3. Effect of Contact Time

The influence of time on the adsorption of CR by the ZnO nanoparticles was investigated, and the findings are presented in Figure 4c. It is evident that the adsorption efficiency increased and reached a plateau after 60 min. This suggests that the adsorption process needs time and, after 60 minutes, the vacant sites become saturated, resulting in an unchanged adsorption efficiency. Therefore, we recommend performing the adsorption of CR for a duration of 60 min.

3.4. *Adsorption Kinetics*

The adsorption kinetics were analyzed using the pseudo-first-order and pseudo-second-order kinetic models. The pseudo-first-order model can be expressed as the following equation [41]:

$$\log(q_e - q_t) = \log q_e - K_1 t \tag{3}$$

where q_t is the adsorption capacity at a certain time, q_e is the adsorption capacity in equilibrium, and K_1 is the rate constant of the pseudo-first-order model. Moreover, the pseudo-second-order model can be expressed as the following equation:

$$\frac{t}{q_t} = \frac{1}{K_2 q_e^2} + \frac{t}{q_e} \tag{4}$$

where K_2 is the rate constant for the pseudo-second-order model. The adsorption data were fitted and the correlation coefficients of the two models were compared (Figure 5). When the correlation coefficient approaches 1, it indicates that the corresponding model is more appropriate for explaining the adsorption mechanism. In this study, the correlation coefficients of the pseudo-first-order and pseudo-second-order models were found to be 0.8051 and 0.9957, respectively. The pseudo-second-order model exhibited a higher correlation coefficient compared to the pseudo-first-order model. These results strongly suggest that the adsorption behavior of CR on the ZnO nanoparticles follows a chemical adsorption process [42]. A similar kinetic process was observed in the studies conducted by Arab and Marahel regarding ZnO nanoparticles [42,43]. The results suggest that the adsorption behavior of CR on ZnO nanoparticles is a chemical adsorption process.

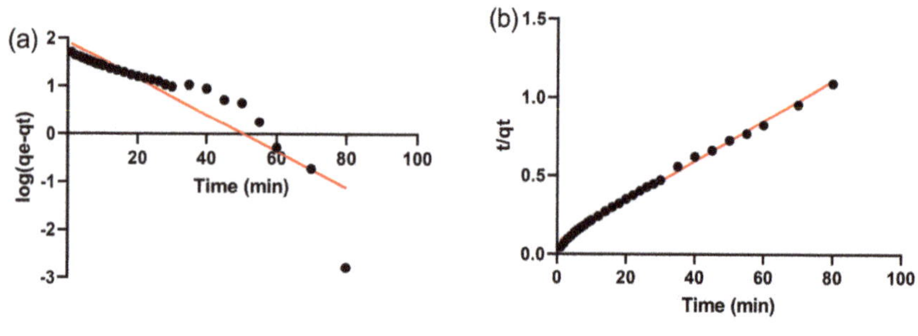

Figure 5. (**a**) Pseudo-first-order kinetics and (**b**) Pseudo-second-order kinetics for ZnO nanoparticles.

3.5. Adsorption Isotherm

The isotherm study was conducted using the Langmuir and Freundlich isotherms. The linear equation for Langmuir and Freundlich can be expressed as [44]:

$$\frac{C_e}{q_e} = \frac{C_e}{q_m} + \frac{1}{K_l q_m} \quad (5)$$

$$\log q_e = \log K_f + \frac{1}{n} \log C_e \quad (6)$$

where C_e is the concentration of the sample in equilibrium, q_m is the maximum adsorption capacity, K_l is the Langmuir coefficient, K_f is the Freundlich coefficient, and $1/n$ is the adsorption intensity. The adsorption data were fitted and the correlation coefficients of the two models were compared, as illustrated in Figure 6. The obtained correlation coefficients for the Langmuir and Freundlich isotherms were 0.9929 and 0.8642, respectively. It is worth noting that the correlation coefficient for the Langmuir model was significantly higher, indicating that the adsorption of CR on ZnO nanoparticles follows a monolayer sorption mechanism with a uniform distribution of active sites on the surface [45].

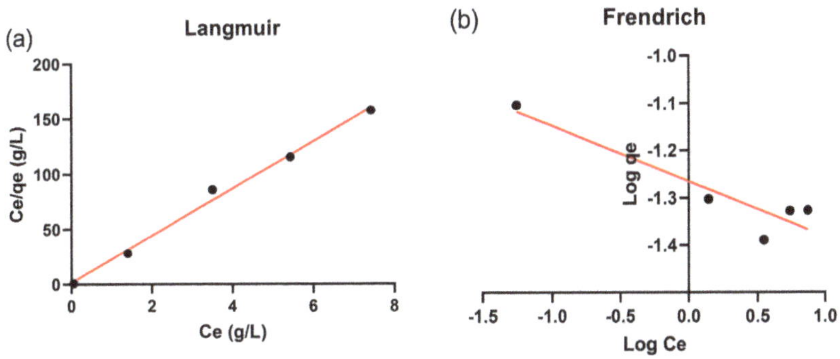

Figure 6. (**a**) Langmuir isotherm and (**b**) Freundlich isotherm for ZnO nanoparticles.

3.6. Reusability Study

After the adsorption of CR, the ZnO nanoparticles were refreshed with ethanol and water. Subsequently, the materials were subjected to a new round of CR adsorption to evaluate their reusability. As shown in Figure 7, the adsorption efficiency gradually decreases over multiple cycles, ranging from 96.2% to 86.1%. This decline in the adsorption efficiency can be attributed to material loss during repeated centrifugation and dispersion, as well as the incomplete desorption of the dye molecules [23]. Nevertheless, the results indicate that the prepared ZnO nanoparticles can be reused for a minimum of three cycles for the effective adsorption and removal of CR.

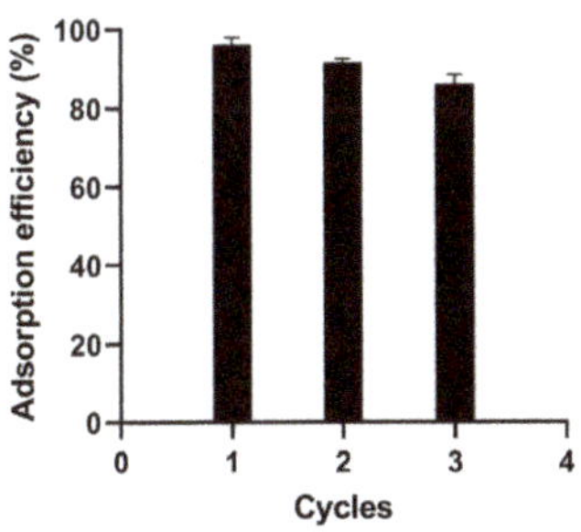

Figure 7. The reusability of the ZnO nanoparticles in the adsorption of CR.

3.7. Antioxidant Activity of ZnO Nanoparticles

Oxidative stress is a kind of chemical reaction that occurs in the body and can lead to the formation of free radicals, which are unstable molecules that can attack lipids, proteins, and DNA, damaging their functions. Antioxidants can protect the body from oxidative stress by scavenging free radicals. ABTS was the utilized hydrogen atom transfer method, and DPPH was the electron transfer method. These were the two typical methods used in this study to evaluate antioxidant activities via colorimetric reactions [46]. DPPH is a relatively stable nitrogen-centered free radical that can accept an electron to become a stable diamagnetic molecule [47]. Meanwhile, ABTS can determine the total antioxidant activity of a hydrogen-donating ability [48]. Assessments of the in vitro antioxidant activities of ZnO nanoparticles and *Hibiscus cannabinus* leaf extracts were carried out to determine their pharmacological activities (Figure 8). Figure 8a,b display the antioxidant activity of the ZnO nanoparticles against DPPH and ABTS. Ascorbic acid was used as a standard in this process. The scavenging activity of the as-synthesized nanoparticles at different concentrations ranged from 8.8 to 75.2% and 38.4 to 100% against DPPH and ABTS, respectively. As a comparison, the scavenging activity of ascorbic acid at 0.25 mg/mL was 21.9% for DPPH and 52.7% for ABTS. Although the antioxidant activity of the ZnO nanoparticles was less than that of standard ascorbic acid, they still exhibited a dose-dependent trend and the values are in a promising range. Moreover, the ZnO nanoparticles showed a better scavenging ability for ABTS than for DPPH. The maximum rate of free-radical scavenging (97.0%) was reached when the concentration of the ZnO nanoparticles was 16.7 mg/mL. As shown in Figure 8c,d, the antioxidant activity of the *Hibiscus cannabinus* leaf extract was also exhibited. The extract showed good scavenging abilities, and the scavenging rate for ABTS was much better than that for DPPH. The same characteristics might indicate that the antioxidant activity of the ZnO nanoparticles was derived from the extract. Sharma reported the ZnO nanoparticles expressed an IC_{50} value of DPPH free-radical-scavenging activity of 34.22 ± 2.52 μg/mL [30]. As many reports show the particles' antioxidant activity against DPPH, it should be noted that, in this study, the antioxidant activity against ABTS of ZnO nanoparticles prepared using *Hibiscus cannabinus* leaves was much better [49–51].

3.8. Antibacterial Activity of ZnO Nanoparticles

The disc diffusion test was used to study the antibacterial activity of the synthesized ZnO nanoparticles. The results showed that apparent inhibition zones were found with diameters around 27.0 mm, according to bacterial strains of *E. coli* and *S. aureus* (Figure 9a,b). The inhibition zones for the two bacterial strains of *Hibiscus cannabinus* leaf extracts were also tested. As Figure 9c–f show, the extract showed moderate inhibition. The inhibition diameters of the extract were 12.34 mm for *E. coli* and 9.74 mm for *S. aureus*. However, when the concentrations were reduced to 15 mg/mL, there was no inhibition, and the MIC values for both were 30 mg/mL. Based on these results, the growth curve of the two bacterial strains was also monitored by measuring the absorbance. As shown in Figure 10, the growth of the control group followed the typical S-shaped bacterial growth rule. With the increasing addition of ZnO nanoparticles into the bacteria, the growth of the bacteria was slowed down. When the concentration of the ZnO nanoparticles reached 0.45 mg/mL, the sample did not show significant growth. This indicates that the treatment with ZnO nanoparticles could delay the growth rate of the strain and even inhibit growth at 0.45 mg/mL, which can be confirmed as the minimal inhibitory concentration. This phenomenon is consistent with the inhibition effects on pathogenic bacteria of other prepared ZnO nanoparticles [30,52–55]. The well-proven antibacterial activities of ZnO nanoparticles ensure the material's wider applications and give them more functions in environmental remediation.

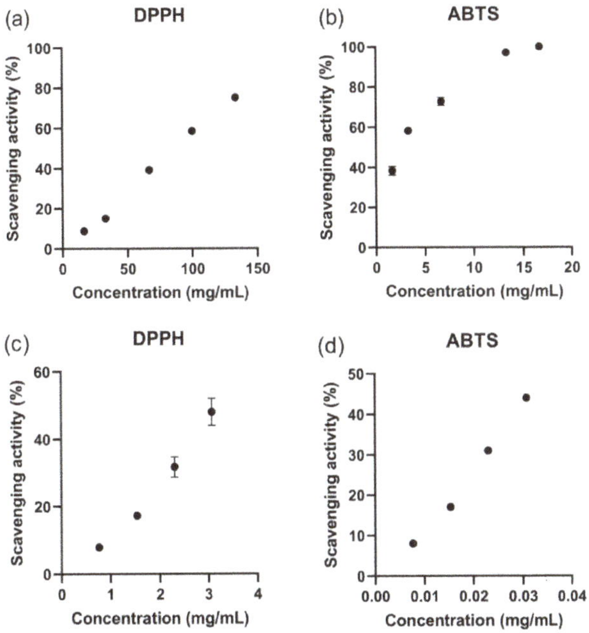

Figure 8. Antioxidant activities of ZnO nanoparticles for (**a**) DPPH free radicals and (**b**) ABTS free radicals; antioxidant activities of the *Hibiscus cannabinus* leaf extract for (**c**) DPPH free radicals and (**d**) ABTS free radicals.

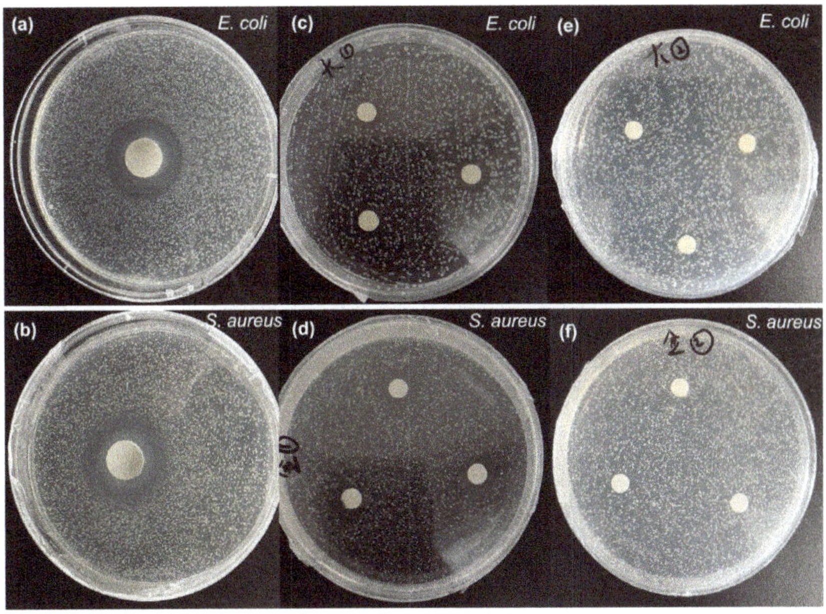

Figure 9. The inhibition zones of ZnO nanoparticles and *Hibiscus cannabinus* leaf extracts on (**a**) *E. coli* and (**b**) *S. aureus*; *Hibiscus cannabinus* leaf extract at 30 mg/mL on (**c**) *E. coli* and (**d**) *S. aureus*; and *Hibiscus cannabinus* leaf extract at 15 mg/mL on (**e**) *E. coli* and (**f**) *S. aureus*.

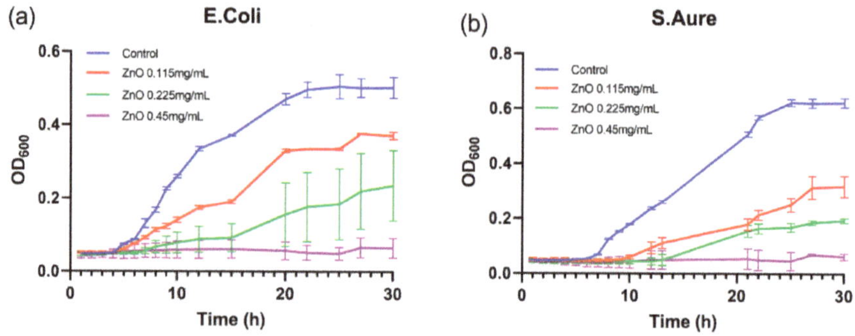

Figure 10. The growth curves of (**a**) *E. coli* and (**b**) *S. aureus* in the presence of different concentrations of ZnO nanoparticles.

4. Conclusions

This study presents a green synthesis approach for the preparation of ZnO nanoparticles, utilizing an extract derived from *Hibiscus cannabinus* leaves. The synthesized ZnO nanoparticles were characterized using UV–Vis, FTIR, XRD, TEM, and SEM techniques, which confirmed their morphology and structure in accordance with previously reported data. The prepared nanoparticles were effectively employed for the adsorption of Congo red in an aqueous solution. The factors influencing the adsorption efficiency were investigated and optimized. The adsorption kinetics were suitably described by the pseudo-second-order model, while the adsorption isotherm followed the Langmuir model. Notably, the as-synthesized materials also exhibited antioxidant and antibacterial properties. The DPPH scavenging activity was found to be moderate, whereas the ABTS scavenging activity was remarkable, displaying a 97.0% radical scavenging rate at a concentration of 16.7 mg/mL. The ZnO nanoparticles also demonstrated strong antibacterial activity against both *E. coli* and *S. aureus*. In conclusion, the environmentally friendly synthesis method employed in this study enabled the production of ZnO nanoparticles that possess excellent adsorption capabilities, as well as antioxidant and antibacterial activities, and which thus have significant potential for use in environmental remediation applications.

Supplementary Materials: The following supporting information can be downloaded at: https://www.mdpi.com/article/10.3390/separations10090466/s1, Figure S1: The size distribution of ZnO nanoparticles.

Author Contributions: Data curation, X.Y.; formal analysis, X.C. and C.C.; investigation, S.Y.; methodology, L.L.; project administration, S.H.; writing—original draft, X.Y.; writing—review and editing, S.H. All authors have read and agreed to the published version of the manuscript.

Funding: This work was supported by the Natural Science Foundation of Changsha, China (Grant No. kq2202323), the Agriculture Research System of China (CARS-19-E22), the Central Public-Interest Scientific Institution Basal Research Fund (1610242023005), the Hunan Provincial Major Science and Technology Project (2021NK1010-3), and the Chinese Agricultural Science and Technology Innovation Project (No. ASTIP-IBFC05).

Institutional Review Board Statement: Not applicable.

Informed Consent Statement: Not applicable.

Data Availability Statement: The data presented in this study are available on request from the corresponding author. The data are not publicly available due to the security requirements of the institute of the corresponding author.

Conflicts of Interest: The authors declare no conflict of interest.

References

1. Majeed, K.; Ambreen, J.; Khan, S.A.; Muhammad, S.; Shah, A.A.; Bhatti, M.A.; Batool, S.S.; Farooq, M.; Shah Bukhari, S.N.; Chandio, A.D.; et al. Effective Removal of Methylene Blue by Mn_3O_4/NiO Nanocomposite under Visible Light. *Separations* **2023**, *10*, 200. [CrossRef]
2. Iqbal, M.; Hanif, M.A.; Rashid, U.; Jilani, M.I.; Alharthi, F.A.; Kazerooni, E.A. Optimization and Kinetic Study of Treating Dye-Contaminated Wastewater Using Bio-Composite Synthesized from Natural Waste. *Separations* **2023**, *10*, 386. [CrossRef]
3. Ahmad, A.; Kamaruddin, M.A.; Hps, A.K.; Yahya, E.B.; Muhammad, S.; Rizal, S.; Ahmad, M.I.; Surya, I.; Abdullah, C.K. Recent Advances in Nanocellulose Aerogels for Efficient Heavy Metal and Dye Removal. *Gels* **2023**, *9*, 416. [CrossRef]
4. Xue, X.; Cheng, R.; Shi, L.; Ma, Z.; Zheng, X. Nanomaterials for Water Pollution Monitoring and Remediation. *Environ. Chem. Lett.* **2017**, *15*, 23–27. [CrossRef]
5. Zelekew, O.A.; Haitosa, H.H.; Chen, X.; Wu, Y.-N. Recent Progress on Plant Extract-Mediated Biosynthesis of ZnO-Based Nanocatalysts for Environmental Remediation: Challenges and Future Outlooks. *Adv. Colloid Interface Sci.* **2023**, *317*, 102931. [CrossRef]
6. Chauhan, A.K.; Kataria, N.; Garg, V.K. Green Fabrication of ZnO Nanoparticles Using *Eucalyptus* spp. Leaves Extract and Their Application in Wastewater Remediation. *Chemosphere* **2020**, *247*, 125803. [CrossRef]
7. Alharthi, M.N.; Ismail, I.; Bellucci, S.; Salam, M.A. Green Synthesis of Zinc Oxide Nanoparticles by Ziziphus Jujuba Leaves Extract: Environmental Application, Kinetic and Thermodynamic Studies. *J. Phys. Chem. Solids* **2021**, *158*, 110237. [CrossRef]
8. Batra, V.; Kaur, I.; Pathania, D.; Sonu; Chaudhary, V. Efficient Dye Degradation Strategies Using Green Synthesized ZnO-Based Nanoplatforms: A Review. *Appl. Surf. Sci. Adv.* **2022**, *11*, 100314. [CrossRef]
9. Dangana, R.S.; George, R.C.; Shittu, U.O.; Agboola, F.K. Facile Biosynthesis, Characterisation and Biotechnological Application of ZnO Nanoparticles Mediated by Leaves of Cnidoscolus Aconitifolius. *Artif. Cells Nanomed. Biotechnol.* **2023**, *51*, 309–317. [CrossRef]
10. Elshafie, H.S.; Osman, A.; El-Saber, M.M.; Camele, I.; Abbas, E. Antifungal Activity of Green and Chemically Synthesized ZnO Nanoparticles against Alternaria Citri, the Causal Agent Citrus Black Rot. *Plant Pathol. J.* **2023**, *39*, 265–274. [CrossRef]
11. Rini, A.; Linda, T.; Hamzah, Y.; Umar, L.; Sari, M.; Rati, Y. Antibacterial Activity of Green Synthesized ZnO Nano-Flower Using Pineapple Peel Extract. *Adv. Nat. Sci.-Nanosci. Nanotechnol.* **2023**, *14*, 025008. [CrossRef]
12. Abid, S.; Munir, R.; Sayed, M.; Nadeem, R.; Muneer, A.; Zahid, M.; Yaseen, M.; Siddiqua, U.; Noreen, S. Synthesis of Zinc Oxide, Ferric, Cu Nano Particles by Almond Shells, Sugar Cane Bagasse, Eggshells, and Their Application as Catalyst for Dye Reactive Red 195 (RR 195) Removal. *Catal. Surv. Asia* **2023**, 1–17. [CrossRef]
13. Loh, X.; Ooi, Z.; Teoh, Y.; Shuit, S. Synthesis and Characterization of Zinc Oxide Nanoparticles Using Peppermint Tea (Mentha Piperita) Dregs Extract and Their Photocatalytic Performance. *Environ. Prog. Sustain. Energy* **2023**, e14202. [CrossRef]
14. Jalali, T.; Ghanavati, F.; Osfouri, S. Green Synthesis of ZnO Nanoparticles Using Marine Brown Algae (Cystoseira) Extract Comprising Sol-Gel, and Combustion Techniques Based on Dye-Sensitized Solar Cells Application. *Int. J. Mod. Phys. B* **2023**, 2450178. [CrossRef]
15. Mushtaq, W.; Ishtiaq, M.; Maqbool, M.; Mazhar, M.W.; Casini, R.; Abd-ElGawad, A.M.; Elansary, H.O. Green Synthesis of Zinc Oxide Nanoparticles Using Viscum Album Extracts: Unveiling Bioactive Compounds, Antibacterial Potential, and Antioxidant Activities. *Plants* **2023**, *12*, 2130. [CrossRef]
16. Al-Mamun, M.; Rafii, M.Y.; Misran, A.B.; Berahim, Z.; Ahmad, Z.; Khan, M.M.H.; Oladosu, Y.; Arolu, F. Kenaf (*Hibiscus cannabinus* L.): A Promising Fiber Crop with Potential for Genetic Improvement Utilizing Both Conventional and Molecular Approaches. *J. Nat. Fibers* **2023**, *20*, 2145410. [CrossRef]
17. Al-Mamun, M.; Rafii, M.Y.; Misran, A.B.; Berahim, Z.; Ahmad, Z.; Khan, M.M.H.; Oladosu, Y. Heterosis and Combining Ability Estimate on Yield and Yield-Related Traits in a Half Diallel Crosses of Kenaf (*Hibiscus cannabinus* L.) in Malaysia. *J. Nat. Fibers* **2023**, *20*, 2192541. [CrossRef]
18. Ayadi, R.; Hanana, M.; Mzid, R.; Hamrouni, L.; Khouja, M.L.; Salhi Hanachi, A. *Hibiscus cannabinus* L.—Kenaf: A Review Paper. *J. Nat. Fibers* **2017**, *14*, 466–484. [CrossRef]
19. Liu, H.; Ying, Q.; Li, C.; Norra, S.; Lichtfouse, E. Enhanced Removal of Antimony in Dyeing Wastewater by Mixing Fe3O4 with Manganese Sand Filter Material. *Water Environ. Res.* **2020**, *92*, 1208–1213. [CrossRef]
20. Nho, S.W.; Cui, X.; Kweon, O.; Jin, J.; Chen, H.; Moon, M.S.; Kim, S.-J.; Cerniglia, C.E. Phylogenetically Diverse Bacteria Isolated from Tattoo Inks, an Azo Dye-Rich Environment, Decolorize a Wide Range of Azo Dyes. *Ann. Microbiol.* **2021**, *71*, 35. [CrossRef]
21. Alshehrei, F. Role of Microorganisms in Biodegradation of Food Additive Azo Dyes: A Review. *Afr. J. Biotechnol.* **2020**, *19*, 799–805. [CrossRef]
22. Sharmila, G.; Thirumarimurugan, M.; Muthukumaran, C. Green Synthesis of ZnO Nanoparticles Using Tecoma Castanifolia Leaf Extract: Characterization and Evaluation of Its Antioxidant, Bactericidal and Anticancer Activities. *Microchem. J.* **2019**, *145*, 578–587. [CrossRef]
23. Zhang, M.; Chang, L.; Zhao, Y.; Yu, Z. Fabrication of Zinc Oxide/Polypyrrole Nanocomposites for Brilliant Green Removal from Aqueous Phase. *Arab. J. Sci. Eng.* **2019**, *44*, 111–121. [CrossRef]
24. Hu, Y.; Liang, P.; Wang, Z.; Jiang, C.; Zeng, Q.; Shen, C.; Wu, Y.; Liu, L.; Yi, Y.; Zhu, H.; et al. Explore the Effect of the Structure-Activity Relationship and Dose-Effect Relationship on the Antioxidant Activity of Licorice Flavonoids. *J. Mol. Struct.* **2023**, *1292*, 136101. [CrossRef]

25. Zhang, L.; Liu, L.; Xiao, A.; Huang, S.; Li, D. Screening and Analysis of Xanthine Oxidase Inhibitors in Jute Leaves and Their Protective Effects against Hydrogen Peroxide-Induced Oxidative Stress in Cells. *Open Chem.* **2020**, *18*, 1481–1494. [CrossRef]
26. Bourhia, M.; Bouothmany, K.; Bakrim, H.; Hadrach, S.; Salamatullah, A.M.; Alzahrani, A.; Khalil Alyahya, H.; Albadr, N.A.; Gmouh, S.; Laglaoui, A.; et al. Chemical Profiling, Antioxidant, Antiproliferative, and Antibacterial Potentials of Chemically Characterized Extract of *Citrullus colocynthis* L. Seeds. *Separations* **2021**, *8*, 114. [CrossRef]
27. Agarwal, H.; Nakara, A.; Menon, S.; Shanmugam, V. Eco-Friendly Synthesis of Zinc Oxide Nanoparticles Using Cinnamomum Tamala Leaf Extract and Its Promising Effect towards the Antibacterial Activity. *J. Drug Deliv. Sci. Technol.* **2019**, *53*, 101212. [CrossRef]
28. Mallikarjunaswamy, C.; Lakshmi Ranganatha, V.; Ramu, R.; Udayabhanu; Nagaraju, G. Facile Microwave-Assisted Green Synthesis of ZnO Nanoparticles: Application to Photodegradation, Antibacterial and Antioxidant. *J. Mater. Sci. Mater. Electron.* **2020**, *31*, 1004–1021. [CrossRef]
29. Abdelbaky, A.S.; Abd El-Mageed, T.A.; Babalghith, A.O.; Selim, S.; Mohamed, A.M.H.A. Green Synthesis and Characterization of ZnO Nanoparticles Using *Pelargonium odoratissimum* (L.) Aqueous Leaf Extract and Their Antioxidant, Antibacterial and Anti-Inflammatory Activities. *Antioxidants* **2022**, *11*, 1444. [CrossRef]
30. Sharma, A.; Nagraik, R.; Venkidasamy, B.; Khan, A.; Dulta, K.; Kumar Chauhan, P.; Kumar, D.; Shin, D.-S. In Vitro Antidiabetic, Antioxidant, Antimicrobial, and Cytotoxic Activity of Murraya Koenigii Leaf Extract Intercedes ZnO Nanoparticles. *Luminescence* **2023**, *38*, 1139–1148. [CrossRef]
31. Saad Algarni, T.; Abduh, N.A.Y.; Al Kahtani, A.; Aouissi, A. Photocatalytic Degradation of Some Dyes under Solar Light Irradiation Using ZnO Nanoparticles Synthesized from Rosmarinus Officinalis Extract. *Green Chem. Lett. Rev.* **2022**, *15*, 460–473. [CrossRef]
32. Li, W.; You, Q.; Zhang, J.; Li, W.; Xu, H. Green Synthesis of Antibacterial LFL-ZnO Using L. Plantarum Fermentation Liquid Assisted by Ultrasound-Microwave. *J. Alloys Compd.* **2023**, *947*, 169697. [CrossRef]
33. Agarwal, H.; Menon, S.; Shanmugam, V.K. Functionalization of Zinc Oxide Nanoparticles Using Mucuna Pruriens and Its Antibacterial Activity. *Surf. Interfaces* **2020**, *19*, 100521. [CrossRef]
34. Fakhari, S.; Jamzad, M.; Kabiri Fard, H. Green Synthesis of Zinc Oxide Nanoparticles: A Comparison. *Green Chem. Lett. Rev.* **2019**, *12*, 19–24. [CrossRef]
35. Hasan, I.M.A.; Tawfik, A.R.; Assaf, F.H. GC/MS Screening of Buckthorn Phytochemicals and Their Use to Synthesize ZnO Nanoparticles for Photocatalytic Degradation of Malachite Green Dye in Water. *Water Sci. Technol.* **2021**, *85*, 664–684. [CrossRef]
36. Ko, G.-A.; Shrestha, S.; Kim Cho, S. Sageretia Thea Fruit Extracts Rich in Methyl Linoleate and Methyl Linolenate Downregulate Melanogenesis via the Akt/GSK3β Signaling Pathway. *Nutr. Res. Pract.* **2018**, *12*, 3–12. [CrossRef]
37. Pejin, B.; Savic, A.; Sokovic, M.; Glamoclija, J.; Ciric, A.; Nikolic, M.; Radotic, K.; Mojovic, M. Further In Vitro Evaluation of Antiradical and Antimicrobial Activities of Phytol. *Nat. Prod. Res.* **2014**, *28*, 372–376. [CrossRef]
38. Saeidnia, S.; Manayi, A.; Gohari, A.R.; Abdollahi, M. The Story of Beta-Sitosterol- A Review. *Eur. J. Med. Plants* **2014**, *4*, 590–609. [CrossRef]
39. Vijayakumar, S.; Divya, M.; Vaseeharan, B.; Ranjan, S.; Kalaiselvi, V.; Dasgupta, N.; Chen, J.; Durán-Lara, E.F. Biogenic Preparation and Characterization of ZnO Nanoparticles from Natural Polysaccharide *Azadirachta indica* L. (Neem Gum) and Its Clinical Implications. *J. Clust. Sci.* **2021**, *32*, 983–993. [CrossRef]
40. Nayak, A.; Sahoo, J.K.; Sahoo, S.K.; Sahu, D. Removal of Congo Red Dye from Aqueous Solution Using Zinc Oxide Nanoparticles Synthesised from *Ocimum sanctum* (Tulsi Leaf): A Green Approach. *Int. J. Environ. Anal. Chem.* **2022**, *102*, 7889–7910. [CrossRef]
41. Ahmad, S.Z.N.; Salleh, W.N.W.; Yusof, N.; Mohd Yusop, M.Z.; Hamdan, R.; Awang, N.A.; Ismail, N.H.; Rosman, N.; Sazali, N.; Ismail, A.F. Pb(II) Removal and Its Adsorption from Aqueous Solution Using Zinc Oxide/Graphene Oxide Composite. *Chem. Eng. Commun.* **2021**, *208*, 646–660. [CrossRef]
42. Arab, C.; El Kurdi, R.; Patra, D. Effect of PH on the Removal of Anionic and Cationic Dyes Using Zinc Curcumin Oxide Nanoparticles as Adsorbent. *Mater. Chem. Phys.* **2022**, *277*, 125504. [CrossRef]
43. Marahel, F.; Ghaedi, M.; Ansari, A. Zinc Oxide Nanoparticles Loaded on Activated Carbon and Its Application for Adsorption Removal of Uric Acid. *Synth. React. Inorg. Met.-Org. Nano-Met. Chem.* **2015**, *45*, 1387–1395. [CrossRef]
44. Eleryan, A.; Aigbe, U.O.; Ukhurebor, K.E.; Onyancha, R.B.; Hassaan, M.A.; Elkatory, M.R.; Ragab, S.; Osibote, O.A.; Kusuma, H.S.; El Nemr, A. Adsorption of Direct Blue 106 Dye Using Zinc Oxide Nanoparticles Prepared via Green Synthesis Technique. *Environ. Sci. Pollut. Res.* **2023**, *30*, 69666–69682. [CrossRef]
45. Phan, D.-N.; Rebia, R.A.; Saito, Y.; Kharaghani, D.; Khatri, M.; Tanaka, T.; Lee, H.; Kim, I.-S. Zinc Oxide Nanoparticles Attached to Polyacrylonitrile Nanofibers with Hinokitiol as Gluing Agent for Synergistic Antibacterial Activities and Effective Dye Removal. *J. Ind. Eng. Chem.* **2020**, *85*, 258–268. [CrossRef]
46. Zare, M.; Namratha, K.; Byrappa, K.; Surendra, D.M.; Yallappa, S.; Hungund, B. Surfactant Assisted Solvothermal Synthesis of ZnO Nanoparticles and Study of Their Antimicrobial and Antioxidant Properties. *J. Mater. Sci. Technol.* **2018**, *34*, 1035–1043. [CrossRef]
47. Ansari, A.; Ali, A.; Khan, N.; Saad Umar, M.; Owais, M.; Shamsuzzaman. Synthesis of Steroidal Dihydropyrazole Derivatives Using Green ZnO NPs and Evaluation of Their Anticancer and Antioxidant Activity. *Steroids* **2022**, *188*, 109113. [CrossRef]

48. Krishnan, B.R.; Ramesh, M.; Selvakumar, M.; Karthick, S.; Sasikumar, A.; Geerthi, D.V.; Senthilkumar, N. A Facile Green Approach of Cone-like ZnO NSs Synthesized Via Jatropha Gossypifolia Leaves Extract for Photocatalytic and Biological Activity. *J. Inorg. Organomet. Polym. Mater.* **2020**, *30*, 4441–4451. [CrossRef]
49. Velsankar, K.; Venkatesan, A.; Muthumari, P.; Suganya, S.; Mohandoss, S.; Sudhahar, S. Green Inspired Synthesis of ZnO Nanoparticles and Its Characterizations with Biofilm, Antioxidant, Anti-Inflammatory, and Anti-Diabetic Activities. *J. Mol. Struct.* **2022**, *1255*, 132420. [CrossRef]
50. Ganesan, V.; Hariram, M.; Vivekanandhan, S.; Muthuramkumar, S. *Periconium* sp. (Endophytic Fungi) Extract Mediated Sol-Gel Synthesis of ZnO Nanoparticles for Antimicrobial and Antioxidant Applications. *Mater. Sci. Semicond. Process.* **2020**, *105*, 104739. [CrossRef]
51. Rahman, A.; Harunsani, M.H.; Tan, A.L.; Ahmad, N.; Khan, M.M. Antioxidant and Antibacterial Studies of Phytogenic Fabricated ZnO Using Aqueous Leaf Extract of *Ziziphus mauritiana* Lam. *Chem. Pap.* **2021**, *75*, 3295–3308. [CrossRef]
52. Vieira, I.R.S.; da Silva, A.A.; da Silva, B.D.; Neto, L.T.; Tessaro, L.; Furtado, C.R.G.; de Sousa, A.M.F.; Carvalho, N.M.F.; Conte-Junior, C.A. Eco-Friendly Synthesis of ZnO Nanomaterial from Green Tea Extract: Photocatalytic, Antibacterial and Antioxidant Potential. *Biomass Convers. Biorefin.* **2023**, 1–15. [CrossRef]
53. Ali, A.H. Experimental Investigations on Effects of ZnO NPS and Annona Muricata Extract for in Vitro and in Vivo Antibacterial Activity. *Mater. Today Proc.* **2022**, *57*, 527–530. [CrossRef]
54. Hameed, H.; Waheed, A.; Sharif, M.S.; Saleem, M.; Afreen, A.; Tariq, M.; Kamal, A.; Al-onazi, W.A.; Al Farraj, D.A.; Ahmad, S.; et al. Green Synthesis of Zinc Oxide (ZnO) Nanoparticles from Green Algae and Their Assessment in Various Biological Applications. *Micromachines* **2023**, *14*, 928. [CrossRef]
55. Dhandapani, K.V.; Anbumani, D.; Gandhi, A.D.; Annamalai, P.; Muthuvenkatachalam, B.S.; Kavitha, P.; Ranganathan, B. Green Route for the Synthesis of Zinc Oxide Nanoparticles from *Melia azedarach* Leaf Extract and Evaluation of Their Antioxidant and Antibacterial Activities. *Biocatal. Agric. Biotechnol.* **2020**, *24*, 101517. [CrossRef]

MDPI AG
Grosspeteranlage 5
4052 Basel
Switzerland
Tel.: +41 61 683 77 34

Separations Editorial Office
E-mail: separations@mdpi.com
www.mdpi.com/journal/separations

www.ingramcontent.com/pod-product-compliance
Lightning Source LLC
Chambersburg PA
CBHW061435210926
51850CB00008B/1523
* 9 7 8 3 7 2 5 8 1 9 8 7 4 *